# INTRODUCTION TO
# QUANTUM FIELD THEORY

### Second Edition

# INTRODUCTION TO
# QUANTUM FIELD THEORY

## Second Edition

## Roberto Casalbuoni
University of Florence, Italy

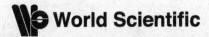

**World Scientific**

NEW JERSEY · LONDON · SINGAPORE · BEIJING · SHANGHAI · HONG KONG · TAIPEI · CHENNAI · TOKYO

*Published by*

World Scientific Publishing Co. Pte. Ltd.
5 Toh Tuck Link, Singapore 596224
*USA office:* 27 Warren Street, Suite 401-402, Hackensack, NJ 07601
*UK office:* 57 Shelton Street, Covent Garden, London WC2H 9HE

**Library of Congress Cataloging-in-Publication Data**
Names: Casalbuoni, R. (Roberto), author.
Title: Introduction to quantum field theory / by Roberto Casalbuoni
    (University of Florence, Italy).
Description: Second edition. | Singapore ; Hackensack, NJ : World Scientific, [2017] |
    Includes bibliographical references and index.
Identifiers: LCCN 2017003584 | ISBN 9789813146662 (hardcover ; alk. paper) |
    ISBN 9813146664 (hardcover ; alk. paper)
Subjects: LCSH: Quantum field theory.
Classification: LCC QC174.45 .C375 2017 | DDC 530.14/3--dc23
LC record available at https://lccn.loc.gov/2017003584

**British Library Cataloguing-in-Publication Data**
A catalogue record for this book is available from the British Library.

Printed in Singapore

To my wife Marta and my daughters Sara and Chiara with love.

# Preface

This book originates from an introductory course in quantum field theory that I have given for many years at the University of Florence. At that time my lecture notes were written in Italian. Then, in the academic year 1997/98, I was on leave at the University of Geneva where I gave the same course. In that occasion I decided to translate the material of my notes in English. These notes in English form the core of the present book. The course was intended to be an introduction to quantum field theory having as a pre-requisite a standard course in non-relativistic quantum mechanics. The language that I have chosen for this course is the one of quantum operators avoiding any recourse to the path integral formalism. My personal feeling is that the path integral approach is a fundamental one, both in quantum mechanics and even more in quantum field theory. However, I also think that this approach is better appreciated after being exposed to the operator formalism.

Today quantum field theory is one of the most important tools in physics. It represents a unified way of approaching many different problems in different areas of physics, as elementary particles, statistical mechanics, critical phenomena, cosmology, etc. For this reason I have tried to keep the book at the level of a non-sophisticated approach pointing more to clarify the relevant aspects.

Entering in some detail, the book begins with a discussion of the quantization of the one-dimensional string. By considering the string as the continuum limit of a discrete chain of harmonic oscillators, it is possible to use ordinary quantum mechanics and show the peculiarities of taking the continuum limit. Then, the book enters in the heart of quantum field theory discussing the free scalar field together with its locality properties. After the scalar field there is a long introduction to the Dirac equation and its properties ending with the quantization of the Dirac field in terms of anticommutators. The part of the book dedicated to the free fields ends with the quantization of the electromagnetic field. For completeness a brief discussion of the massive vector fields is presented. With respect to my original lecture notes I have inserted at the beginning of the book a presentation of the Lorentz group and of his spinor representations, together with a very brief discussion of the Poincaré group. Although the derivation of the Dirac equation is done following the traditional way, the introduction of the spinor representations of the Lorentz group offers the possibility of obtaining the same results in a much simpler way, both for the massive and the massless case. In Chapter 6 the reader will find an extensive discussion of the symmetry properties in field theory, ranging from exact to spontaneously broken symmetries. The Goldstone theorem is discussed and, after introducing the local (or gauge) symmetry for the electromagnetic field, I present the generalization to non-abelian symmetries. The Chapter ends with the Higgs mechanism, the key to the present understanding of weak interactions. This Chapter should be useful also to students interested in critical phenomena. The next two Chapters are dedicated to perturbation theory and its applications to the scattering of elementary particles and to the mechanism of emission and absorption of the electromagnetic radiation by an atomic system. In the last Chapter the reader can find a discussion of the infinities in quantum field theory and how to take care of them. By taking spinor electrodynamics as an example I give a detailed discussion of regularization and renormalization processes at one-loop level.

I would like to thank particularly, Daniele Dominici, Giulio Pettini and Matthias Steinhauser for a careful reading of the manuscript and for several suggestions.

A particular thank goes to my wife Marta, for her patience and love in allowing me to spend many weekends to write this book.

## Preface for the Second Edition

The path integral formalism has become an essential theoretical tool in field theory, since the role of gauge theories, in describing strong, weak and electromagnetic interactions among elementary particles, has been recognized. Therefore, when World Scientific asked me for a second edition of this book, I decided to introduce the path integral quantization of gauge theories. In order to maintain the introductory style of the book, I have added three new Chapters. Chapter 11 deals with the path integral quantization in quantum mechanics. In this Chapter, it is shown how to derive this formalism from the standard formulation of quantum mechanics. But some space has been reserved to the physical interpretation of this new form of quantization and how, ordinary quantum mechanics follows from the path integral. In Chapter 12, the extension of the path integral quantization to field theory is discussed. Among other subjects, the perturbative expansion is illustrated for the case of the $\lambda\phi^4$ theory. Furthermore, the formalism for fermions is discussed thoroughly. The last Chapter is dedicated to an accurate discussion of the quantiztion of the gauge fields, both in the abelian and the non-abelian cases. In the last part of this Chapter, the derivation of the Ward identities from path integral is illustrated. Finally, we illustrate the case of classical symmetries broken at quamum level, the so-called anomalies, for the chiral $U(1)$ symmetry. Following what was done in the first edition, a series of problems has been added to the new Chapters.

I would like to thank Matthias Steinhauser for a careful reading of the additional material and for several useful suggestions.

# Contents

# Chapter 1

# Introduction

## 1.1 Notation and units

In relativistic quantum theories the two fundamental constants, the velocity of light, $c$, and the Planck constant, $\hbar$, appear almost everywhere. Therefore it is convenient to choose a unit system (natural units) where their numerical value is one

$$c = \hbar = 1. \tag{1.1}$$

Our conventions are as follows: the Minkowski space-time metric tensor $g_{\mu\nu}, \mu, \nu = 0, 1, 2, 3$, is diagonal with eigenvalues $(+1, -1, -1, -1)$. In general we say that a metric in $m + n$ dimensions has a signature $(m, n)$ if it has $m$ negative and $n$ positive eigenvalues. Therefore $g_{\mu\nu}$ has a signature $(3, 1)$. The position and momentum four-vectors are given by

$$x^\mu = (t, \vec{x}), \qquad p^\mu = (E, \vec{p}), \quad \mu = 0, 1, 2, 3, \tag{1.2}$$

where $\vec{x}$ and $\vec{p}$ are the three-dimensional position and momentum. We will adopt the convention of summing over repeated indices unless stated otherwise. Then the scalar product between two four-vectors can be written as

$$a \cdot b = a^\mu b^\nu g_{\mu\nu} = a_\mu b^\mu = a^\mu b_\mu = a^0 b^0 - \vec{a} \cdot \vec{b}, \tag{1.3}$$

where the indices have been lowered using the metric tensor

$$a_\mu = g_{\mu\nu} a^\nu \tag{1.4}$$

and can be raised by using the inverse metric tensor $g^{\mu\nu} \equiv g_{\mu\nu}$. The four-gradient is defined as

$$\partial_\mu = \frac{\partial}{\partial x^\mu} = \left( \frac{\partial}{\partial t}, \frac{\partial}{\partial \vec{x}} \right) = \left( \partial_t, \vec{\nabla} \right). \tag{1.5}$$

1

The four-momentum operator in position space is

$$p^\mu \to i\frac{\partial}{\partial x_\mu} = (i\partial_t, -i\vec{\nabla}).$$ (1.6)

The following relations will be useful

$$p^2 = p_\mu p^\mu \to -\frac{\partial}{\partial x^\mu}\frac{\partial}{\partial x_\mu} \equiv -\Box,$$ (1.7)

$$x \cdot p = Et - \vec{p} \cdot \vec{x}.$$ (1.8)

The Ricci tensor in 4 dimensions, $\epsilon^{\mu\nu\rho\sigma}$ is completely antisymmetric and such that

$$\epsilon^{0123} = +1.$$ (1.9)

For the electromagnetism we will use the Heaviside-Lorentz system, where the vacuum dielectric constant is fixed to one

$$\epsilon_0 = 1.$$ (1.10)

From the relation $\epsilon_0 \mu_0 = 1/c^2$ it follows

$$\mu_0 = 1.$$ (1.11)

In these units the Coulomb force is given by

$$|\vec{F}| = \frac{e_1 e_2}{4\pi}\frac{1}{|\vec{x}_1 - \vec{x}_2|^2}$$ (1.12)

and there are no visible constants in the Maxwell equations. For instance, the Gauss law is

$$\vec{\nabla} \cdot \vec{E} = \rho,$$ (1.13)

where $\rho$ is the charge density. The dimensionless fine structure constant

$$\alpha = \frac{e^2}{4\pi\epsilon_0 \hbar c}$$ (1.14)

is given by

$$\alpha = \frac{e^2}{4\pi}.$$ (1.15)

Any physical quantity can be expressed by using as fundamental unit energy, mass, length or time in an equivalent fashion. In fact, from our choice the following equivalence relations follow

$$ct \approx \ell \Longrightarrow \text{time} \approx \text{length},$$
$$E \approx mc^2 \Longrightarrow \text{energy} \approx \text{mass},$$
$$E \approx pv \Longrightarrow \text{energy} \approx \text{momentum},$$
$$Et \approx \hbar \Longrightarrow \text{energy} \approx (\text{time})^{-1} \approx (\text{length})^{-1}.$$ (1.16)

In practice, it is enough to notice that the product $c\hbar$ has dimensions $[E \cdot \ell]$. Therefore

$$c\hbar = 3 \cdot 10^8 \text{ mt} \cdot \sec^{-1} \cdot 1.05 \cdot 10^{-34} \text{ J} \cdot \sec = 3.15 \cdot 10^{-26} \text{ J} \cdot \text{mt.} \quad (1.17)$$

Recalling that

$$1 \text{ eV} = e \cdot 1 = 1.602 \cdot 10^{-19} \text{ J}, \quad (1.18)$$

where $e$ is the electric charge of the proton expressed in Coulomb, it follows

$$c\hbar = \frac{3.15 \cdot 10^{-26}}{1.6 \cdot 10^{-13}} \text{ MeV} \cdot \text{mt} = 197 \text{ MeV} \cdot \text{fm}, \quad (1.19)$$

where $1 \text{ fm} = 10^{-13}$ cm. From which

$$1 \text{ MeV}^{-1} = 197 \text{ fm.} \quad (1.20)$$

Using this relation we can easily convert a quantity given in MeV (the typical unit used in elementary particle physics) in fermi. For instance, since the elementary particle masses are usually given in MeV, the Compton wave length of an electron is simply given by

$$\lambda^e_{\text{Compton}} = \frac{1}{m_e} \approx \frac{1}{0.5 \text{ MeV}} \approx \frac{200 \text{ MeV} \cdot \text{fermi}}{0.5 \text{ MeV}} \approx 400 \text{ fermi.} \quad (1.21)$$

Therefore the approximate relation to keep in mind is $1 \approx 200 \text{ MeV} \cdot \text{fermi}$. Furthermore, using

$$c = 3 \cdot 10^{23} \text{ fermi} \cdot \sec^{-1}, \quad (1.22)$$

we get

$$1 \text{ fermi} = 3.3 \cdot 10^{-24} \text{ sec} \quad (1.23)$$

and

$$1 \text{ MeV}^{-1} = 6.58 \cdot 10^{-22} \text{ sec.} \quad (1.24)$$

Also, using

$$1 \text{ barn} = 10^{-24} \text{ cm}^2, \quad (1.25)$$

it follows from (1.20)

$$1 \text{ GeV}^{-2} = 0.389 \text{ mbarn.} \quad (1.26)$$

## 1.2   Major steps in quantum field theory

For a more complete history of quantum field theory see, for example
[Pais (1986)] and [Weinberg (1995a)].

**1924**

- Bose and Einstein introduce a new statistics for light-quanta
  (photons) [Bose (1924); Einstein (1924)].

**1925**

- January - Pauli formulates the exclusion principle [Pauli
  (1925)].
- July - Heisenberg's first paper on quantum mechanics (matrix
  mechanics) [Heisenberg (1925)].
- September - Born and Jordan extend Heisenberg's formulation
  of quantum mechanics to electrodynamics [Born and Jordan
  (1925)]. A more complete treatment was given in the famous
  three men's paper by Born, Heisenberg and Jordan [Born *et al.*
  (1925)].

**1926**

- January - Schrödinger's formulation of the wave equation
  [Schroedinger (1926a)].
- February - Fermi introduces a new statistics, known today as
  the Fermi-Dirac statistics [Fermi (1926)].
- August - Dirac relates statistics and symmetry properties of
  the wave function, and shows that the quantized electromagnetic field is equivalent to a set of harmonic oscillators satisfying the Bose-Einstein statistics [Dirac (1926)].

**1927**

- March - Davisson and Germer detect the electron diffraction
  by a crystal [Davisson and Germer (1927)].
- October - Jordan and Klein show that the electromagnetic
  field satisfies commutation rules [Jordan and Klein (1927)].

**1928**

- January - Dirac's generalization of quantum mechanics to the
  relativistic case (Dirac equation) [Dirac (1928a,b)].

- January - Jordan and Wigner introduce anticommuting fields for describing particles satisfying Fermi-Dirac statistics [Jordan and Wigner (1928)].

**1929**

- January - Pauli and Heisenberg develop the analog for fields of Lagrangian and Hamiltonian methods of analytical mechanics [Heisenberg and Pauli (1929, 1930)].
- March - Weyl formulates gauge invariance and its relation to charge conservation [Weyl (1929a)].
- Klein and Nishina complete the theory of the scattering Compton based on the Dirac equation [Klein and Nishina (1929)].

**1931**

- Dirac introduces the idea of anti-electron (positron) to interpret the negative energy solutions of his equation [Dirac (1931)].

**1932**

- Anderson detects the positron [Anderson (1932)].

**1934**

- Dirac and Heisenberg evaluate the vacuum polarization of the photon.
- First battle with infinities in quantum field theory [Dirac (1934); Heisenberg (1934)].

**1936**

- Serber introduces the concept of renormalized charge [Serber (1936)].

**1946**

- Tomonaga begins the program of renormalization but, due to the war, his work is completely ignored by the western physicists [Tomonaga (1946)] (see also the next papers [Koba *et al.* (1947)]).

**1947**

- Lamb and Rutherford discover the "Lamb shift" [Lamb and Retherford (1947)]

- Bethe performs the first calculation of the Lamb shift [Bethe (1947)].

**1948**

- Schwinger evaluates the corrections to the $g-2$ of the electron [Schwinger (1948a)].
- The renormalization program starts in the west [Schwinger (1948b, 1949a,b, 1951a,b,c,d)], [Kanesawa and Tomonaga (1948a,b); Tomonaga (1948); Ito *et al.* (1948); Koba and Tomonaga (1948)], [Feynman (1948a,b,c, 1949a,b, 1950)], [Dyson (1949a,b)].

## 1.3   The Lorentz group

Since the main subject of this book is quantum relativistic field theory, we will provide a brief introduction to some of the properties of the Lorentz group.

The Lorentz transformations are defined as the ones leaving invariant the distance between two points infinitesimally close in Minkowski space $dx^\mu = (dx^0, d\vec{x}) = (dt, d\vec{x})$

$$ds^2 = dx^\mu dx^\nu g_{\mu\nu} = dx^\mu dx_\mu. \tag{1.27}$$

Then, a Lorentz transformation

$$dx'^\mu = \Lambda^\mu_\nu dx^\nu \tag{1.28}$$

is such that

$$dx'^\mu dx'_\mu = dx^\mu dx_\mu, \tag{1.29}$$

that is

$$\Lambda^\mu_\rho g_{\mu\nu} \Lambda^\nu_\sigma = g_{\rho\sigma}. \tag{1.30}$$

By lowering the upper index of $\Lambda^\mu_\nu$

$$\Lambda_{\mu\nu} = g_{\mu\rho} \Lambda^\rho_\nu, \tag{1.31}$$

we can write (1.30) as

$$\Lambda_{\mu\rho} g^{\mu\nu} \Lambda_{\nu\sigma} = g_{\rho\sigma}, \tag{1.32}$$

or, in matrix terms

$$\Lambda^T g \Lambda = g, \tag{1.33}$$

where we have introduced the matrices $\Lambda$ and $g$ with matrix elements $\Lambda_{\mu\nu}$ and $g_{\mu\nu} = g^{\mu\nu}$ respectively. These matrices are orthogonal with respect to a metric with signature $(3, 1)$. They form a group called $O(3, 1)$. From the previous relation it follows

$$\det|\Lambda| = \pm 1. \tag{1.34}$$

Lorentz transformation with determinant equal to $+1$ are said to be proper[1], whereas the ones with determinant equal to $-1$ are said improper. From eq. (1.32) we have

$$1 = g_{00} = \Lambda_{00}^2 - \sum_1^3 \Lambda_{i0}^2, \tag{1.35}$$

implying

$$\Lambda_{00} \geq 1 \quad \text{or} \quad \Lambda_{00} \leq -1. \tag{1.36}$$

From this it follows that the Lorentz group can be divided into four parts, with only one, $L_+^\uparrow$, connected to the identity.

Table 1.1   The four different parts of the Lorentz group

| Symbol | $\Lambda_{00}$ | $\det|\Lambda|$ | Type of transformation |
|--------|--------|--------|------------------------|
| $L_+^\uparrow$ | $\geq +1$ | $+1$ | proper orthochronous |
| $L_+^\downarrow$ | $\leq -1$ | $+1$ | proper nonorthochronous |
| $L_-^\uparrow$ | $\geq +1$ | $-1$ | improper orthochronous |
| $L_-^\downarrow$ | $\leq -1$ | $-1$ | improper nonorthochronous |

The other parts can be reached, starting from the identity, through discrete transformations, namely parity, $P$ and time-reversal, $T$ (see Table 1.1) as shown in Fig. 1.1

$$P = \begin{pmatrix} 1 & & & \\ & -1 & & \\ & & -1 & \\ & & & -1 \end{pmatrix}, \quad T = \begin{pmatrix} -1 & & & \\ & 1 & & \\ & & 1 & \\ & & & 1 \end{pmatrix}. \tag{1.37}$$

We will denote by $g(\Lambda)$ the abstract group element associated to the transformation $\Lambda$. It follows from eq. (1.28) that the group composition law for two transformations $\Lambda_1$ and $\Lambda_2$ is

$$g(\Lambda_1)g(\Lambda_2) = g(\Lambda_1\Lambda_2). \tag{1.38}$$

---

[1] These matrices form a subgroup of $O(3, 1)$ called $SO(3, 1)$ because it consists of special matrices with unit determinant.

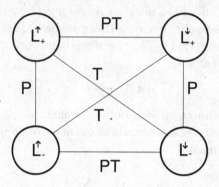

Fig. 1.1   The different parts of the Lorentz group as connected via parity, $P$, and time-reversal, $T$.

It is worth to notice that due to the condition (1.33) a generic Lorentz transformation depends on six parameters, corresponding to three rotations and three special Lorentz transformations along the three axes.

   Let us now study infinitesimal Lorentz transformations. Since they are defined as the transformations leaving invariant the norm of a four-vector

$$x^2 = x'^2, \tag{1.39}$$

for an infinitesimal transformation

$$x' = x + \delta x, \tag{1.40}$$

we get

$$x^2 \approx x^2 + 2x \cdot \delta x \implies x \cdot \delta x = 0. \tag{1.41}$$

Since Lorentz transformations are linear

$$x'_\mu = \Lambda_{\mu\nu} x^\nu \approx x_\mu + \epsilon_{\mu\nu} x^\nu, \tag{1.42}$$

we get

$$x \cdot \delta x = 0 \implies x^\mu \epsilon_{\mu\nu} x^\nu = 0. \tag{1.43}$$

The most general solution for the parameters $\epsilon_{\mu\nu}$ of the transformation is that they form a second order antisymmetric tensor

$$\epsilon_{\mu\nu} = -\epsilon_{\nu\mu}. \tag{1.44}$$

We recover, in this way, the result that the number of independent parameters characterizing a Lorentz transformation is six.

From these considerations it follows that the transformations connected to the identity can be written as

$$g(\Lambda) = e^{-\frac{i}{2}\omega_{\mu\nu}J^{\mu\nu}}, \tag{1.45}$$

with $\omega_{\mu\nu} = -\omega_{\nu\mu}$ the parameters of the transformation ($\Lambda \approx 1 + \omega$) and $J_{\mu\nu} = -J_{\nu\mu}$ the infinitesimal generators. The latter satisfy the following commutation relations

$$[J_{\mu\nu}, J_{\rho\sigma}] = i[J_{\nu\rho}g_{\mu\sigma} - J_{\mu\rho}g_{\nu\sigma} + J_{\mu\sigma}g_{\nu\rho} - J_{\nu\sigma}g_{\mu\rho}]. \tag{1.46}$$

The generators of rotations and boosts can be expressed in terms of $J_{\mu\nu}$ as

$$J_i = \frac{1}{2}\epsilon_{ijk}J_{jk}, \quad K_i = J_{i0}, \tag{1.47}$$

with commutation relations

$$[J_i, J_j] = -[K_i, K_j] = i\epsilon_{ijk}J_k,$$
$$[J_i, K_j] = i\epsilon_{ijk}K_k. \tag{1.48}$$

A crucial observation in the study of the Lorentz group is that it is homomorphic to the group of complex $2 \times 2$ matrices with determinant equal to one, the group $SL(2,C)^2$. To prove this statement let us introduce

$$\sigma^\mu = (1, \vec{\tau}), \quad \tilde{\sigma}^\mu = (1, -\vec{\tau}), \tag{1.49}$$

with $\vec{\tau}$ the Pauli matrices and 1 the identity matrix in the 2-dimensional complex space. Then we define

$$\hat{x} = x^\mu \tilde{\sigma}_\mu = x^0 + \vec{x} \cdot \vec{\tau} = \begin{pmatrix} x^0 + x^3 & x^1 - ix^2 \\ x^1 + ix^2 & x^0 - x^3 \end{pmatrix}, \tag{1.50}$$

with $x^\mu$ a four-vector. Let us notice that

$$\text{Tr}(\hat{x}) = 2x^0, \quad \det|\hat{x}| = x^\mu x_\mu. \tag{1.51}$$

On this space we can define the following linear transformations

$$\hat{x}' = U\hat{x}U^\dagger, \tag{1.52}$$

with $\det|U| = 1$. Clearly this leaves invariant the determinant of $\hat{x}$ and therefore the norm of the four-vector $x^\mu$. Furthermore $\hat{x}'$ is hermitian as $\hat{x}$. That is to say, we are doing a linear transformation on the real four-vector $x_\mu$ preserving its norm and its reality properties. In other words the linear transformation induced by $U$ is nothing but a Lorentz transformation

$$x'^\mu = \Lambda^\mu_\nu x. \tag{1.53}$$

---

[2]Here $S$ stay for special, that is determinant equal to one, $L$ for linear, 2 for $2 \times 2$ matrices and $C$ for complex valued.

From this it follows

$$\hat{x}' = x'^{\mu}\tilde{\sigma}_{\mu} = \Lambda_{\nu}^{\mu}x^{\nu}\tilde{\sigma}_{\mu} = Ux^{\nu}\tilde{\sigma}_{\nu}U^{\dagger}. \tag{1.54}$$

Using

$$\text{Tr}[\sigma_{\mu}\tilde{\sigma}_{\nu}] = 2g_{\mu\nu}, \tag{1.55}$$

we find

$$\Lambda_{\mu\nu} = \frac{1}{2}\text{Tr}[\sigma_{\mu}U\tilde{\sigma}_{\nu}U^{\dagger}]. \tag{1.56}$$

This relation shows that there is a 2 to 1 correspondence between $SL(2,C)$ and $O(3,1)$, since both $U$ and $-U$ correspond to the same Lorentz transformation, $\Lambda$. It is not difficult to evaluate the matrices $U$ giving rise to rotations around the axis $i$, and boosts (special Lorentz transformations) along the direction $i$

$$U_i(\alpha) = e^{i\alpha_i J_i} = \cos(\alpha_i/2) + i\tau_i\sin(\alpha_i/2),$$
$$U_i(\chi) = e^{i\chi_i K_i} = \cosh(\chi_i/2) - \tau_i\sinh(\chi_i/2). \tag{1.57}$$

In particular, we see that the generators of rotations and boosts, in this $2 \times 2$ representation, are respectively

$$J_i = \frac{\tau_i}{2}, \qquad K_i = i\frac{\tau_i}{2}. \tag{1.58}$$

Notice that the $J_i$'s are hermitian, whereas the $K_i$'s are antihermitian. Therefore this representation is not unitary. The same happens[3] for all the finite-dimensional representations of $SL(2,C)$. In fact, due to the non compact nature of the Lorentz group, it is possible to show that the unitary representations are infinite-dimensional. Let us also observe that the relation between $SL(2,C)$ and $O(3,1)$ is the same existing between $SU(2)$ and $O(3)$. In fact, if we restrict the $SL(2,C)$ transformations to be unitary, we see from the first of eqs. (1.51) that they do not change the fourth component of the four-vector and therefore they are pure rotations. It follows from these considerations that instead of studying the representations of $O(3,1)$ we can study the representations of $SL(2,C)$. However, the representations of $O(3,1)$, or tensor representations, will be characterized by the condition $T(U) = T(-U)$, where $T$ is the representation and $U \in SL(2,C)$.

In the case of finite-dimensional representations one can define two hermitian combinations of the generators, namely

$$\vec{J}^{(1)} = \frac{1}{2}(\vec{J} - i\vec{K}), \qquad \vec{J}^{(2)} = \frac{1}{2}(\vec{J} + i\vec{K}), \tag{1.59}$$

---

[3]In particular the hermiticity of $\vec{J}$ and the anti-hermiticity of $\vec{K}$.

with commutation rules

$$[J_i^{(1)}, J_j^{(1)}] = i\epsilon_{ijk} J_k^{(1)}, \quad [J_i^{(2)}, J_j^{(2)}] = i\epsilon_{ijk} J_k^{(2)}$$

$$[J_i^{(1)}, J_j^{(2)}] = 0. \tag{1.60}$$

This is the algebra of $SU(2) \otimes SU(2)$. Therefore a generic finite-dimensional irreducible representation of the Lorentz group (or $SL(2,C)$) is characterized by assigning the pair $(s_1, s_2)$ with $s_1$ and $s_2$ the spin corresponding to the two groups $SU(2)$. The lowest dimensional representations are

$$\zeta^\alpha \in \left(\frac{1}{2}, 0\right), \quad \zeta^{\dot\alpha} \in \left(0, \frac{1}{2}\right). \tag{1.61}$$

The elements of these representations are called undotted and dotted spinors respectively. The following relation is also important

$$(s_1, s_2) = (s_1, 0) \otimes (0, s_2). \tag{1.62}$$

It should be noticed that the physical angular momentum $\vec{J}$ is the sum of the two $SU(2)$ generators. This means that the physical spin, $j$, contained in a given representation $(s_1, s_2)$ is obtained by summing the two angular momenta, and it will get one of the following values

$$|s_1 - s_2| \le j \le (s_1 + s_2). \tag{1.63}$$

For instance, the representation $(1/2, 1/2)$ has a spin content 0 and 1 and it corresponds to a four-vector. Another relevant observation is that the infinitesimal generators $\vec{J}$ and $\vec{K}$ transform under parity as a pseudo-vector and a proper vector respectively

$$P\vec{J}P^{-1} = \vec{J}, \quad P\vec{K}P^{-1} = -\vec{K}, \tag{1.64}$$

showing that parity interchanges the two $SU(2)$. Said in a different way, parity transforms undotted spinors in dotted ones and vice versa. From this it follows that a representation invariant under parity should contain pairs of representations connected by parity. The simplest example is the representation

$$\left(\frac{1}{2}, 0\right) \oplus \left(0, \frac{1}{2}\right). \tag{1.65}$$

This is a four-dimensional representation of the Lorentz group and we shall see that it corresponds to the Dirac representation.

Analogously the representation

$$(1, 0) \oplus (0, 1) \tag{1.66}$$

describes two spin 1 and corresponds to an antisymmetric tensor of rank 2, as the electromagnetic field strength tensor.

In the next section we will show how to construct all the finite-dimensional representations of $SL(2, C)$.

## 1.4   Spinor representations of the Lorentz group

We can regard $SL(2,C)$ as a representation of itself. In this case we introduce a two-dimensional complex vector space (spinor space) and define transformations according to

$$\zeta'^{\alpha} = a^{\alpha}_{\beta}\zeta^{\beta}, \quad a \in SL(2,C), \quad \alpha, \beta = 1, 2. \tag{1.67}$$

We see from eqs. (1.58) and (1.59) that

$$\vec{J}^{(1)} = \frac{\vec{\tau}}{2}, \quad \vec{J}^{(2)} = 0. \tag{1.68}$$

Therefore the spinors transform according to

$$\zeta^{\alpha} \in \left(\frac{1}{2}, 0\right). \tag{1.69}$$

It is convenient to introduce the following special $2 \times 2$ matrix

$$\epsilon^{\alpha\beta} = \epsilon_{\alpha\beta} = i(\tau_2)_{\alpha\beta} = \begin{pmatrix} 0 & 1 \\ -1 & 0 \end{pmatrix}, \tag{1.70}$$

with the following properties

$$\epsilon^{-1} = \epsilon^{T} = \epsilon^{\dagger} = -\epsilon. \tag{1.71}$$

Furthermore, for any matrix $a \in SL(2,C)$ we have

$$a = \epsilon^{-1}a^{T-1}\epsilon. \tag{1.72}$$

Together with the representation specified by $a$ we can introduce the representations defined by $a^{T-1}$, $a^*$ and $a^{\dagger-1}$. In fact we can easily verify that they satisfy the same product rule as the one defined by $a$. For instance,

$$a_1^{T-1}a_2^{T-1} = (a_1 a_2)^{T-1} \tag{1.73}$$

and analogously for the other cases. It follows from (1.72) that the representation specified by $a^{T-1}$ is equivalent to the one specified by $a$. In the same way $a^*$ and $a^{\dagger-1}$ are equivalent. On the other hand, $a$ and $a^*$ are not equivalent[4]. Then, it is convenient to introduce explicitly a second spinor space transforming according to $a^*$. We will use a special notation for the components of the elements of this space

$$\zeta'^{\dot{\alpha}} = (a^{\alpha}_{\beta})^*\zeta^{\dot{\beta}} \equiv a^{\dot{\alpha}}_{\dot{\beta}}\zeta^{\dot{\beta}}. \tag{1.74}$$

---

[4]Notice that if we restrict the matrices to $SU(2)$ then, due to the unitarity condition, it follows that $a^* = a^{T-1}$ and correspondingly $a^*$ is equivalent to $a$.

The first spinor space is said to be the space of the undotted spinors, whereas the second one the space of the dotted spinors. By using the hermiticity of $\vec{J}$ and the anti-hermiticity of $\vec{K}$ we have

$$\left(e^{i\vec{\alpha}\cdot\vec{J}}\right)^{\dagger-1} = e^{i\vec{\alpha}\cdot\vec{J}}, \quad \left(e^{i\vec{\alpha}\cdot\vec{K}}\right)^{\dagger-1} = e^{-i\vec{\alpha}\cdot\vec{K}}. \tag{1.75}$$

Since $a^*$ and $a^{\dagger-1}$ are equivalent representations, the previous relations show that going from the $a$ representation to the $a^*$ is equivalent to send $\vec{J} \to \vec{J}$ and $\vec{K} \to -\vec{K}$, that is to say to exchange $\vec{J}^{(1)}$ and $\vec{J}^{(2)}$. But this means that the complex conjugate of an undotted spinor transforms as a dotted spinor and vice versa. Therefore

$$\zeta^{\dot{\alpha}} \in \left(0, \frac{1}{2}\right). \tag{1.76}$$

We can make use of the matrix $\epsilon$ to lower and raise the spinor indices

$$\zeta_\alpha = \zeta^\beta \epsilon_{\beta\alpha}, \quad \zeta_{\dot{\alpha}} = \zeta^{\dot{\beta}} \epsilon_{\dot{\beta}\dot{\alpha}}, \tag{1.77}$$

with

$$\epsilon^{\dot{\alpha}\dot{\beta}} = \epsilon_{\dot{\alpha}\dot{\beta}} = \epsilon^{\alpha\beta} = \epsilon_{\alpha\beta}. \tag{1.78}$$

The dotted and undotted spinors with lower indices transform respectively as $a^{T-1}$ and $a^{\dagger-1}$. Inverting eqs. (1.77) we find

$$\zeta^\alpha = \epsilon^{\alpha\beta} \zeta_\beta, \quad \zeta^{\dot{\alpha}} = \epsilon^{\dot{\alpha}\dot{\beta}} \zeta_{\dot{\beta}}. \tag{1.79}$$

Since spinors with upper and lower indices transform according to $a$ and $a^{T-1}$ we see that the expression

$$\zeta_\alpha \eta^\alpha = \zeta_\alpha \epsilon^{\alpha\beta} \eta_\beta = -\zeta^\alpha \eta_\alpha \tag{1.80}$$

is invariant under $SL(2, C)$ transformations. Analogous result is found for dotted spinors.

In agreement with eq. (1.56) we define spinor indices for the matrices $\sigma^\mu$ and $\tilde{\sigma}^\mu$ as

$$(\sigma^\mu)_{\dot{\alpha}\beta} = (1, \vec{\tau}), \quad (\tilde{\sigma}^\mu)^{\alpha\dot{\beta}} = (1, -\vec{\tau}). \tag{1.81}$$

We can define higher order spinors via tensor products. For instance, the following rank two spinors

$$\zeta_{\alpha\beta}, \quad \zeta^\beta_\alpha, \quad \zeta^{\alpha\beta}, \quad \zeta_{\alpha\dot{\beta}} \tag{1.82}$$

transform respectively as

$$a^{T-1} \otimes a^{T-1}, \quad a \otimes a^{T-1}, \quad a \otimes a, \quad a^{T-1} \otimes a^{\dagger-1}. \tag{1.83}$$

These representations can be generally expressed as direct sum of irreducible representations. The latter are given by spinors of the type

$$\zeta^{\alpha_1\alpha_2\cdots\alpha_m;\dot{\beta}_1\dot{\beta}_2\cdots\dot{\beta}_n} \tag{1.84}$$

transforming as

$$\zeta'^{\alpha_1\alpha_2\cdots\alpha_m;\dot{\beta}_1\dot{\beta}_2\cdots\dot{\beta}_n} = a^{\alpha_1}_{\gamma_1}a^{\alpha_2}_{\gamma_2}\cdots a^{\alpha_m}_{\gamma_m}a^{\dot{\beta}_1}_{\dot{\gamma}_1}a^{\dot{\beta}_2}_{\dot{\gamma}_2}\cdots a^{\dot{\beta}_m}_{\dot{\gamma}_m}\zeta^{\alpha_1\alpha_2\cdots\alpha_m;\dot{\beta}_1\dot{\beta}_2\cdots\dot{\beta}_n} \tag{1.85}$$

symmetrized in both the $m$ undotted and the $n$ dotted indices. Here

$$a^{\dot{\beta}}_{\dot{\alpha}} = (a^{\beta}_{\alpha})^*. \tag{1.86}$$

The number of undotted indices, $m$, and the number of dotted indices, $n$, are respectively twice the spin $s_1$ and $s_2$ characterizing the representation of $SU(2)\otimes SU(2)$. We say also that these spinors are of type $[m,n]$.

As an example, let us consider the representation $(1/2, 1/2)$. This has angular momentum 0 and 1 and corresponds to a four-vector. This follows from the observation that combining one undotted and one dotted spinors via the $\sigma_\mu$ matrices, it is possible to construct the combination

$$\eta^{\dot{\alpha}}(\sigma^\mu)_{\dot{\alpha}\beta}\zeta^\beta \tag{1.87}$$

transforming as a four-vector. The generators of $SL(2,C)$ for the two inequivalent spinor representations are given by $J_{\mu\nu} = i\sigma_{\mu\nu}/2$ and $J_{\mu\nu} = i\tilde{\sigma}_{\mu\nu}/2$, where

$$(\sigma_{\mu\nu})^\alpha_\beta = \frac{1}{2}(\tilde{\sigma}_\mu)^{\alpha\dot{\gamma}}(\sigma_\nu)_{\dot{\gamma}\beta} - (\tilde{\sigma}_\nu)^{\alpha\dot{\gamma}}(\sigma_\mu)_{\dot{\gamma}\beta} \tag{1.88}$$

and

$$(\tilde{\sigma}_{\mu\nu})^{\dot{\beta}}_{\dot{\alpha}} = \frac{1}{2}(\sigma_\mu)_{\dot{\alpha}\gamma}(\tilde{\sigma}_\nu)^{\gamma\dot{\beta}} - (\sigma_\nu)_{\dot{\alpha}\gamma}(\tilde{\sigma}_\mu)^{\gamma\dot{\beta}}. \tag{1.89}$$

One can show that lowering one spinor index to $\sigma_{\mu\nu}$

$$(\sigma_{\mu\nu})_{\alpha\beta} = (\sigma_{\mu\nu})^\gamma_\alpha\epsilon_{\gamma\beta}, \tag{1.90}$$

the resulting matrices are symmetric in the spinor indices. Now, let us consider the tensor product of two undotted spinors. According to the rule for adding two angular momenta one finds

$$\left(\frac{1}{2},0\right)\otimes\left(\frac{1}{2},0\right) = (0,0)\oplus(1,0). \tag{1.91}$$

This corresponds to taking the antisymmetric and the symmetric parts of the product of two undotted spinors with

$$\zeta^\alpha\epsilon_{\alpha\beta}\eta^\beta \in (0,0), \quad \zeta^\alpha(\sigma_{\mu\nu})_{\alpha\beta}\eta^\beta \in (1,0). \tag{1.92}$$

By doing the same operation with two dotted spinors one gets the representations $(0,0)$ and $(0,1)$. The combination of $(1,0)$ and $(0,1)$ gives rise to a generic antisymmetric tensor containing two spin 1. As an example think of the electromagnetic field strength consisting of $\vec{E}$ and $\vec{B}$. Finally notice that, in order to have tensor representations we need $s_1 + s_2$ to be integer or, $m + n$ even. Since in this case the transformation $a$ appears an even number of times, the condition $T(a) = T(-a)$ is automatically satisfied.

## 1.5 The Poincaré group

The Poincaré group is defined as the set of transformations leaving invariant the distance between two points in Minkowski space

$$(y - x)^2 = (y - x)_\mu (y - x)_\nu g^{\mu\nu}. \tag{1.93}$$

Therefore it consists of Lorentz transformations and translations

$$x'^\mu = \Lambda^\mu_\nu x^\nu + a^\mu. \tag{1.94}$$

As such it is characterized by the pair $(\Lambda, a)$. Denoting the corresponding group element by $g(\Lambda, a)$ and applying two Poincaré transformations one can easily derive the multiplication rule for two group elements

$$g(\Lambda_1, a_1) g(\Lambda_2, a_2) = g(\Lambda_1 \Lambda_2, \Lambda_1 a_2 + a_1). \tag{1.95}$$

The number of parameters associated to a generic Poincaré transformation is ten. Six of them correspond to Lorentz transformations and four to translations. By taking again the component of the Poincaré group connected to the identity we will write

$$g(\Lambda, a) = e^{-\frac{i}{2}\omega_{\mu\nu}J^{\mu\nu} + i\epsilon_\mu P^\mu}. \tag{1.96}$$

The commutation rules involving the 4 new generators $P_\mu$ are

$$[J_{\rho\sigma}, P_\mu] = i(g_{\sigma\mu}P_\rho - g_{\rho\mu}P_\sigma) \tag{1.97}$$

and

$$[P_\mu, P_\nu] = 0. \tag{1.98}$$

We can define the Pauli-Lubanski four-vector as

$$W_\mu = \frac{1}{2}\epsilon_{\mu\nu\rho\sigma}J^{\nu\rho}P^\sigma. \tag{1.99}$$

One can easily show that $W_\mu$ commutes with the momentum operator

$$[W_\mu, P_\nu] = 0 \tag{1.100}$$

and that it has commutation relations of the same type of the momentum with the Lorentz generators

$$[J_{\rho\sigma}, W_\mu] = i(g_{\sigma\mu}W_\rho - g_{\rho\mu}W_\sigma). \qquad (1.101)$$

It is possible to show that $P^2$ and $W^2$ are Casimir operators of the Poincaré group, that is they commute with all Poincaré generators and, in particular, with the Lorentz ones (see the exercises in this Chapter)

$$[J_{\mu\nu}, P^2] = [J_{\mu\nu}, W^2] = 0. \qquad (1.102)$$

On a state vector for a particle of mass $m$, momentum $\vec{p} = \vec{0}$, spin $s$ and $z$-component $s_z$, we have

$$W^0|\vec{p} = \vec{0}, m, s, s_z\rangle = 0, \quad W^i|\vec{p} = \vec{0}, m, s, s_z\rangle = -mJ_i|\vec{p} = \vec{0}, m, s, s_z\rangle \qquad (1.103)$$

and

$$W^2|\vec{p} = \vec{0}, m, s, s_z\rangle = -m^2 s(s+1)|\vec{p} = \vec{0}, m, s, s_z\rangle. \qquad (1.104)$$

## 1.6   Exercises

(1) Convert 1/gram in units of length (cm.) and time (sec.).
(2) Show that in the units used for writing eq. (1.12) the electric charge is dimensionless. Hint: use eq. (1.14).
(3) Evaluate the masses of the electron and the proton in GeV.
(4) Using the commutation relations for the generators of the Poincaré generators show that $P^2$ and $W^2$ are Casimir operators for this algebra, that is they commute with both $J_{\mu\nu}$ and $P_\mu$. Show that $J_{\mu\nu}J^{\mu\nu}$ is a Casimir of the Lorentz group, but it is not a Casimir of the Poincaré group and give an argument to explain the last sentence on physical grounds.
(5) Using the 2×2 representation for a boost along the third axis, show that this corresponds to the special Lorentz transformation

$$x'^0 = x^0 \cosh\chi + x^3 \sinh\chi, \quad x'^3 = x^0 \sinh\chi + x^3 \cosh\chi. \qquad (1.105)$$

(6) Associate to a four-vector in Minkowski space, $x^\mu$, the 5-dimensional column vector

$$\begin{pmatrix} x \\ 1 \end{pmatrix}. \qquad (1.106)$$

Show that the $5 \times 5$ matrix

$$\begin{pmatrix} \Lambda & a \\ 0 & 1 \end{pmatrix} \qquad (1.107)$$

gives a representation of the Poincaré group.

# Chapter 2

# Lagrangian formalism for continuum systems and quantization

## 2.1 Many degrees of freedom

The aim of this book is to extend ordinary quantum mechanics, describing non relativistic particles in interaction with a given force, to the relativistic case where forces are described by fields, as in the case of the electromagnetism. The most relevant differences between the two cases are that the forces become dynamical degrees of freedom, and that one needs a relativistic treatment of the problem. In order to get a consistent description we will need to quantize the field degrees of freedom.

The concept of field is a very general one. A field represents a physical quantity depending on the space-time point. Examples are the distribution of the temperature in a room, the distribution of the pressure in the atmosphere, the particle velocities inside a fluid, the electric and magnetic fields, $\vec{E}$ and $\vec{B}$, in a given region of space. The common physical feature of these systems is the existence of a fundamental or ground state. For some of the examples listed above, the ground states are given in Table 2.1:

Table 2.1 Examples of fields and ground states.

| Field | Ground state |
|---|---|
| Temperature | T = constant |
| Pressure | P = constant |
| $\vec{E}, \vec{B}$ | Vacuum |

Then, the field is defined in terms of fluctuations around the ground state. If these deviations are small one gets linear equations for the fields. Of course the approximation can be improved by using a perturbative approach. If the linear approximation is considered one gets, in general, dynamical equations which are very similar for many different physical systems. Often the fluctuations satisfy a second order differential equation describing wave propagation. The quantization of such systems leads to a description in terms of particles corresponding to various classical excitations.

It happens that many physical systems, both with finite or infinite number of degrees of freedom, can be put in Hamiltonian form. This is true also for the case of systems described by a wave equation. The Hamiltonian description is what we need to proceed to the canonical quantization of the dynamical system.

We will start the discussion of field quantization with a simple one-dimensional system, a vibrating string. We will consider $N$ linear oscillators (for instance a one-dimensional string of atoms). In this way we get a vibrating string in the continuum limit. The advantage is to start with a finite number of degrees of freedom, a problem well known in ordinary quantum mechanics. After that we will take the continuum limit by sending $N \to \infty$ with a separation among the atoms going to zero, obtaining a vibrating string and its quantized properties.

## 2.2   Linear atomic string

Let us consider a string of $N + 1$ harmonic oscillators, or $N + 1$ atoms (of unit mass) interacting through a harmonic force, as in Fig. 2.1. The length

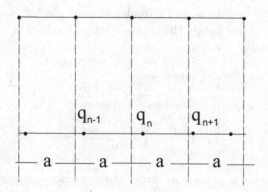

Fig. 2.1   In the upper line the atoms are in their equilibrium position, whereas in the lower line they are displaced by the quantities $q_n$.

of the string is $L$ and the interatomic distance is $a$. Therefore $L = Na$. The equations of motion for the displacements of the atoms with respect to the equilibrium configurations are

$$\ddot{q}_n = \omega^2 \left[(q_{n+1} - q_n) + (q_{n-1} - q_n)\right] = \omega^2 \left[q_{n+1} + q_{n-1} - 2q_n\right]. \quad (2.1)$$

Here $\omega$ is the common frequency of the oscillation of the atoms. The potential energy of the system is

$$U = \frac{1}{2}\omega^2 \sum_{n=1}^{N} (q_n - q_{n+1})^2. \quad (2.2)$$

In order to define the problem, one has to specify the boundary conditions. Usually one considers two possibilities:

- Periodic boundary conditions, that is $q_{N+1} = q_1$.
- Fixed boundary conditions, that is $q_{N+1} = q_1 = 0$.

To quantize the problem it is convenient to go to the Hamiltonian formulation. The Hamiltonian is given by $(p_n = \dot{q}_n)$

$$H = T + U = \frac{1}{2}\sum_{n=1}^{N} \left(p_n^2 + \omega^2 (q_n - q_{n+1})^2\right), \quad p_n = \dot{q}_n. \quad (2.3)$$

The equations of motion can be diagonalized by looking for the eigenmodes. Let us put

$$q_n^{(j)} = A_j e^{ik_j an} e^{-i\omega_j t}, \quad (2.4)$$

where the index $j$ enumerates the possible eigenvalues. Notice that in this equation the dependence on the original equilibrium position has been made explicit through

$$q_n^{(j)} = q^{(j)}(x_n) = q^{(j)}(na) \approx e^{ik_j x_n}. \quad (2.5)$$

Here $x_n = na$ is the equilibrium position of the $n^{th}$ atom. By substituting eq. (2.4) into the equations of motion we get

$$-\omega_j^2 q_n^{(j)} = -4\omega^2 q_n^{(j)} \sin^2\left(\frac{k_j a}{2}\right), \quad (2.6)$$

from which

$$\omega_j^2 = 4\omega^2 \sin^2\left(\frac{k_j a}{2}\right). \quad (2.7)$$

The relation between $\omega_j$ and $k_j$ (frequency of oscillation), shown graphically in Fig. 2.2, is called a dispersion relation. Although $k_j$ is a scalar, it will be

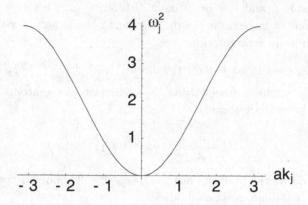

Fig. 2.2   The first Brillouin zone in arbitrary units.

called wave vector, since in more than one dimension it is indeed a vectorial quantity. We may notice that wave vectors differing by multiples integer of $2\pi/a$, that is, such that

$$k'_j = k_j + 2m\frac{\pi}{a}, \qquad m = \pm 1, \pm 2, ..., \tag{2.8}$$

correspond to the same $\omega_j$. This allows us to restrict $k_j$ to be in the so called first Brillouin zone, defined by $|k_j| \leq \pi/a$. Let us now take into account the boundary conditions. Here we will choose periodic boundary conditions, that is $q_{N+1} = q_1$, or, more generally, $q_{n+N} = q_n$. This gives us

$$q_{n+N}^{(j)} = A_j e^{ik_j a(n+N)} e^{-i\omega_j t} = q_n^{(j)} = A_j e^{ik_j an} e^{-i\omega_j t}, \tag{2.9}$$

from which

$$k_j aN = 2\pi j, \qquad (j = \text{integer}). \tag{2.10}$$

Since $aN = L$ ($L$ is the length of the string)

$$k_j = \frac{2\pi}{aN} j = \frac{2\pi}{L} j, \qquad j = 0, \pm 1, \pm 2, \cdots \pm \frac{N}{2}, \tag{2.11}$$

where we have assumed $N$ to be even. The restriction on $j$ follows from considering the first Brillouin zone ($|k_j| \leq \pi/a$). Notice that the possible values of $k_j$ are $2(N/2)+1 = N+1$, and that $j = 0$ corresponds to a uniform translation of the string (with zero frequency). Since we are interested only in the oscillatory motions, we will omit this solution in the following. It follows that we have $N$ non trivial independent solutions

$$q_n^{(j)} = A_j e^{-i\omega_j t} e^{iak_j n}. \tag{2.12}$$

The most general solution is obtained by a linear superposition (we include the time dependence in $Q_j$ and $P_j$)

$$q_n = \sum_j e^{i\frac{2\pi}{N}jn} \frac{Q_j}{\sqrt{N}}, \tag{2.13}$$

$$p_n = \sum_j e^{-i\frac{2\pi}{N}jn} \frac{P_j}{\sqrt{N}}. \tag{2.14}$$

From the reality of $q_n$ and $p_n$ we get

$$Q_j^\star = Q_{-j}, \qquad P_j^\star = P_{-j}. \tag{2.15}$$

In the following we will make use of the following relation (see Exercise (1) in this Chapter)

$$\sum_{n=1}^{N} e^{i\frac{2\pi}{N}(j'-j)n} = N\delta_{jj'}. \tag{2.16}$$

By using this equation we can invert the previous expansions

$$\sum_{n=1}^{N} q_n e^{-i\frac{2\pi}{N}j'n} = \sum_j \sum_n e^{i\frac{2\pi}{N}(j-j')n} \frac{Q_j}{\sqrt{N}} = \sqrt{N}Q_{j'}, \tag{2.17}$$

obtaining

$$Q_j = \frac{1}{\sqrt{N}} \sum_{n=1}^{N} q_n e^{-i\frac{2\pi}{N}jn}, \tag{2.18}$$

$$P_j = \frac{1}{\sqrt{N}} \sum_{n=1}^{N} p_n e^{i\frac{2\pi}{N}jn}. \tag{2.19}$$

Notice that (see eq. (2.3))

$$P_j = \dot{Q}_{-j} = \dot{Q}_j^\dagger. \tag{2.20}$$

Substituting inside the Hamiltonian we find

$$H = \sum_{j=1}^{N/2} \left( |P_j|^2 + \omega_j^2 |Q_j|^2 \right). \tag{2.21}$$

This is nothing but the Hamiltonian of $N$ decoupled harmonic oscillators each having a frequency $\omega_j$, as it can be seen by putting

$$P_j = \frac{1}{\sqrt{2}}(X_j + iY_j), \quad Q_j = \frac{1}{\sqrt{2}}(Z_j + iT_j). \tag{2.22}$$

The result we have obtained so far shows that the string of $N$ atoms is equivalent to $N$ decoupled harmonic oscillators. The oscillator modes have been obtained through the expansion of the displacements from the equilibrium in normal modes. We are now in the position to introduce the concept of displacement field. Let us define a function of the equilibrium position of the atoms

$$x_n = na, \qquad L = Na, \qquad (2.23)$$

as the displacement of the $n^{th}$ atom from its equilibrium position

$$u(x_n, t) = q_n(t). \qquad (2.24)$$

The field $u(x_n, t)$ satisfies the following equation of motion

$$\ddot{u}(x_n, t) = \omega^2 \left[ u(x_{n+1}, t) + u(x_{n-1}, t) - 2u(x_n, t) \right]$$
$$= \omega^2 \left[ (u(x_{n+1}, t) - u(x_n, t)) - (u(x_n, t) - u(x_{n-1}, t)) \right]. (2.25)$$

Let us now consider the continuum limit of this system. Physically this is equivalent to say that we are looking at the system at a scale much bigger than the interatomic distance. We will define the limit by taking $a \to 0$ and keeping fixed the length of the string, that is to say

$$a \to 0, \qquad N \to \infty, \qquad aN = L \quad \text{fixed.} \qquad (2.26)$$

In the limit, $u(x_n, t)$ defines a function of the variable $x$ varying in the interval $(0, L)$. Furthermore

$$\frac{u(x_n, t) - u(x_{n-1}, t)}{a} \to u'(x, t) \qquad (2.27)$$

and

$$(u(x_{n+1}, t) - u(x_n, t)) - (u(x_n, t) - u(x_{n-1}, t)) \to a(u'(x_{n+1}, t) - u(x_n, t))$$
$$\to a^2 u''(x, t). \qquad (2.28)$$

The equation of motion becomes

$$\ddot{u}(x, t) = a^2 \omega^2 u''(x, t). \qquad (2.29)$$

In order to give a sense to this equation we let $\omega$ to diverge in such a way that

$$\lim_{a \to 0} a\omega = v, \quad v \text{ finite,} \qquad (2.30)$$

where $v$ has the dimension of a velocity. What we get is the equation for the string as the propagation of waves with velocity $v$

$$\ddot{u}(x, t) = v^2 u''(x, t). \qquad (2.31)$$

In the limit we also have

$$\sum_{n=1}^{N+1} a \to \int_0^L dx, \tag{2.32}$$

from which

$$H = \frac{1}{2a} \int_0^L dx \left[ \left( \dot{u}(x,t) \right)^2 + v^2 \left( u'(x,t) \right)^2 \right]. \tag{2.33}$$

To get finite energy we need also a redefinition of the field variable

$$u(x,t) = \sqrt{a}\phi(x,t), \tag{2.34}$$

from which

$$H = \frac{1}{2} \int_0^L dx \left[ \left( \dot{\phi}(x,t) \right)^2 + v^2 \left( \phi'(x,t) \right)^2 \right]. \tag{2.35}$$

The normal modes decomposition for the field $\phi(x,t)$ becomes

$$\phi(x,t) = \frac{u(x,t)}{\sqrt{a}} \approx \frac{q_n(t)}{\sqrt{a}} \approx \sum_j e^{ik_j an} \frac{Q_j}{\sqrt{aN}}, \tag{2.36}$$

$$k_j = \frac{2\pi}{L} j, \qquad -\infty < j < +\infty, \tag{2.37}$$

that is

$$\phi(x,t) = \frac{1}{\sqrt{L}} \sum_{j=-\infty}^{+\infty} e^{i\frac{2\pi}{L}jx} Q_j(t). \tag{2.38}$$

The eigenfrequencies are given by

$$\omega_j^2 = 4\omega^2 \sin^2\left(\frac{\pi a}{L}j\right) \to 4\omega^2 \left(\frac{\pi j}{L}\right)^2 a^2 = (a\omega k_j)^2 \to v^2 k_j^2. \tag{2.39}$$

In the continuum limit the frequency is proportional to the wave vector $k_j$. The relation between the normal modes $Q_j(t)$ and the field $\phi(x,t)$ can be inverted by using the following relation

$$\int_0^L dx \, e^{ix(k-k')} = L\delta_{k,k'}, \tag{2.40}$$

which holds for $k$ and $k'$ of the form (2.37). The Hamiltonian is easily obtained as

$$H = \sum_{j=1}^{\infty} \left( |\dot{Q}_j|^2 + v^2 k_j^2 |Q_j|^2 \right). \tag{2.41}$$

The main result here is that in the continuum limit the Hamiltonian of the system describes an infinite set of decoupled harmonic oscillators. In the following we will show that the quantization of field theories of the type described in this Section gives rise, naturally, to a description in terms of particles.

## 2.3   String quantization

We have shown that a string of $N$ atoms can be described in terms of a set of decoupled harmonic oscillators, and this property holds true also in the continuum limit $(N \to \infty)$. In the discrete case we have shown that the Hamiltonian of the system can be written as

$$H = \sum_{j=1}^{N/2} \left( |P_j|^2 + \omega_j^2 |Q_j|^2 \right),  \tag{2.42}$$

where

$$P_j = \dot{Q}_j^\dagger, \quad Q_j^\dagger = Q_{-j}, \quad P_j^\dagger = P_{-j},  \tag{2.43}$$

$$\omega_j^2 = 4\omega^2 \sin^2 \frac{k_j a}{2}, \quad k_j = \frac{2\pi}{L} j, \quad |j| = 1, 2, \ldots, \frac{N}{2},  \tag{2.44}$$

whereas in the continuum limit

$$H = \sum_{j=1}^{\infty} \left( |P_j|^2 + \omega_j^2 |Q_j|^2 \right),  \tag{2.45}$$

with

$$\omega_j = v|k_j|, \quad |j| = 1, 2, \ldots, \infty  \tag{2.46}$$

and $k_j$ given by eq. (2.44). In both cases the quantization is trivially done through the canonical commutation relations

$$[Q_j, P_k] = i\delta_{jk}, \quad [Q_j^\dagger, P_k^\dagger] = i\delta_{jk}.  \tag{2.47}$$

Having to do with harmonic oscillators it is convenient to introduce annihilation and creation operators

$$a_j = \sqrt{\frac{\omega_j}{2}} Q_j + i \frac{1}{\sqrt{2\omega_j}} P_j^\dagger, \quad a_j^\dagger = \sqrt{\frac{\omega_j}{2}} Q_j^\dagger - i \frac{1}{\sqrt{2\omega_j}} P_j,  \tag{2.48}$$

with $j$ assuming a finite or an infinite number of values according to the system being discrete or continuum. In both cases we have

$$[a_j, a_k^\dagger] = \delta_{jk}  \tag{2.49}$$

and

$$[a_j, a_k] = [a_j^\dagger, a_k^\dagger] = 0.  \tag{2.50}$$

Notice that

$$a_{-j} = \sqrt{\frac{\omega_j}{2}} Q_j^\dagger + i \frac{1}{\sqrt{2\omega_j}} P_j \neq a_j^\dagger  \tag{2.51}$$

and

$$a^\dagger_{-j} = \sqrt{\frac{\omega_j}{2}} Q_j - i \frac{1}{\sqrt{2\omega_j}} P^\dagger_j \neq a_j. \tag{2.52}$$

Therefore $a_j$ and $a^\dagger_j$ are $2N$ (in the discrete case) independent operators as $Q_j$ and $P_j$. The previous relations can be inverted to give

$$Q_j = \frac{1}{\sqrt{2\omega_j}} (a_j + a^\dagger_{-j}), \qquad P_j = -i\sqrt{\frac{\omega_j}{2}} (a_{-j} - a^\dagger_j). \tag{2.53}$$

The Hamiltonian expressed in terms of $a_j$ and $a^\dagger_j$ is written as

$$H = \sum_{j=1}^{N/2} \omega_j [a^\dagger_j a_j + a^\dagger_{-j} a_{-j} + 1] = \sum_{j=-N/2}^{N/2} \omega_j \left[ a^\dagger_j a_j + \frac{1}{2} \right]. \tag{2.54}$$

The ground state is characterized by the equation

$$a_j |0\rangle = 0, \tag{2.55}$$

with a corresponding energy

$$E_0 = \sum_j \frac{\omega_j}{2}. \tag{2.56}$$

In the continuum limit the energy of the fundamental state is infinite (we will come back to this point later on). The generic energy eigenstate is obtained by applying creation operators to the ground state

$$|n_{-N/2}, \cdots, n_{N/2}\rangle = \frac{1}{(n_{-N/2}! \cdots n_{N/2}!)^{1/2}} (a^\dagger_{-N/2})^{n_{-N/2}} \cdots (a^\dagger_{N/2})^{n_{N/2}} |0\rangle. \tag{2.57}$$

The Hilbert space spanned by these vectors is called a <u>Fock</u> space. The state given above can be thought of being formed by $n_{-N/2}$ quanta with energy $\omega_{-N/2}$, up to $n_{N/2}$ quanta with energy $\omega_{N/2}$. In this interpretation the $n_j$ quanta (or particles) of energy $\omega_j$ are indistinguishable one from the other. Furthermore, in a given state, we can put as many particles as we want. Therefore the particles we are describing here satisfy the Bose-Einstein statistics. Formally this follows from the commutation relation

$$[a^\dagger_i, a^\dagger_j] = 0, \tag{2.58}$$

implying the symmetry of the states with respect to the exchange of two quanta. For instance a two-particle state is given by

$$|i, j\rangle = a^\dagger_i a^\dagger_j |0\rangle = a^\dagger_j a^\dagger_i |0\rangle = |j, i\rangle. \tag{2.59}$$

As we have already noticed the energy of the fundamental state becomes infinite in the continuum limit. This is perhaps the most simple of the infinities that we will encounter in our study of field quantization. We will learn much later in this course about the possibility of keeping them under control. At this moment let us notice that normally only relative energies are important and the value of $E_0$ (see eq. (2.56)) is not physically relevant. However, there are situations, as in the Casimir effect (see later), where it is indeed relevant. Forgetting momentarily about these special situations we can define a new Hamiltonian by subtracting $E_0$. This can be done in a rather formal way by defining the concept of **normal ordering**. Given an operator which is a monomial in the creation and annihilation operators, we define its normal ordered form by pushing all the annihilation operators to the right of the creation operators. We then extend the definition to polynomials by linearity. For instance, in the case of the Hamiltonian (2.45) we have

$$: H : \ \equiv N(H) = N\left(\sum_j \frac{\omega_j}{2}\left(a_j^\dagger a_j + a_j a_j^\dagger\right)\right) = \sum_j \omega_j a_j^\dagger a_j. \qquad (2.60)$$

Coming back to the discrete case, recalling eqs. (2.13) and (2.14) and using the canonical commutators (2.47), we get

$$[q_n, p_m] = \sum_{jk} e^{i\frac{2\pi}{N}(jn-km)} \frac{1}{N}[Q_j, P_k] = \frac{i}{N}\sum_j e^{i\frac{2\pi}{N}j(n-m)} = i\delta_{nm}. \qquad (2.61)$$

In the continuum case we use the expansion (2.38).

$$\phi(x,t) = \frac{1}{\sqrt{L}}\sum_{j=-\infty}^{+\infty} e^{i\frac{2\pi}{L}jx}Q_j, \quad \dot\phi(x,t) = \frac{1}{\sqrt{L}}\sum_{j=-\infty}^{+\infty} e^{-i\frac{2\pi}{L}jx}P_j, \qquad (2.62)$$

from which

$$[\phi(x,t), \dot\phi(y,t)] = \frac{1}{L}\sum_{jk} e^{i\frac{2\pi}{L}(jx-ky)}i\delta_{jk}$$

$$= \frac{i}{L}\sum_{j=-\infty}^{+\infty} e^{i\frac{2\pi}{L}j(x-y)} = i\delta(x-y). \qquad (2.63)$$

This relation could have been obtained starting from the discrete case and going to the continuum limit by recalling that

$$\phi(x,t) \approx \frac{u(x,t)}{\sqrt{a}} \approx \frac{q_n}{\sqrt{a}}, \qquad (2.64)$$

implying

$$[\phi(x_n, y), \dot{\phi}(x_m, t)] = i\frac{\delta_{nm}}{a}. \tag{2.65}$$

In the continuum limit

$$\lim_{a \to 0} \frac{\delta_{nm}}{a} = \delta(x - y), \tag{2.66}$$

if for $a \to 0$, $x_n \to x$, and $x_m \to y$. In fact,

$$1 = \sum_n a \left( \frac{\delta_{nm}}{a} \right) \to \int dx \left( \lim_{a \to 0} \frac{\delta_{nm}}{a} \right), \tag{2.67}$$

showing that the properties of a delta-approximation are indeed satisfied. It follows that we have the following correspondence between the discrete and the continuum case

$$q_n \to \phi(x, t), \qquad p_n \to \dot{\phi}(x, t). \tag{2.68}$$

In other words, $\phi(x, t)$ and $\dot{\phi}(x, t)$ appear to be the analog of a pair of canonical variables. This remark suggests a way to approach the quantization of a field theory different from the one followed so far. The way we have used is based upon the construction of the normal modes of the oscillators, but in the discrete case this is not necessary at all. In fact, the quantization can be made starting from the commutation relations among the canonical variables, $[q, p] = i$, without having an *a priori* knowledge of the dynamics of the system. This suggests that one could start directly by the field operators $\phi(x, t)$, $\dot{\phi}(x, t)$, and quantize the theory by requiring $[\phi(x, t), \dot{\phi}(y, t)] = i\delta(x - y)$. To make this approach a consistent one, we need to extend the Hamiltonian and Lagrangian descriptions to a continuum system. Let us recall how we proceed in the discrete case. We start by giving a Lagrangian function $L(q_n, \dot{q}_n, t)$. Then we define the conjugated momenta by the equation

$$p_n = \frac{\partial L}{\partial \dot{q}_n}. \tag{2.69}$$

We then go to the Hamiltonian formalism by taking the conjugated momenta, $p_n$, as independent variables. The previous equation is used in order to solve the velocities in terms of $q_n$ and $p_n$. Next we define the Hamiltonian as

$$H(q_n, p_n) = \sum_n p_n \dot{q}_n - L. \tag{2.70}$$

At the classical level the time evolution of the observables is obtained through the equation

$$\dot{A} = \{A, H\}, \qquad (2.71)$$

where the Poisson brackets can be defined starting form the brackets between the canonical variables $\{q_n, p_m\} = \delta_{nm}$. The theory is then quantized through the "correspondence" rule

$$\{.,.\} \to -i[.,.]. \qquad (2.72)$$

with $\{.,.\}$ the classical Poisson bracket and $[.,.]$ the commutator. In the next Section we will learn how to extend the Lagrangian and Hamiltonian formalisms to the continuum case.

## 2.4 The canonical formalism for continuum systems

We can evaluate the Lagrangian for the string using the kinetic and the potential energy. Let us start with the discrete case. The kinetic energy is given by

$$T = \frac{1}{2} \sum_{n=1}^{N} p_n^2 = \frac{1}{2} \sum_{n=1}^{N} \dot{u}^2(x_n, t) = \frac{1}{2} \sum_{n=1}^{N} a\dot{\phi}^2(x_n, t) \to \frac{1}{2} \int_0^L \dot{\phi}^2(x, t) dx \qquad (2.73)$$

and the potential energy by

$$U = \frac{1}{2} \sum_{n=1}^{N} \omega^2 (q_n - q_{n+1})^2 = \frac{1}{2} \sum_{n=1}^{N} \omega^2 a(\phi(x_n, t) - \phi(x_{n+1}, t))^2 \qquad (2.74)$$

and using that for $a \to 0$, $v = a\omega$, is finite, it follows

$$U = \frac{1}{2} \sum_{n=1}^{N} av^2 \left( \frac{\phi(x_n, t) - \phi(x_{n+1}, t)}{a} \right)^2 \to \frac{v^2}{2} \int_0^L \phi'^2(x, t) dx. \qquad (2.75)$$

Therefore the total energy and the Lagrangian are

$$E = T + U = \frac{1}{2} \int_0^L dx \left[ \dot{\phi}^2(x, t) + v^2 \phi'^2(x, t) \right] \qquad (2.76)$$

and

$$L = T - U = \frac{1}{2} \int_0^L dx \left[ \dot{\phi}^2(x, t) - v^2 \phi'^2(x, t) \right], \qquad (2.77)$$

respectively. The important result is that in the continuum limit, the Lagrangian can be written as a spatial integral of a function of the field $\phi$ and of its first derivative, $\dot{\phi}$, called Lagrangian density

$$\mathcal{L} = \frac{1}{2}\left(\dot{\phi}^2 - v^2\phi'^2\right) \tag{2.78}$$

with the total Lagrangian given by

$$L = \int_0^L \mathcal{L}\,dx. \tag{2.79}$$

Of course, this is not the most general situation that can be imagined, but we will consider only the case in which the Lagrangian density depends only on the field and on its first derivatives

$$L = \int \mathcal{L}(\phi, \dot{\phi}, \phi', x, t)dx. \tag{2.80}$$

We will restrict ourselves to theories in which the Lagrangian contains at most first order time derivatives of the fields, since theories with higher order derivatives have potential problems with probability conservation.

Given the Lagrangian, the next step is to build up the action functional. The extrema of the action give rise to the equations of motion. The action is given by

$$S = \int_{t_1}^{t_2} L\,dt = \int_{t_1}^{t_2} dt \int dx \mathcal{L}(\phi, \dot{\phi}, \phi', x, t). \tag{2.81}$$

We require $S$ to be stationary with respect to those variations that are consistent with the boundary conditions satisfied by the fields. If $\Sigma$ is the spatial surface delimiting the region of spatial integration (for the string it reduces to the end points), we will require

$$\delta\phi(x, t) = 0 \quad \text{on } \Sigma. \tag{2.82}$$

Furthermore we will require the variations at the times $t_1$ and $t_2$ to be zero for any space point $x$

$$\delta\phi(x, t_1) = \delta\phi(x, t_2) = 0, \quad \text{for any } x. \tag{2.83}$$

In the discrete case we have only boundary conditions of the second type, but here the first ones are necessary in order to be consistent with the boundary conditions chosen for the field. Let us now require $S$ to be stationary against variations satisfying the boundary conditions (2.82) and (2.83)

$$0 = \delta S = \int_{t_1}^{t_2} dt \int dx \delta\mathcal{L} = \int_{t_1}^{t_2} dt \int dx \left(\frac{\partial \mathcal{L}}{\partial \phi}\delta\phi + \frac{\partial \mathcal{L}}{\partial \dot{\phi}}\delta\dot{\phi} + \frac{\partial \mathcal{L}}{\partial \phi'}\delta\phi'\right). \tag{2.84}$$

Integrating by parts

$$0 = \int \, dx \left[ \frac{\partial \mathcal{L}}{\partial \dot{\phi}} \delta\phi \right]_{t_1}^{t_2} + \int_{t_1}^{t_2} dt \left[ \frac{\partial \mathcal{L}}{\partial \phi'} \delta\phi \right]_0^L$$
$$+ \int_{t_1}^{t_2} dt \int \, dx \left[ \frac{\partial \mathcal{L}}{\partial \phi} - \frac{\partial}{\partial t} \frac{\partial \mathcal{L}}{\partial \dot{\phi}} - \frac{\partial}{\partial x} \frac{\partial \mathcal{L}}{\partial \phi'} \right] \delta\phi. \qquad (2.85)$$

The boundary terms are zero due to eqs. (2.82) and (2.83). Then from the arbitrariness of $\delta\phi$ within the region of integration, we get the Euler-Lagrange equations

$$\frac{\partial \mathcal{L}}{\partial \phi} - \frac{\partial}{\partial t} \frac{\partial \mathcal{L}}{\partial \dot{\phi}} - \frac{\partial}{\partial x} \frac{\partial \mathcal{L}}{\partial \phi'} = 0. \qquad (2.86)$$

In fact $\delta\phi$ can be chosen to be zero everywhere except for a small region around any given point $x$ (see Fig. 2.3).

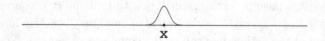

**Fig. 2.3** Here the arbitrary variation $\delta\phi(x)$ is chosen to be zero all along the string, except for a small region around the point $x$.

This discussion can be easily extended to the case of $N$ fields $\phi_i$, $i = 1, \ldots, N$ (think, as an example, to the electromagnetic field), and to the case of $n$ spatial dimensions with points labelled by $x_\alpha$, $\alpha = 1, \ldots, n$. In this case the structure of the action will be

$$S = \int_{t_1}^{t_2} dt \int_V d^n x \mathcal{L} \left( \phi_i, \dot{\phi}_i, \frac{\partial \phi_i}{\partial x_\alpha} \right). \qquad (2.87)$$

Here $V$ is the integration space volume. We will require again the stationarity of the action with respect to variations of the fields satisfying the boundary conditions

$$\delta\phi_i(x_\alpha, t) = 0, \qquad \text{on } \Sigma, \text{ for any } t, \ \ t_1 \le t \le t_2, \qquad (2.88)$$

where $\Sigma$ is the boundary of $V$, and

$$\delta\phi_i(x_\alpha, t_1) = \delta\phi_i(x_\alpha, t_2) = 0, \qquad \text{for any } x_\alpha \in V. \qquad (2.89)$$

As before, the first boundary conditions are required in order to be consistent with the boundary conditions for the fields as, for instance, in the case of spatial regions extended up to infinity, where the fields are required to go to zero. The Euler-Lagrange equations one gets in this case are

$$\frac{\partial \mathcal{L}}{\partial \phi_i} - \frac{\partial}{\partial t}\frac{\partial \mathcal{L}}{\partial \dot{\phi}_i} - \frac{\partial}{\partial x_\alpha}\frac{\partial \mathcal{L}}{\partial \phi_i^{;\alpha}} = 0, \quad i = 1, \ldots, N, \quad \alpha = 1, \ldots, n, \quad (2.90)$$

where

$$\phi_i^{;\alpha} \equiv \frac{\partial \phi_i}{\partial x_\alpha}. \tag{2.91}$$

To go to the Hamiltonian description, one introduces the momentum densities conjugated to the fields $\phi_i$:

$$\Pi_i = \frac{\partial \mathcal{L}}{\partial \dot{\phi}_i} \tag{2.92}$$

and the Hamiltonian density

$$\mathcal{H} = \sum_i \Pi_i \dot{\phi}_i - \mathcal{L}. \tag{2.93}$$

In the case of the string, one gets from eq. (2.77)

$$\frac{\partial \mathcal{L}}{\partial \dot{\phi}} = \dot{\phi}, \quad \frac{\partial \mathcal{L}}{\partial \phi'} = -v^2 \phi', \quad \frac{\partial \mathcal{L}}{\partial \phi} = 0. \tag{2.94}$$

From which one recovers the equations of motion for the field $\phi$. Furthermore

$$\Pi = \dot{\phi}, \tag{2.95}$$

implying

$$\mathcal{H} = \Pi\dot{\phi} - \mathcal{L} = \Pi^2 - \left(\frac{1}{2}\Pi^2 - \frac{1}{2}v^2\phi'^2\right) = \frac{1}{2}\left(\Pi^2 + v^2\phi'^2\right), \tag{2.96}$$

which coincides with the energy density given in eq. (2.76), after using eq. (2.95).

The greatest advantage of using the Lagrangian formalism instead of the Hamiltonian one is the possibility to formulate in a simple way the symmetry properties of the theory. We shall see that this is due to the first theorem of Emmy Noether [Noether (1918)] which makes possible to correlate in a simple way the symmetry properties of the Lagrangian with the existence of conservation laws[1]. Due to this correspondence it is possible to use the Noether theorem in a constructive way. That is, to restrict the form of the Lagrangian requiring a particular type of symmetry. We will

---

[1] The first Noether's theorem deals with global symmetries, whereas the second one has to do with local symmetries which will be discussed in Chapter 6.

discuss the theorem later on. Here we will show how the equations of motion of the string generate the conservation laws.

The energy contained in the segment $[a, b]$ of the string, with $0 \le a \le b \le L$ is given by

$$E(a,b) = \frac{1}{2} \int_a^b dx \left[ \dot{\phi}^2 + v^2 \phi'^2 \right]. \qquad (2.97)$$

We can evaluate its time variation

$$\frac{dE(a,b)}{dt} = \int_a^b dx \left[ \dot{\phi}\ddot{\phi} + v^2 \phi' \dot{\phi}' \right] = v^2 \int_a^b dx \left[ \dot{\phi}\phi'' + \phi'\dot{\phi}' \right]$$

$$= v^2 \int_a^b dx \frac{\partial}{\partial x} \left[ \dot{\phi}\phi' \right] = v^2 \left[ \dot{\phi}\phi' \right]_a^b, \qquad (2.98)$$

where we have made use of the equations of motion in the second step. Defining the local quantity

$$\mathcal{P}(x,t) = -v^2 \dot{\phi}\phi', \qquad (2.99)$$

which is the analogous of the Poynting's vector in electrodynamics, we get

$$-\frac{dE(a,b)}{dt} = [\mathcal{P}(b,t) - \mathcal{P}(a,t)]. \qquad (2.100)$$

This is the classical energy conservation law, expressing the fact that if the energy decreases in the segment $[a, b]$, then there must be a flux of energy at the end points $a$ and $b$. The total energy is conserved due to the boundary conditions, $P(0,t) = P(L,t)$. But the previous law says something more, because it gives us a local conservation law, as it follows by taking the limit $b \to a$. In fact, in this limit

$$E(a,b) \to (b-a)\mathcal{H}, \qquad (2.101)$$

with $\mathcal{H}$ given by (2.96), and

$$\frac{\partial \mathcal{H}}{\partial t} + \frac{\partial \mathcal{P}}{\partial x} = 0. \qquad (2.102)$$

This conservation law can be checked by using the explicit expressions of $\mathcal{H}$ and $P$, and the equations of motion.

## 2.5   The canonical quantization of a continuum system

As we have seen in Section 2.4, in field theory the momentum densities are defined as

$$\Pi_i = \frac{\partial \mathcal{L}}{\partial \dot{\phi}_i}. \qquad (2.103)$$

It is then natural to assume the following commutation relations

$$[\phi_i(x_\alpha, t), \Pi_j(y_\alpha, t)] = i\delta_{ij}\delta^n(x_\alpha - y_\alpha) \quad \alpha = 1, \ldots, n, \quad i, j = 1, \ldots, N. \tag{2.104}$$

and

$$[\phi_i(x_\alpha, t), \phi_j(y_\alpha, t)] = 0, \qquad [\Pi_i(x_\alpha, t), \Pi_j(y_\alpha, t)] = 0. \tag{2.105}$$

In the string case we have $\Pi = \dot{\phi}$ and we reproduce eq. (2.63). Starting from the previous commutation relations and expanding the fields in terms of normal modes one gets back the commutation relations for the creation and annihilation operators. Therefore we reconstruct the particle interpretation. Using the Heisenberg representation (but omitting from now on the corresponding index for the operators), the expansion of the string field in terms of creation and annihilation operators is obtained using eqs. (2.62) and (2.53).

$$\phi(x, t) = \frac{1}{\sqrt{L}} \sum_j e^{i\frac{2\pi}{L}jx} Q_j$$

$$= \frac{1}{\sqrt{L}} \sum_j \frac{1}{\sqrt{2\omega_j}} \left[ e^{i\frac{2\pi}{L}jx} a_j(t) + e^{-i\frac{2\pi}{L}jx} a_j^\dagger(t) \right]. \tag{2.106}$$

From the equations of motion of the string

$$\ddot{\phi} - v^2\phi'' = 0, \tag{2.107}$$

we find the equations of motion for $Q_j$

$$\ddot{Q}_j + \omega_j^2 Q_j = 0. \tag{2.108}$$

These equations are consistent with the Hamilton equations:

$$\dot{Q}_j = P_j^\dagger, \quad \dot{P}_j = -\omega^2 Q_j^\dagger. \tag{2.109}$$

Then, using the decomposition (2.48) of $a_j$ in terms of $Q_j$ and $P_j = \dot{Q}_j^\dagger$, we get easily

$$\dot{a}_j + i\omega_j a_j = 0, \tag{2.110}$$

from which

$$a_j(t) = a_j(0)e^{-i\omega_j t} \equiv a_j e^{-i\omega_j t}, \quad a_j^\dagger(t) = a_j^\dagger(0)e^{i\omega_j t} \equiv a_j^\dagger e^{i\omega_j t} \tag{2.111}$$

and

$$\phi(x, t) = \frac{1}{\sqrt{L}} \sum_j \frac{1}{\sqrt{2\omega_j}} \left[ e^{i\left(\frac{2\pi}{L}jx - \omega_j t\right)} a_j + e^{-i\left(\frac{2\pi}{L}jx - \omega_j t\right)} a_j^\dagger \right]. \tag{2.112}$$

Let us investigate the structure of this expansion. It can be written in the following way

$$\phi(x,t) = \sum_j \left[ f_j(x,t)a_j + f_j^*(x,t)a_j^\dagger \right], \qquad (2.113)$$

with

$$f_j(x,t) = \frac{1}{\sqrt{2\omega_j L}} e^{i\left(\frac{2\pi}{L}jx - i\omega_j t\right)} = \frac{1}{\sqrt{2\omega_j L}} e^{i(k_j x - i\omega_j t)}, \qquad (2.114)$$

where we have made use of the definition of $k_j$ (see eq. (2.37)). The functions $f_j(x,t)$ and their complex conjugated satisfy the wave equation

$$\frac{\partial^2 f_j(x,t)}{\partial t^2} - v^2 \frac{\partial^2 f_j(x,t)}{\partial x^2} = 0 \qquad (2.115)$$

with periodic boundary conditions

$$f_j(0,t) = f_j(L,t). \qquad (2.116)$$

It is immediate to verify that they are a complete set of orthonormal functions

$$\sum_j f_j^*(x,t) i\partial_t^{(-)} f_j(y,t) = \delta(x-y), \qquad (2.117)$$

$$\int_0^L dx f_j^*(x,t) i\partial_t^{(-)} f_l(x,t) = \delta_{jl}, \qquad (2.118)$$

where

$$A\partial_t^{(-)}B = A(\partial_t B) - (\partial_t A)B. \qquad (2.119)$$

Let us consider the first relation. We have

$$\sum_j f_j^*(x,t) i\partial_t^{(-)} f_j(y,t) = \sum_j 2\omega_j f_j^*(x,t) f_j(y,t)$$

$$= \sum_j \frac{1}{L} e^{-ik_j(x-y)} = \delta(x-y). \qquad (2.120)$$

Evaluating this expression with two $f_j(x,t)$'s or two $f_j^*(x,t)$'s one gets zero. As far as the second relation is concerned we get

$$\int_0^L dx f_j^*(x,t) i\partial_t^{(-)} f_l(x,t) = \int_0^L dx[\omega_l + \omega_j] f_j^*(x,t) f_l(x,t)$$

$$= \frac{1}{L} \frac{\omega_l + \omega_j}{2\sqrt{\omega_j \omega_l}} \int_0^L dx e^{ix(k_l - k_j)} e^{i(\omega_j - \omega_l)t}$$

$$= \frac{1}{L} \frac{\omega_l + \omega_j}{2\sqrt{\omega_j \omega_l}} e^{i(\omega_j - \omega_l)t} L\delta_{jl} = \delta_{jl}. \qquad (2.121)$$

Also in this case, by taking two $f_j(x,t)$'s or two $f_j^*(x,t)$'s, the result is zero due to the appearance of the factor $(\omega_l - \omega_j)$.

An interesting question is why the operator $\partial_t^{(-)}$ appears in these relations. The reason is that the scalar product should be time independent, otherwise the statement of orthonormality of the solutions would be time dependent. For instance, in the case of the Schrödinger equation, we define the scalar product as

$$\int d^3x \, \psi^*(\vec{x},t)\psi(\vec{x},t), \tag{2.122}$$

because for hermitian Hamiltonians this is indeed time independent, as it can be checked by differentiating the scalar product with respect to time. In the present case, we can define a time independent scalar product, by considering two solutions $f$ and $\tilde{f}$ of the wave equation, and evaluating the following two expressions

$$\int_0^L dx \tilde{f}\left[\frac{\partial^2 f}{\partial t^2} - v^2\frac{\partial^2 f}{\partial x^2}\right] = 0, \tag{2.123}$$

$$\int_0^L dx\left[\frac{\partial^2 \tilde{f}}{\partial t^2} - v^2\frac{\partial^2 \tilde{f}}{\partial x^2}\right] f = 0. \tag{2.124}$$

Subtracting these two expressions one from the other we get

$$\int_0^L dx\left[\frac{\partial}{\partial t}\left(\tilde{f}\frac{\partial f}{\partial t} - \frac{\partial \tilde{f}}{\partial t}f\right) - v^2\frac{\partial}{\partial x}\left(\tilde{f}\frac{\partial f}{\partial x} - \frac{\partial \tilde{f}}{\partial x}f\right)\right] = 0. \tag{2.125}$$

If both $f$ and $\tilde{f}$ satisfy periodic boundary conditions, the second term is zero, and it follows that the quantity

$$\int_0^L dx\tilde{f}\partial_t^{(-)}f \tag{2.126}$$

is a constant of motion. Using eq. (2.118), we can invert the relation between the field, and the creation and annihilation operators. We get

$$a_j = \int_0^L dxf_j^*(x,t)i\partial_t^{(-)}\phi(x,t) \tag{2.127}$$

and

$$a_j^\dagger = \int_0^L dx\phi(x,t)i\partial_t^{(-)}f_j(x,t). \tag{2.128}$$

From the field commutation relations we find

$$[a_j, a_k^\dagger] = \delta_{jk}. \tag{2.129}$$

In analogous way we get

$$[a_j, a_k] = [a_j^\dagger, a_k^\dagger] = 0. \tag{2.130}$$

We have seen that the total energy of the string is a constant of motion. There is another constant of motion corresponding to the total momentum of the string

$$P = \int_0^L dx\, \mathcal{P} = -\int_0^L dx\, \dot\phi\phi', \tag{2.131}$$

where $\mathcal{P}$ has been defined in eq. (2.99). We will show in the following that the conservation of the total momentum of the string is derived from the invariance of the theory under spatial translations. Here let us check that this is in fact a conserved quantity:

$$\frac{dP}{dt} = -\int_0^L dx(\ddot\phi\phi' + \dot\phi\dot\phi') = -\int_0^L dx\, \frac{1}{2}\frac{\partial}{\partial x}(\dot\phi^2 + v^2\phi'^2) = 0, \tag{2.132}$$

where we have used the equations of motion of the string and the boundary conditions. By using the field expansion, we find

$$P = -\sum_l \frac{1}{2}k_l[a_{-l}a_l e^{-2i\omega_l t} + a_{-l}^\dagger a_l^\dagger e^{2i\omega_l t} - a_l a_l^\dagger - a_l^\dagger a_l]. \tag{2.133}$$

The first two terms give zero contribution because they are antisymmetric in the summation index ($k_l \approx l$). Therefore

$$P = \frac{1}{2}\sum_j k_j[a_j a_j^\dagger + a_j^\dagger a_j] = \sum_j k_j a_j^\dagger a_j, \tag{2.134}$$

where we have used

$$\sum_j k_j = 0, \tag{2.135}$$

since $k_j \to -k_j$ when $j \to -j$. We see that $P$ has an expression similar to that of $H$ (see eq. (2.60)). Therefore the states

$$(a_{-N/2}^\dagger)^{n_{-N/2}} \cdots (a_j^\dagger)^{n_j} \cdots |0\rangle = |\psi\rangle \tag{2.136}$$

have energy

$$H|\psi\rangle = (n_{-N/2}\omega_{-N/2} + \cdots + n_j\omega_j + \cdots)|\psi\rangle \tag{2.137}$$

and a momentum

$$P|\psi\rangle = (n_{-N/2}k_{-N/2} + \cdots + n_j k_j + \cdots)|\psi\rangle, \tag{2.138}$$

as it follows from

$$[H, a_j^\dagger] = \omega_j a_j^\dagger, \qquad [P, a_j^\dagger] = k_j a_j^\dagger. \tag{2.139}$$

## 2.6 Exercises

(1) Prove Eq. (2.16). (Hint: add and subtract 1 to the equation.)
(2) Evaluate the expressions (2.127), (2.128), for the annihilation and creation operators and their commutation relations given in eqs. (2.129) and (2.130).
(3) Derive eq. (2.133) for the momentum operator.
(4) Show that the addition of a divergence to the Lagrangian density does not change the equations of motion.
(5) Given an operator $A$ of the form

$$A = \sum_j \alpha_j a_j^\dagger a_j, \tag{2.140}$$

with

$$[a_j, a_k] = [a_j^\dagger, a_k^\dagger] = 0, \qquad [a_j, a_k^\dagger] = \delta_{ik}, \tag{2.141}$$

and an eigenstate of $A$, $|\rho\rangle$, corresponding to the eigenvalue $\rho$. Show that $a_j^\dagger|\rho\rangle$ is an eigenstate of $A$ with eigenvalue $\rho + \alpha_j$ and that $a_j|\rho\rangle$ corresponds to the eigenvalue $\rho - \alpha_j$.

# Chapter 3

# The Klein-Gordon field

## 3.1 The problems in relativistic quantum mechanics

The extension of quantum mechanics to the relativistic case gives rise to several problems. The difficulties originate from the relativistic dispersion relation

$$E^2 = |\vec{p}|^2 + m^2. \tag{3.1}$$

This relation gives rise to two solutions

$$E = \pm\sqrt{|\vec{p}|^2 + m^2}. \tag{3.2}$$

It is not difficult to convince himself that the solutions with negative energy have unphysical behavior. For instance, increasing the momentum, the energy decreases! But these solutions do not create a real problem at the classical level. In fact, we see from eq. (3.2) that there is a gap of $2m$ between the energies corresponding to the two types of solutions. At the classical level, the energy is always transferred in a continuous fashion. So there is no way to start with a positive energy particle and end up with one of negative energy. On the contrary, in quantum mechanics, through the emission of a quantum of energy $E > 2m$, a positive energy state may lose energy up to the point of becoming a state with negative energy. Since a physical system has the tendency to decrease its energy, the positive energy states would migrate to the states of negative energy, causing a collapse of the usual matter. However, we shall see that although it is not possible to ignore these solutions, they can be interpreted in terms of antiparticles. In this way one can get rid of the problems connected with the negative energy solutions. From this new interpretation, new phenomena emerge as annihilation of particles and antiparticles and creation of particle-antiparticle pairs. These new effects have deep consequences on

39

the interpretation of the theory. Suppose that we wish to localize a particle on a distance smaller than its Compton wavelength, that is smaller than $1/m$. Correspondingly, due to the uncertainty principle, we will have an uncertainty on the momentum greater than $m$. This means that the momentum (and the energy) of the particle could well reach values of order $2m$, enough to create a particle-antiparticle pair. Of course, this is possible only if we violate the conservation of energy and momentum. But this is indeed possible since the uncertainty on both energy and momentum will be greater than $1/\Delta x \approx m$. Therefore the attempt of localization at the Compton scale has the result of pair creating particles and antiparticles, meaning that we will be unable to define the concept of a localized single particle. At the Compton scale there is no such thing as a particle, but the picture we get from the previous considerations is the one of a cloud of particles and antiparticles surrounding our initial particle, and there is no way to distinguish the original particle from the many around it.

These considerations imply that the relativistic theories cannot be seen as theories with a fixed number of particles, as in the case of ordinary quantum mechanics. In this sense a field theory looks as the most natural way to describe such systems. In fact, since the Fock space has no definite number of particles, it offers the possibility of describing the physics of a variable number of particles.

There are other justifications for using field theories. For instance, by looking at the quantization of the electromagnetic field, physicists realized that one obtains a natural explanation of the particle-wave duality, and that in the particle description one has to do with a variable number of photons. On the contrary, physical entities as the electrons, were always described in particle terms till 1924 when De Broglie, in his PhD thesis [De Broglie (1924)], made the hypothesis that particles would exhibit a wave-like character, correlating the particle momentum with the wavelength of the wave associated to it. This hypothesis was verified experimentally by [Davisson and Germer (1927)]. This suggested that the particle-wave duality would be a feature valid for any type of waves and/or particles. Therefore, based on the analogy with the electromagnetic field, it became natural to introduce a field for any kind of particle.

Historically, the attempt of making quantum mechanics a relativistic theory was pursued by looking for relativistic generalizations of the Schrödinger equation. Later it was realized that these equations should be rather used as equations for the fields describing the corresponding particles. As we shall see, these equations describe correctly the energy dispersion relation and the spin of the various particles. Therefore their solutions can be used as a basis for the expansion of the fields in terms of creation and annihilation operators. In order to illustrate this procedure, let us start considering the Schrödinger equation for a free particle

$$i\frac{\partial \psi}{\partial t} = H\psi, \tag{3.3}$$

where $H$ is the Hamiltonian operator

$$H = \frac{|\vec{p}|^2}{2m} = -\frac{1}{2m}|\vec{\nabla}|^2. \tag{3.4}$$

If we take a wave function describing an eigenstate of energy and momentum

$$\psi \approx e^{-iEt+i\vec{p}\cdot\vec{x}}, \tag{3.5}$$

all the informations in the equation amount to reproduce correctly the energy-momentum relation

$$E = \frac{|\vec{p}|^2}{2m}. \tag{3.6}$$

In the relativistic case one could try to reproduce the positive energy branch of the dispersion relation (3.1). In that case one could start from the Hamiltonian

$$H = \sqrt{|\vec{p}|^2 + m^2}, \tag{3.7}$$

giving rise to the wave equation

$$i\frac{\partial \psi}{\partial t} = \left(\sqrt{-|\vec{\nabla}|^2 + m^2}\right)\psi. \tag{3.8}$$

There are two obvious problems with this equation:

- spatial and time derivatives appear in a non-symmetric way;
- the equation is non-local in space since it depends on an infinite number of spatial derivatives

$$\left(\sqrt{-|\vec{\nabla}|^2 + m^2}\right)\psi = m\left(\sqrt{1 - \frac{|\vec{\nabla}|^2}{m^2}}\right)\psi = m\sum_{k=0}^{\infty} c_k \left(|\vec{\nabla}|^2\right)^k \psi. \tag{3.9}$$

Both these difficulties are eliminated by iterating eq. (3.8)

$$-\frac{\partial^2 \psi}{\partial t^2} = \left(-|\vec{\nabla}|^2 + m^2\right)\psi. \tag{3.10}$$

This equation is both local and invariant under Lorentz transformations, in fact we can write it in the following form

$$\left(\Box + m^2\right)\psi = 0,\tag{3.11}$$

where

$$\Box = \frac{\partial^2}{\partial t^2} - |\vec{\nabla}|^2\tag{3.12}$$

is the d'Alembert operator in $(3+1)$ dimensions defined in eq. (1.7). Notice that in order to solve the difficulties we have mentioned above we have been obliged to consider both types of solutions: positive energy, $E = \sqrt{|\vec{p}|^2 + m^2}$, and negative energy $E = -\sqrt{|\vec{p}|^2 + m^2}$. The equation we have obtained in this way is known as the Klein-Gordon equation [Klein (1926); Gordon (1926); Schroedinger (1926b)]. As a relativistic extension of the Schrödinger theory it was initially discarded because it gives rise to a non-definite positive probability. In fact, if $\psi$ and $\psi^\star$ are two solutions of such an equation, we can write the following identity

$$0 = \psi^\star \left(\Box + m^2\right)\psi - \psi\left(\Box + m^2\right)\psi^\star = \partial_\mu \left[\psi^\star \partial^\mu \psi - (\partial^\mu \psi^\star)\psi\right].\tag{3.13}$$

Therefore the current

$$J_\mu = \psi^\star \partial_\mu \psi - (\partial_\mu \psi^\star)\psi\tag{3.14}$$

has zero four-divergence and the quantity

$$\int d^3x\, J_0 = \int d^3x(\psi^\star \dot{\psi} - \dot{\psi}^\star \psi)\tag{3.15}$$

is a constant of motion. But we cannot interpret the time-component of the current as a probability density, as we do in the Schrödinger case, because it is not positive definite.

## 3.2 Quantization of the Klein-Gordon field

In this Section we will discuss the quantization of the Klein-Gordon field, that is a field satisfying eq. (3.11). The quantization will be performed by following the steps we have previously outlined, that is

- construction of the Lagrangian density and determination of the canonical momentum density $\Pi(x)$;
- quantization through the requirement of canonical commutation relations

$$[\phi(x,t),\Pi(y,t)] = i\delta^3(x-y),$$
$$[\phi(x,t),\phi(y,t)] = [\Pi(x,t),\Pi(y,t)] = 0\,,\tag{3.16}$$

- expansion of $\phi(x,t)$ in terms of a complete set of solutions of the Klein-Gordon equation. This in order to define the appropriate creation and annihilation operators;
- construction of the Fock space.

We start by looking for the Lagrangian density, requiring that the related Euler-Lagrangian equation gives rise to the Klein-Gordon equation. To this end let us recall how one proceeds in the discrete case. Suppose to have a system of $N$ degrees of freedom satisfying the following equations of motion

$$m_i \ddot{q}_i = -\frac{\partial V}{\partial q_i}. \tag{3.17}$$

Multiplying these equations by some arbitrary variations $\delta q_i$, satisfying the following boundary conditions

$$\delta q_i(t_1) = \delta q_i(t_2) = 0, \tag{3.18}$$

summing over $i$, and integrating in time between $t_1$ and $t_2$, we get

$$\int_{t_1}^{t_2} dt \left[ \sum_{i=1}^N m_i \ddot{q}_i \delta q_i \right] = - \int_{t_1}^{t_2} dt \sum_{i=1}^N \delta q_i \frac{\partial V}{\partial q_i}. \tag{3.19}$$

Integration by parts leads to

$$\delta \left\{ \int_{t_1}^{t_2} dt \left[ \frac{1}{2} \sum_{i=1}^N m_i \dot{q}_i^2 - V \right] \right\} - \left[ \sum_{i=1}^N m_i \dot{q}_i \delta q_i \right]_{t_1}^{t_2} = 0. \tag{3.20}$$

Using the boundary conditions we see that, if the equations of motion are satisfied, then the action

$$S = \int_{t_1}^{t_2} \left[ \frac{1}{2} \sum_{i=1}^N m_i \dot{q}_i^2 - V \right] dt \tag{3.21}$$

is stationary. Conversely from the requirement that the action is stationary under variations satisfying eq. (3.18), the equations of motion follow. Analogously, in the Klein-Gordon case, we multiply the equation by arbitrary local variations of the field $\delta\phi(x) = \tilde{\phi}(x) - \phi(x)$, with boundary conditions for the fields

$$\delta\phi(x,t_1) = \delta\phi(x,t_2) = 0, \quad \lim_{\vec{x}\to\infty} \delta\phi(x,t) = 0. \tag{3.22}$$

Then we integrate over space and time. After integrating by parts we find

$$0 = \int_{t_1}^{t_2} dt \int d^3x \left[ \frac{\partial}{\partial t} \left( \dot{\phi}\delta\phi \right) - \dot{\phi}\dot{\delta\phi} - \vec{\nabla} \cdot \left( \vec{\nabla}\phi\delta\phi \right) + \vec{\nabla}\phi \cdot \vec{\nabla}\delta\phi + m^2\phi\delta\phi \right]. \tag{3.23}$$

Using the boundary conditions we get

$$0 = \delta \int_{t_1}^{t_2} dt \int d^3x \left[ \frac{1}{2}\dot{\phi}^2 - \frac{1}{2}\vec{\nabla}\phi \cdot \vec{\nabla}\phi - \frac{1}{2}m^2\phi^2 \right]. \tag{3.24}$$

Therefore the Lagrangian will be given by

$$L = \int d^3x \mathcal{L}, \tag{3.25}$$

with

$$\mathcal{L} = \frac{1}{2} \left[ \partial_\mu\phi\partial^\mu\phi - m^2\phi^2 \right]. \tag{3.26}$$

We have just shown that the quantity (the action)

$$S = \int_{t_1}^{t_2} dt L \tag{3.27}$$

is stationary along the path where the equations of motion are satisfied. We can now write down the canonical momentum density

$$\Pi = \frac{\partial \mathcal{L}}{\partial \dot{\phi}} = \dot{\phi} \tag{3.28}$$

and the canonical commutation relations

$$[\phi(x,t), \dot{\phi}(y,t)] = i\delta^3(x-y), \quad [\phi(x,t), \phi(y,t)] = [\dot{\phi}(x,t), \dot{\phi}(y,t)] = 0. \tag{3.29}$$

Let us now construct a complete set of solutions of the Klein-Gordon equation. First of all we need a scalar product. But we have already shown in the previous Section that the Klein-Gordon equation admits a conserved quantity (see eq. (3.15)). Therefore, if $f$ and $g$ are two solutions, the scalar product can be defined as

$$\langle f|g \rangle = i \int d^3x f^* \partial_t^{(-)} g, \tag{3.30}$$

where the operator $\partial_t^{(-)}$ has been defined in eq. (2.119). Let us now look for plane-wave solutions

$$f(x) = A(k)e^{-ikx} = A(k)e^{-i(k_0 x_0 - \vec{k}\cdot\vec{x})}. \tag{3.31}$$

From the wave equation we get

$$(\Box + m^2)f = (-k^2 + m^2)f = 0, \tag{3.32}$$

implying

$$k^2 = m^2 \implies k_0^2 = |\vec{k}|^2 + m^2. \tag{3.33}$$

To fix the normalization, we proceed as in the one-dimensional case by taking a finite volume and requiring periodic boundary conditions (normalization in the box). By taking a cube of side $L$ we require

$$\phi(x + L, y, z, t) = \phi(x, y + L, z, t) = \phi(x, y, z + L, t) = \phi(x, y, z, t), \quad (3.34)$$

implying

$$\vec{k} = \frac{2\pi}{L}\vec{n}, \quad (3.35)$$

where

$$\vec{n} = n_1\vec{i}_1 + n_2\vec{i}_2 + n_3\vec{i}_3, \quad (3.36)$$

is a vector with integer components $(n_1, n_2, n_3)$. The normalization condition is

$$\langle f_{\vec{k}}|f_{\vec{k}'}\rangle = i\int_V d^3x f_{\vec{k}}^* \partial_t^{(-)} f_{\vec{k}'} = \delta_{\vec{k},\vec{k}'}, \quad (3.37)$$

where the **delta** is a Kronecker symbol defined as

$$\delta_{\vec{k},\vec{k}'} = \prod_{i=1}^{3}\delta_{n_i,n_i'}, \quad (3.38)$$

with $\vec{n}$ and $\vec{n}'$ two vectors with integer components, related to $\vec{k}$ and $\vec{k}'$, by the relation (3.35). It follows

$$\int_V d^3x A_{\vec{k}}^* A_{\vec{k}'} e^{i(k_0 - k_0')x_0 - i(\vec{k} - \vec{k}')\cdot\vec{x}}(k_0' + k_0) = \delta_{\vec{k},\vec{k}'}. \quad (3.39)$$

Using

$$\int_L dx e^{i\frac{2\pi}{L}(n_1 - n_1')x} = L\delta_{n_1,n_1'}, \quad (3.40)$$

we get

$$i\int_V d^3x f_{\vec{k}}^* \partial_t^{(-)} f_{\vec{k}'} = |A_{\vec{k}}|^2 2k_0 L^3 \delta_{\vec{k},\vec{k}'}, \quad (3.41)$$

where

$$k_0^2 = \left(\frac{2\pi}{L}\right)^2 |\vec{n}|^2 + m^2. \quad (3.42)$$

By considering the positive solutions of this equation, we obtain

$$A_{\vec{k}} = \frac{1}{L^{3/2}}\frac{1}{\sqrt{2\omega_k}}, \quad \omega_k = \sqrt{\left(\frac{2\pi}{L}\right)^2 |\vec{n}|^2 + m^2} = \sqrt{|\vec{k}|^2 + m^2} \quad (3.43)$$

and

$$f_{\vec{k}}(x) = \frac{1}{L^{3/2}} \frac{1}{\sqrt{2\omega_k}} e^{-ikx}. \qquad (3.44)$$

Often we will make use of the so-called normalization in the continuum. The space integration is then extended to all of $R^3$ and we require

$$\langle f_{\vec{k}} | f_{\vec{k}'} \rangle = i \int d^3x f_{\vec{k}}^* \partial_t^{(-)} f_{\vec{k}'} = \delta^3(\vec{k} - \vec{k}'). \qquad (3.45)$$

In this case the spatial momentum can assume all the possible values in $R^3$. It follows

$$\int d^3x A_{\vec{k}}^* A_{\vec{k}'} e^{ikx - ik'x} (k_0 + k_0') = (2\pi)^3 \delta^3(\vec{k} - \vec{k}') |A_{\vec{k}}|^2 2k_0 \qquad (3.46)$$

and the corresponding normalization is

$$A_{\vec{k}} = \frac{1}{\sqrt{(2\pi)^3}} \frac{1}{\sqrt{2\omega_k}}, \qquad (3.47)$$

where

$$\omega_k = \sqrt{|\vec{k}|^2 + m^2}. \qquad (3.48)$$

We see that one goes from the normalization in the box to the normalization in the continuum through the formal substitution

$$\frac{1}{\sqrt{V}} \rightarrow \frac{1}{\sqrt{(2\pi)^3}}. \qquad (3.49)$$

The wave function in the continuum is

$$f_{\vec{k}}(x) = \frac{1}{\sqrt{(2\pi)^3}} \frac{1}{\sqrt{2\omega_k}} e^{-ikx}. \qquad (3.50)$$

In both cases the dispersion relation

$$k_0^2 = |\vec{k}|^2 + m^2 \qquad (3.51)$$

is obviously satisfied, but we have to remember that there are two independent solutions

$$k_0 = \pm\sqrt{|\vec{k}|^2 + m^2} = \pm\omega_k. \qquad (3.52)$$

As a consequence we get two different types of wave functions with positive and negative energy behaving respectively as $e^{-i\omega_k x_0}$ and $e^{i\omega_k x_0}$, $\omega_k > 0$. The second kind of solutions has negative norm in the scalar product we have defined. This would be a big problem if this equation had the same interpretation as the Schrödinger equation. However in field theory this problem does not arise. In fact, the physical Hilbert space is the Fock space, where the scalar product is between the states built up in terms of creation and annihilation operators. Correspondingly, for the negative energy solutions we will make use of the same normalization coefficient found for the positive one.

Having two types of solutions the most general expansion for the field operator (in the Heisenberg representation) is

$$\phi(x) = \frac{1}{\sqrt{(2\pi)^3}} \int d^3k \frac{1}{\sqrt{2\omega_k}} \left[ a(\vec{k})e^{-i\omega_k x_0 + i\vec{k}\cdot\vec{x}} + \tilde{a}(\vec{k})e^{i\omega_k x_0 + i\vec{k}\cdot\vec{x}} \right]. \quad (3.53)$$

Exchanging $\vec{k} \to -\vec{k}$ in the second term we get

$$\phi(x) = \frac{1}{\sqrt{(2\pi)^3}} \int d^3k \frac{1}{\sqrt{2\omega_k}} \left[ a(\vec{k})e^{-ikx} + \tilde{a}(-\vec{k})e^{ikx} \right]$$

$$= \int d^3k [f_{\vec{k}} a(\vec{k}) + f_{\vec{k}}^* \tilde{a}(-\vec{k})]. \quad (3.54)$$

Notice that the positive and negative energy solutions are orthogonal (remember the one-dimensional case discussed in Section 2.5). We can then invert the previous expansion with the result

$$a(\vec{k}) = i \int d^3x f_{\vec{k}}^*(x) \partial_t^{(-)} \phi(x), \quad \tilde{a}(-\vec{k}) = i \int d^3x \phi(x) \partial_t^{(-)} f_{\vec{k}}(x). \quad (3.55)$$

If $\phi(x)$ is hermitian, we have

$$\tilde{a}(-\vec{k}) = a^\dagger(\vec{k}) \quad (3.56)$$

and the expansion becomes

$$\phi(x) = \int d^3k [f_{\vec{k}}(x) a(\vec{k}) + f_{\vec{k}}^*(x) a^\dagger(\vec{k})]. \quad (3.57)$$

From the equations (3.55) one can easily evaluate the commutators among the operators $a(\vec{k})$ and $a^\dagger(\vec{k})$, obtaining

$$[a(\vec{k}), a^\dagger(\vec{k}')] = \delta^3(\vec{k} - \vec{k}'), \quad (3.58)$$

$$[a(\vec{k}), a(\vec{k}')] = [a^\dagger(\vec{k}), a^\dagger(\vec{k}')] = 0. \quad (3.59)$$

These commutation relations depend on the normalization defined for the $f_{\vec{k}}$'s. For instance, if we change this normalization by a factor $N_{\vec{k}}$

$$\langle f_{\vec{k}} | f_{\vec{k}'} \rangle = i \int d^3x f_{\vec{k}}^* \partial_t^{(-)} f_{\vec{k}'} = N_{\vec{k}} \delta^3(\vec{k} - \vec{k}'), \quad (3.60)$$

leaving unchanged the expansion for the field

$$\phi = \int d^3k [f_{\vec{k}} a(\vec{k}) + f_{\vec{k}}^* a^\dagger(\vec{k})], \quad (3.61)$$

we get

$$a(\vec{k}) = \frac{i}{N_{\vec{k}}} \int d^3x f_{\vec{k}}^* \partial_t^{(-)} \phi, \qquad a^\dagger(\vec{k}) = \frac{i}{N_{\vec{k}}} \int d^3x \phi \partial_t^{(-)} f_{\vec{k}} \qquad (3.62)$$

and therefore

$$[a(\vec{k}), a^\dagger(\vec{k}')] = \frac{i}{N_{\vec{k}} N_{\vec{k}'}} \int d^3x f_{\vec{k}}^* \partial_t^{(-)} f_{\vec{k}'} = \frac{1}{N_{\vec{k}}} \delta^3(\vec{k} - \vec{k}'). \qquad (3.63)$$

For instance, a normalization which is used very often is the covariant one

$$\phi(x) = \frac{1}{(2\pi)^3} \int d^3k \frac{1}{2\omega_k} [A(\vec{k}) e^{-ikx} + A^\dagger(\vec{k}) e^{ikx}]. \qquad (3.64)$$

The name of this normalization is because the factor $1/2\omega_k$ makes the integration over the three-momentum Lorentz invariant. In fact, one has

$$\frac{1}{(2\pi)^3} \int d^3k \frac{1}{2\omega_k} = \frac{1}{(2\pi)^4} \int d^4k (2\pi) \delta(k^2 - m^2) \theta(k_0), \qquad (3.65)$$

as it follows by noticing that for $k_0 \approx \omega_k$

$$k^2 - m^2 \approx 2\omega_k \left( k_0 - \sqrt{|\vec{k}|^2 + m^2} \right). \qquad (3.66)$$

In this case the basis functions for the expansion are

$$f_{\vec{k}}(x) = \frac{1}{(2\pi)^3} \frac{1}{2\omega_k} e^{-ikx}, \qquad (3.67)$$

with normalization

$$i \int d^3x f_{\vec{k}}^* \partial_t^{(-)} f_{\vec{k}'} = \frac{1}{(2\pi)^3} \frac{1}{2\omega_k} \delta^3(\vec{k} - \vec{k}') \qquad (3.68)$$

and therefore

$$[A(\vec{k}), A^\dagger(\vec{k}')] = (2\pi)^3 2\omega_k \delta^3(\vec{k} - \vec{k}'). \qquad (3.69)$$

## 3.3 Noether's theorem for relativistic fields

We will now review Noether's theorem [Noether (1918)]. This theorem relates symmetries of the action with conserved quantities. More precisely, for any continuous transformation of fields and/or coordinates leaving invariant the action, there exists a conserved quantity. Transformations, that do not involve the coordinates, are called **internal transformations**. Now consider the case of a generic transformation. The variation of a local quantity $F(x)$ (that is a function of the space-time point) is given, at first order, by

$$\Delta F(x) = \tilde{F}(x') - F(x) = \tilde{F}(x + \delta x) - F(x)$$

$$\cong \tilde{F}(x) - F(x) + \delta x^\mu \frac{\partial F(x)}{\partial x^\mu}. \tag{3.70}$$

The total variation $\Delta$ takes into account both the variation of the coordinates and the form variation of $F$ ($F \to \tilde{F}$). It is then convenient to define a **local** variation $\delta F$, depending only on the variation in form of $F(x)$

$$\delta F(x) = \tilde{F}(x) - F(x). \tag{3.71}$$

Therefore the total variation can be written as the sum of two transformations

$$\Delta F(x) = \delta F(x) + \delta x^\mu \frac{\partial F(x)}{\partial x^\mu}. \tag{3.72}$$

Let us now start from a generic four-dimensional action

$$S = \int_V d^4x \, \mathcal{L}(\phi^i, x), \qquad i = 1, \dots, N \tag{3.73}$$

and let us consider an arbitrary variation of fields and coordinates,

$$x'^\mu = x^\mu + \delta x^\mu, \tag{3.74}$$

$$\Delta\phi^i(x) = \tilde{\phi}^i(x') - \phi^i(x) \approx \delta\phi^i(x) + \delta x^\mu \frac{\partial \phi^i}{\partial x^\mu}. \tag{3.75}$$

If the action is invariant under the transformation, then

$$\tilde{S}_{V'} = S_V. \tag{3.76}$$

The variation of $S_V$ under (3.74) and (3.75) is given by (here $\phi^i_{,\mu} = \partial\phi^i/\partial x^\mu$),

$$\delta S_V = \int_{V'} d^4x' \tilde{\mathcal{L}}(\tilde{\phi}^i, x') - \int_V d^4x \mathcal{L}(\phi^i, x)$$

$$= \int_V d^4x \tilde{\mathcal{L}}(\tilde{\phi}^i, x + \delta x) \frac{\partial(x')}{\partial(x)} - \int_V d^4x \mathcal{L}(\phi^i, x), \tag{3.77}$$

where we have performed a change of variables $x' = x + \delta x$. Then

$$\delta S_V \approx \int_V d^4x \tilde{\mathcal{L}}(\tilde{\phi}^i, x + \delta x)(1 + \partial_\mu \delta x^\mu) - \int_V d^4x \mathcal{L}(\phi^i, x)$$

$$= \int_V d^4x [\tilde{\mathcal{L}}(\tilde{\phi}^i, x + \delta x) - \mathcal{L}(\phi^i, x)] + \int_V d^4x \mathcal{L}(\phi^i, x)\partial_\mu \delta x^\mu$$

$$\approx \int_V d^4x \left[ \frac{\partial \mathcal{L}}{\partial \phi^i} \delta\phi^i + \frac{\partial \mathcal{L}}{\partial \phi^i_{,\mu}} \delta\phi^i_{,\mu} + \frac{\partial \mathcal{L}}{\partial x^\mu} \delta x^\mu \right] + \int_V d^4x \mathcal{L}\partial_\mu \delta x^\mu$$

$$= \int_V d^4x \left[ \frac{\partial \mathcal{L}}{\partial \phi^i} - \partial_\mu \frac{\partial \mathcal{L}}{\partial \phi^i_{,\mu}} \right] \delta\phi^i$$

$$+ \int_V d^4x \partial_\mu \left[ \mathcal{L}\delta x^\mu + \frac{\partial \mathcal{L}}{\partial \phi^i_{,\mu}} \delta\phi^i \right]. \tag{3.78}$$

When we choose variations such that the surface term is vanishing (see eqs. (2.82) and (2.83)) we get the Euler-Lagrange equations of motion

$$\frac{\partial \mathcal{L}}{\partial \phi^i} - \partial_\mu \frac{\partial \mathcal{L}}{\partial \phi^i_{,\mu}} = 0. \tag{3.79}$$

Considering transformations of fields satisfying the equations of motion and leaving invariant $S_V$, we find, using eq. (3.75)

$$\int_V d^4x\, \partial_\mu \left[ \mathcal{L}\delta x^\mu + \frac{\partial \mathcal{L}}{\partial \phi^i_{,\mu}} \Delta\phi^i - \delta x^\nu \frac{\partial \mathcal{L}}{\partial \phi^i_{,\mu}} \phi^i_{,\nu} \right] = 0. \tag{3.80}$$

This proves Noether's theorem, asserting that for any infinitesimal transformation of fields and coordinates leaving invariant the action, there exists a current with zero four-divergence and a corresponding conserved quantity, the space integral of the zero component of the current. According to the choice one does for the variations $\delta x^\mu$ and $\Delta\phi^i$, and for the corresponding symmetries of the action, one gets different conserved quantities.

Let us start with an action invariant under space and time translations. In this case we have $\delta x^\mu = a^\mu$ with $a^\mu$ independent of $x$ and $\Delta\phi^i = 0$[1]. From the general result in eq. (3.80) we get the following local conservation law

$$T^\mu_\nu = \frac{\partial \mathcal{L}}{\partial \phi^i_{,\mu}} \phi^i_{,\nu} - \mathcal{L}g^\mu_\nu, \quad \partial_\mu T^\mu_\nu = 0. \tag{3.81}$$

$T_{\mu\nu}$ is called the energy-momentum tensor of the system. From its local conservation we get four constants of motion

$$P_\nu = \int d^3x\, T^0_\nu. \tag{3.82}$$

$P_\mu$ is the four-momentum of the system. In the case of internal symmetries we take $\delta x^\mu = 0$. The conserved current is

$$J^\mu = \frac{\partial \mathcal{L}}{\partial \phi^i_{,\mu}} \Delta\phi^i = \frac{\partial \mathcal{L}}{\partial \phi^i_{,\mu}} \delta\phi^i, \quad \partial_\mu J^\mu = 0, \tag{3.83}$$

with an associated constant of motion given by

$$Q = \int d^3x\, J^0. \tag{3.84}$$

In general, if the system has more that one internal symmetry, we may have more than one conserved charge $Q$, that is we have a conserved charge for each independent infinitesimal transformation $\Delta\phi^i$.

---

[1]There is no variation in form for a field under a space-time translation.

The last case we will consider is the invariance with respect to Lorentz transformations. For an infinitesimal transformation (see eq. (1.42))

$$x'_\mu = x_\mu + \epsilon_{\mu\nu}x^\nu. \tag{3.85}$$

In general, the relativistic fields are chosen to belong to a representation of the Lorentz group (for instance the Klein-Gordon field belongs to the scalar representation). This means that under a Lorentz transformation the components of the field mix together as, for instance, the components of a vector field under rotations. Therefore, the transformation law of the fields $\phi^i$ under an infinitesimal Lorentz transformation can be written in general as

$$\Delta\phi^i = -\frac{1}{2}\Sigma^{ij}_{\mu\nu}\epsilon^{\mu\nu}\phi^j, \tag{3.86}$$

where we have required that the transformation of the fields is of first order in the Lorentz parameters $\epsilon_{\mu\nu}$. The coefficients $\Sigma_{\mu\nu}$ (antisymmetric in the indices $(\mu, \nu)$) define a matrix in the indices $(i, j)$ which can be shown to be the representative of the infinitesimal generators of the Lorentz group in the field representation. Using this equation and the expression for $\delta x_\mu$ we get the local conservation law

$$0 = \partial_\mu\left[\left(\frac{\partial\mathcal{L}}{\partial\phi^i_{,\mu}}\phi^i_{,\nu} - \mathcal{L}g^\mu_\nu\right)\epsilon^{\nu\rho}x_\rho + \frac{1}{2}\frac{\partial\mathcal{L}}{\partial\phi^i_{,\mu}}\Sigma^{ij}_{\nu\rho}\epsilon^{\nu\rho}\phi^j\right]$$
$$= \frac{1}{2}\epsilon^{\nu\rho}\partial_\mu\left[\left(T^\mu_\nu x_\rho - T^\mu_\rho x_\nu\right) + \frac{\partial\mathcal{L}}{\partial\phi^i_{,\mu}}\Sigma^{ij}_{\nu\rho}\phi^j\right] \tag{3.87}$$

and defining

$$\mathcal{M}^\mu_{\rho\nu} = x_\rho T^\mu_\nu - x_\nu T^\mu_\rho - \frac{\partial\mathcal{L}}{\partial\phi^i_{,\mu}}\Sigma^{ij}_{\rho\nu}\phi^j, \tag{3.88}$$

we obtain six locally conserved currents (one for each Lorentz transformation)

$$\partial_\mu\mathcal{M}^\mu_{\nu\rho} = 0 \tag{3.89}$$

and consequently six constants of motion (notice that the lower indices are antisymmetric)

$$M_{\nu\rho} = \int d^3x\,\mathcal{M}^0_{\nu\rho}. \tag{3.90}$$

Three of these constants (the ones with $\nu$ and $\rho$ assuming spatial values) are nothing but the components of the angular momentum of the field.

In the Klein-Gordon case

$$T_{\mu\nu} = \partial_\mu \phi \partial_\nu \phi - \frac{1}{2} \left( \partial_\rho \phi \partial^\rho \phi - m^2 \phi^2 \right) g_{\mu\nu}, \tag{3.91}$$

from which

$$T^0_0 = \frac{1}{2} \dot{\phi}^2 + \frac{1}{2} \left( |\vec{\nabla}\phi|^2 + m^2 \phi^2 \right). \tag{3.92}$$

This current corresponds to the invariance under time translations, and it must be identified with the energy density of the field (compare with eq. (2.96) for the one-dimensional case). In analogous way

$$T^0_i = \dot{\phi} \frac{\partial \phi}{\partial x^i} \tag{3.93}$$

is the momentum density of the field. Using $\dot{\phi} = \Pi$, the energy and momentum of the Klein-Gordon field can be written in the form

$$P^0 = H = \int d^3x T^{00} = \frac{1}{2} \int d^3x \left( \Pi^2 + |\vec{\nabla}\phi|^2 + m^2 \phi^2 \right), \tag{3.94}$$

$$P^i = \int d^3x T^{0i} = - \int d^3x \Pi \frac{\partial \phi}{\partial x^i}, \quad (\vec{P} = - \int d^3x \Pi \vec{\nabla}\phi). \tag{3.95}$$

## 3.4    Energy and momentum of the Klein-Gordon field

It is very easy to verify that the energy density found previously coincides with the Hamiltonian density evaluated in the canonical way through the Legendre transformation of the Lagrangian density

$$\mathcal{H} = \Pi \dot{\phi} - \mathcal{L}. \tag{3.96}$$

Then, we will show that the four-momentum $P^\mu$ is the generator, as it should be, of space-time translations. This amounts to saying that it satisfies the following commutation relation with the field

$$[\phi(x), P^\mu] = i \frac{\partial \phi}{\partial x_\mu}. \tag{3.97}$$

In fact,

$$[\phi(\vec{y}, t), H] = \frac{1}{2} \int d^3x [\phi(\vec{y}, t), \Pi^2(\vec{x}, t)] = i\Pi(\vec{y}, t) = i\dot{\phi}(\vec{y}, t). \tag{3.98}$$

Analogously

$$[\phi(\vec{y}, t), P^i] = - \int d^3x [\phi(\vec{y}, t), \Pi(\vec{x}, t) \frac{\partial \phi(\vec{x}, t)}{\partial x^i}] = -i \frac{\partial \phi(\vec{y}, t)}{\partial y^i} = i \frac{\partial \phi(\vec{y}, t)}{\partial y_i}. \tag{3.99}$$

Therefore the operator

$$U = e^{ia^\mu P_\mu} \tag{3.100}$$

generates translations in $x$. In fact, by looking at the first order in $a^\mu$, it follows

$$e^{ia\cdot P}\phi(x)e^{-ia\cdot P} \approx \phi(x) + ia^\mu[P_\mu,\phi(x)] = \phi(x) + a^\mu\frac{\partial\phi(x)}{\partial x^\mu} \approx \phi(x+a). \tag{3.101}$$

With a calculation completely analogous to the one done in Section 2.5 we can evaluate the Hamiltonian and the momentum in terms of creation and annihilation operators

$$H = \frac{1}{2}\int d^3k\,\omega_k[a^\dagger(\vec{k})a(\vec{k}) + a(\vec{k})a^\dagger(\vec{k})], \tag{3.102}$$

$$\vec{P} = \int d^3k\,\vec{k}\,a^\dagger(\vec{k})a(\vec{k}). \tag{3.103}$$

They satisfy the following commutation relations with $a^\dagger(\vec{k})$

$$[H,a^\dagger(\vec{k})] = \omega_k a^\dagger(\vec{k}), \qquad [\vec{P},a^\dagger(\vec{k})] = \vec{k}a^\dagger(\vec{k}). \tag{3.104}$$

This shows that the operators $a^\dagger(\vec{k})$, acting on the vacuum, create states of momentum $\vec{k}$ and energy $\omega_k = (|\vec{k}|^2 + m^2)^{1/2}$, whereas the annihilation operators $a(\vec{k})$ annihilate the vacuum states corresponding to the same energy and momentum. In the case of the box normalization, for any $\vec{k} = (2\pi/L)\vec{n}$ (that is for any choice of the three integer components of the vector $\vec{n}$), one can build up a state $|n_{\vec{k}}\rangle$

$$|n_{\vec{k}}\rangle = \frac{1}{\sqrt{n_{\vec{k}}!}}\left(a^\dagger(\vec{k})\right)^{n_{\vec{k}}}|0\rangle, \tag{3.105}$$

containing $n_{\vec{k}}$ particles of momentum $\vec{k}$ and energy $\omega_k$. The most general state is obtained by tensor product of states similar to the previous one. Any of these states is characterized by a triple of integers defining the momentum $\vec{k}$, that is

$$|n_{\vec{k}_1}\dots n_{\vec{k}_\alpha}\rangle = \prod_\otimes |n_{\vec{k}_i}\rangle = \frac{1}{\sqrt{n_{\vec{k}_1}!\dots n_{\vec{k}_\alpha}!}}(a^\dagger(\vec{k}_1))^{n_{\vec{k}_1}}\dots(a^\dagger(\vec{k}_\alpha))^{n_{\vec{k}_\alpha}}|0\rangle. \tag{3.106}$$

The fundamental state is the one with zero particles in any cell of the momentum space (vacuum state)

$$|0\rangle = \prod_\otimes |0\rangle_i, \tag{3.107}$$

where $|0\rangle_i$ is the fundamental state for the momentum in the cell $i$. That is

$$a_{\vec{k}_i}|0\rangle_i = 0. \tag{3.108}$$

In this normalization the Hamiltonian is given by

$$H = \frac{1}{2}\sum_{\vec{k}} \omega_k [a^\dagger(\vec{k})a(\vec{k}) + a(\vec{k})a^\dagger(\vec{k})] \tag{3.109}$$

and therefore

$$H|0\rangle = \frac{1}{2}\sum_{\vec{k}} \omega_k |0\rangle. \tag{3.110}$$

This sum is infinite. Recalling that $\vec{k} = (2\pi/L)\vec{n}$, it follows that the cell in the $\vec{k}$-space has a volume

$$\Delta V_{\vec{k}} = \frac{(2\pi)^3}{L^3}, \tag{3.111}$$

from which

$$\frac{1}{2}\sum_{\vec{k}} \omega_k = \frac{1}{2}\sum_{\vec{k}} \frac{\Delta V_{\vec{k}}\omega_k}{\Delta V_{\vec{k}}} \implies \frac{1}{2}\frac{L^3}{(2\pi)^3}\int d^3k \sqrt{|\vec{k}|^2 + m^2}, \tag{3.112}$$

which is divergent.

Let us recall that this problem can be formally avoided through the use of the normal product. In other words by subtracting the infinite energy of the vacuum from the Hamiltonian. In this case

$$: H := \sum_{\vec{k}} \omega_k \, a^\dagger(\vec{k})a(\vec{k}), \tag{3.113}$$

whereas in the continuum

$$: H := \int d^3k \, \omega_k \, a^\dagger(\vec{k})a(\vec{k}). \tag{3.114}$$

As we see, the energy of the vacuum depends on the quantization volume. This implies that it depends on the boundary conditions of the problem. In the real vacuum this is not a difficulty, but this point should be carefully considered when quantizing fields living inside a finite volume. In this case the dependence on the boundary produces measurable effects, as it was pointed out theoretically by [Casimir (1948)], and then proved experimentally by [Sparnaay (1958)]. We will discuss very briefly the Casimir effect arising when we have an electromagnetic field confined between two large perfectly conducting plates. We idealize the two plates as two large parallel

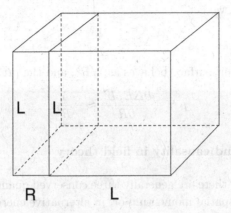

Fig. 3.1 The Casimir effect.

squares of side $L$ at a distance $R \ll L$. The theory shows that there is an attractive force per unit surface, $p$, between the two plates given by

$$p = -\frac{\pi^2}{240} \frac{\hbar c}{R^4}. \tag{3.115}$$

We can understand the origin of this force in a very qualitative way, in particular its dependence on the plate distance, by quantizing the electromagnetic field (that we will take here as a Klein-Gordon field with zero mass, $m = 0$) in a box of side $L$. In this case the vacuum energy is

$$E_0 \approx L^3 \int_{1/L}^{k_{max}} k \, d^3k, \tag{3.116}$$

with the integration between a lower momentum of order $1/L$ and an arbitrary upper momentum which is necessary in order to make the integral finite. The energy contained in the region bounded by the two surfaces of side $L$ (at the left in Fig. 3.1) is given by

$$E \approx L^2 R \int_{1/L}^{k_{max}} k \, d^3k. \tag{3.117}$$

If we now insert two plates of side $L$, as shown in Fig. 3.1, at a distance $R$, we get an analogous result, but with a lower momentum of order $1/R$. Therefore the variation of energy results to be

$$\Delta E \approx L^2 R \int_{1/R}^{1/L} k \, d^3k = L^2 R \int_{1/R}^{1/L} k^3 \, dk = \frac{L^2 R}{4} \left[ \left( \frac{1}{L} \right)^4 - \left( \frac{1}{R} \right)^4 \right]. \tag{3.118}$$

Then, for $R \ll L$, we get

$$\Delta E \approx -\frac{L^2}{R^3}. \tag{3.119}$$

The energy per unit surface behaves as $1/R^3$, and the pressure is given by

$$p \approx -\frac{\partial \Delta E/L^2}{\partial R} \approx -\frac{1}{R^4}. \tag{3.120}$$

## 3.5 Locality and causality in field theory

For a free particle there are generally three conserved quantum numbers as, for instance, the spatial momentum or, in alternative energy, the square of the angular momentum and its third component. All these quantities can be expressed as space integrals of **local** functions of the fields. The locality property is crucial and it is strictly connected to causality. To understand this point, let us consider the example of a free Klein-Gordon field. In this case there is a further constant of motion, the number of particles

$$N = \int d^3 k\, a^\dagger(\vec{k}) a(\vec{k}). \tag{3.121}$$

We will show that this cannot be written as the spatial integral of a local quantity, and that this implies the non-observability of the number of particles. We know that for the free Klein-Gordon theory there is the following conserved current

$$J_\mu = \phi^\dagger(\partial_\mu \phi) - (\partial_\mu \phi^\dagger)\phi. \tag{3.122}$$

This expression vanishes for a hermitian field but, nevertheless, the number of particles, $N$, can be expressed in terms of the positive and negative energy components of the field

$$\phi^{(+)}(x) = \int d^3 k \frac{1}{\sqrt{2\omega_k (2\pi)^3}} e^{-i(\omega_k t - \vec{k}\cdot\vec{x})} a(\vec{k}), \quad \phi^{(-)}(x) = \phi^{(+)\dagger}(x). \tag{3.123}$$

In fact, it is not difficult to show that

$$N = \int d^3 x\, \phi^{(-)}\, i\partial_t^{(-)}\, \phi^{(+)}. \tag{3.124}$$

This is a constant of motion, because $\phi^{(+)}$ and $\phi^{(-)}$ are both solutions of the equation of motion, and therefore

$$j_\mu = i\phi^{(-)}(\partial_\mu \phi^{(+)}) - (\partial_\mu \phi^{(-)})\phi^{(+)} \tag{3.125}$$

is a conserved current. However, $\phi^{(+)}$ and $\phi^{(-)}$ are not local functions of $\phi$ and therefore also the conserved current is not local. In fact, to relate the negative and positive frequency components to the field we need a time integration. Let us define

$$\phi(x) = \int d^4k \phi(k) e^{-ikx}, \qquad (3.126)$$

with

$$\phi(k) = \sqrt{\frac{2\omega_k}{(2\pi)^3}} \delta(k^2 - m^2) \left(a(k)\theta(k_0) + \tilde{a}(k)\theta(-k_0)\right), \qquad (3.127)$$

one has

$$\phi^{(+)}(x) = \int d^4k \phi^{(+)}(k) e^{-ikx}, \qquad (3.128)$$

with

$$\phi^{(+)}(k) = \theta(k_0)\phi(k). \qquad (3.129)$$

Using the convolution theorem for the Fourier transform we get

$$\phi^{(+)}(x) = \int d^4x' \tilde{\theta}(x - x')\phi(x'). \qquad (3.130)$$

But

$$\begin{aligned}
\tilde{\theta}(x - x') &= \frac{1}{(2\pi)^4} \int d^4k\, e^{ik(x-x')} \theta(k_0) \\
&= \delta^3(\vec{x} - \vec{x}') \int \frac{dk_0}{2\pi} e^{ik_0(x_0 - x_0')} \theta(k_0) \\
&= \delta^3(\vec{x} - \vec{x}') \tilde{\theta}(x_0 - x_0').
\end{aligned} \qquad (3.131)$$

Therefore

$$\phi^{(+)}(\vec{x}, x_0) = \int dx_0' \tilde{\theta}(x_0 - x_0')\phi(\vec{x}, x_0'). \qquad (3.132)$$

To show the implications of having to do with a non-local current, let us define a particle density operator

$$\mathcal{N}(x) = i\phi^{(-)} \partial_t^{(-)} \phi^{(+)}. \qquad (3.133)$$

This operator does not commute with itself at equal times and different space points

$$[\mathcal{N}(\vec{x}, t), \mathcal{N}(\vec{y}, t)] \neq 0, \qquad \vec{x} \neq \vec{y}. \qquad (3.134)$$

However, for local operators[2], $\mathcal{O}(\phi)$, this commutator is automatically zero, due to the canonical commutation relations

$$[\mathcal{O}(\phi(\vec{x}, t)), \mathcal{O}(\phi(\vec{y}, t))] = 0, \qquad \forall \vec{x} \neq \vec{y}. \qquad (3.135)$$

We want to argue that the vanishing of this commutator is just the necessary condition in order that $\mathcal{O}$ represents an observable quantity. In fact, if the commutator of a local operator with itself is not zero at space-like distances, the measurement of the observable at some point, $x$, would influence the measurements done at points with space-like separation from $x$, because we cannot diagonalize the operator simultaneously at two such points. But this would imply the propagation of a signal at a velocity greater than the light velocity, in contrast with the causality principle. We see that the vanishing of the commutator of a local observable with itself at space-like distances is a necessary condition in order to satisfy the causality principle. We show now that this is automatically satisfied if the operator under consideration is a local function of the fields. We will start showing that the commutator of the field with itself is a Lorentz invariant function. Therefore, from the vanishing of the commutator for separations between points of the type $x^{\mu} = (t, \vec{x})$, and $y^{\mu} = (t, \vec{y})$, it follows the vanishing for arbitrary space-like separations. Let us evaluate the commutator

$$\begin{aligned}
[\phi(x), \phi(y)] &= \int \frac{d^3 k_1 d^3 k_2}{(2\pi)^3 \sqrt{2\omega_{k_1} 2\omega_{k_2}}} \Big[ [a(\vec{k}_1), a^{\dagger}(\vec{k}_2)] e^{-ik_1 x + ik_2 y} \\
&\quad + [a^{\dagger}(\vec{k}_1), a(\vec{k}_2)] e^{ik_1 x - ik_2 y} \Big] \\
&= \int \frac{d^3 k}{(2\pi)^3 2\omega_k} \Big[ e^{-ik(x-y)} - e^{ik(x-y)} \Big] \\
&= -2i \int \frac{d^3 k}{(2\pi)^3 2\omega_k} \sin(\omega_k(x_0 - y_0)) e^{i\vec{k}(\vec{x}-\vec{y})}. \qquad (3.136)
\end{aligned}$$

Using eq. (3.65), this expression can be written in invariant form

$$\begin{aligned}
[\phi(x), \phi(y)] &= \int \frac{d^4 k}{(2\pi)^3} \theta(k_0) \delta(k^2 - m^2) \Big[ e^{-ik(x-y)} - e^{ik(x-y)} \Big] \\
&= \int \frac{d^4 k}{(2\pi)^3} \epsilon(k_0) \delta(k^2 - m^2) e^{-ik(x-y)}. \qquad (3.137)
\end{aligned}$$

Since the sign of the fourth component of a time-like four-vector is invariant under proper Lorentz transformations, we see that defining

$$[\phi(x), \phi(y)] = i\Delta(x - y), \qquad (3.138)$$

---

[2] A local operator is a function of the fields and of a finite number of derivatives of the fields.

the function

$$\Delta(x - y) = -i \int \frac{d^4 k}{(2\pi)^3} \epsilon(k_0) \delta(k^2 - m^2) e^{-ik(x-y)} \qquad (3.139)$$

is Lorentz invariant and, as such, it depends only on $(x - y)^2$. Since $\Delta(x - y)$ vanishes at equal times, it follows that it is zero for arbitrary space-like separations. Therefore the canonical commutation relations ensure the observability of the Klein-Gordon field. For the negative and positive energy components we get

$$\Delta^{(+)}(x - y) \equiv [\phi^{(+)}(x), \phi^{(-)}(y)] = \int \frac{d^3 k}{(2\pi)^3 2\omega_k} e^{-ik(x-y)}$$

$$= \int \frac{d^4 k}{(2\pi)^3} \theta(k_0) \delta(k^2 - m^2) e^{-ik(x-y)}. \qquad (3.140)$$

Also in this case we have a Lorentz invariant function, and therefore it is enough to study its equal times behavior:

$$\Delta^{(+)}(0, \vec{x}) = \int \frac{d^3 k}{(2\pi)^3 2\omega_k} e^{i\vec{k} \cdot \vec{x}} = \int \frac{k^2 dk d(\cos\theta) d\varphi}{(2\pi)^3 2\omega_k} e^{ikr \cos\theta}$$

$$= -i \frac{1}{4\pi^2 r} \int_0^\infty \frac{k dk}{2\omega_k} \left[ e^{ikr} - e^{-ikr} \right]$$

$$= -\frac{1}{8\pi^2 r} \frac{d}{dr} \int_{-\infty}^{+\infty} dk \frac{e^{ikr}}{\sqrt{|\vec{k}|^2 + m^2}}. \qquad (3.141)$$

By putting $k = m \sinh\theta$, we get

$$\Delta^{(+)}(0, \vec{x}) = -\frac{1}{8\pi^2 r} \frac{d}{dr} \int_{-\infty}^{+\infty} d\theta e^{imr \sinh\theta}. \qquad (3.142)$$

The integral defines a Hankel function $H_0^{(1)}$

$$\int_{-\infty}^{+\infty} d\theta e^{imr \sinh\theta} = i\pi H_0^{(1)}(imr). \qquad (3.143)$$

Using the following relation between $H_0^{(1)}$ and $H_1^{(1)}$

$$\frac{d}{dr} H_0^{(1)}(imr) = -im H_1^{(1)}(imr), \qquad (3.144)$$

we obtain

$$\Delta^{(+)}(0, \vec{x}) = -\frac{m}{8\pi r} H_1^{(1)}(imr). \qquad (3.145)$$

The asymptotic behavior of the Hankel function $H_1^{(1)}(imr)$ for large and small values of $r$ is given by

$$\lim_{r \to \infty} H_1^{(1)}(imr) \approx -\sqrt{\frac{2}{\pi mr}} e^{-mr}, \qquad \lim_{r \to 0} H_1^{(1)}(imr) \approx -\frac{2}{mr}, \qquad (3.146)$$

from which

$$\lim_{r\to\infty} \Delta^{(+)}(0,\vec{x}) \approx \frac{m}{8\pi r}\sqrt{\frac{2}{\pi m r}}e^{-mr}, \qquad \lim_{r\to 0}\Delta^{(+)}(0,\vec{x}) \approx \frac{1}{4\pi r^2}. \qquad (3.147)$$

We see that for space-like separations this commutator does not vanish. But for space separations larger than the Compton wavelength $1/m$, $\Delta^{(+)}$ is practically zero. Remember that for an electron the Compton wavelength is about $3.9 \cdot 10^{-11}$ cm. Clearly, an analogous result is obtained for the commutator of the particle density operator. From this we can derive the impossibility of localize a Klein-Gordon particle (but the result can be extended to any relativistic particle) over distances of the order of $1/m$. To prove this let us introduce the following operators

$$N(V) = \int_V d^3x \mathcal{N}(x), \qquad (3.148)$$

where $V$ is a sphere. Suppose we want to localize the particle around a

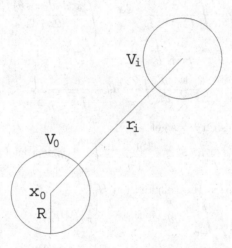

Fig. 3.2  In order to localize a particle inside $V_0$, there should be no other particles within a distance $r_i \approx R$.

point $x_0$ with an uncertainty $R$. We consider a sphere $V_0$ centered at $x_0$ with radius $R$. We then take other spheres $V_i$ not connected to $V_0$, that is with centers separated by $x_0$ by a distance $r_i > 2R$, as shown in Fig. 3.2. The requirement to localize the particle within $V_0$, with a radius $R < 1/m$, is equivalent to ask for the existence of a state with eigenvalue 1 for the operator $N(V_0)$ and eigenvalues 0 for all the $N(V_i)$ with $r_i \approx 1/m$. But such an eigenstate does not exist because, as we have shown previously

$[N(V_0), N(V_i)] \neq 0$. On the contrary, if we take volumes $V_i$ at distances $r_i$ much bigger than $1/m$, the corresponding operators $N(V_i)$ commute, and we can construct the desired state. Therefore it is possible to localize the particle only over distances much bigger than the Compton wavelength. The physical explanation is that to realize the localization over distances much smaller than $1/m$, we need energies much bigger than $m$. But in this case there is a non-zero probability to create particle antiparticle pairs.

To summarize, a local quantity can be an observable if and only if the commutator with itself vanishes for space-like distances, otherwise there would be a violation of the causality principle. If the quantity is a local function of the fields, the previous condition is automatically satisfied due to the canonical commutation relations. As a consequence, a relativistic particle described by a hermitian Klein-Gordon field, cannot be localized over distances of the order of $1/m$, because the particle density is not a local function of the fields. From these considerations we see also that the negative energy components of the fields are essential for the internal consistency of the theory. Otherwise the particle interpretation of the field (that is the commutation relations among creation and annihilation operators) and the locality properties (vanishing of the commutators at space-like distances) would not be compatible.

## 3.6 The charged scalar field

We have shown that a hermitian Klein-Gordon field describes a set of identical scalar particles. If we want to describe different kind of particles we need to introduce different kind of fields. As an example, consider two different hermitian scalar fields. The free Lagrangian is simply the sum of the two free Lagrangian densities describing separately the fields

$$\mathcal{L} = \frac{1}{2} \sum_{i=1}^{2} \left[ (\partial_\mu \phi_i)(\partial^\mu \phi_i) - m_i^2 \phi_i^2 \right] \tag{3.149}$$

and we can write immediately the canonical commutation relations, since the two fields refer to different degrees of freedom,

$$[\phi_i(\vec{x}, t), \dot{\phi}_j(\vec{y}, t)] = i\delta_{ij}\delta^3(\vec{x} - \vec{y}), \tag{3.150}$$

$$[\phi_i(\vec{x}, t), \phi_j(\vec{y}, t)] = [\dot{\phi}_i(\vec{x}, t), \dot{\phi}_j(\vec{y}, t)] = 0. \tag{3.151}$$

All the considerations made for a single field can be trivially extended to the case of two fields. In particular there will be two kinds of creation and annihilation operators, $a^\dagger{}_i(\vec{k})$, $i = 1, 2$, which we will assume mutually commuting. The two-particle state made with different creation operators is not any more symmetric

$$a^\dagger{}_1(\vec{k}_1)a^\dagger{}_2(\vec{k}_2)|0\rangle \neq a^\dagger{}_2(\vec{k}_1)a^\dagger{}_1(\vec{k}_2)|0\rangle. \tag{3.152}$$

This means that the two fields correspond to distinguishable particles.

Something new happens when the mass term for the two fields is the same ($m_1 = m_2 = m$)

$$\mathcal{L} = \frac{1}{2}\left[(\partial_\mu\phi_1)(\partial^\mu\phi_1) + (\partial_\mu\phi_2)(\partial^\mu\phi_2)\right] - \frac{1}{2}m^2\left[\phi_1^2 + \phi_2^2\right]. \tag{3.153}$$

Then, the theory acquires a symmetry under rotations in the plane of the two fields $\phi_1$ and $\phi_2$

$$\phi_1' = \phi_1\cos\theta + \phi_2\sin\theta,$$
$$\phi_2' = -\phi_1\sin\theta + \phi_2\cos\theta. \tag{3.154}$$

In fact, the Lagrangian is a function of the norm of the following two-dimensional vectors

$$\vec{\phi} = (\phi_1, \phi_2), \qquad \partial_\mu\vec{\phi} = (\partial_\mu\phi_1, \partial_\mu\phi_2) \tag{3.155}$$

and their norm is invariant under rotations in the plane. For infinitesimal transformations we have

$$\delta\phi_1 = \phi_2\theta, \qquad \delta\phi_2 = -\phi_1\theta, \tag{3.156}$$

or, in a more compact form

$$\delta\phi_i = \epsilon_{ij}\phi_j\theta, \tag{3.157}$$

where $\epsilon_{ij}$ is the two-dimensional antisymmetric Ricci tensor, defined by

$$\epsilon_{12} = -\epsilon_{21} = 1. \tag{3.158}$$

From the Noether theorem, we have a conserved current, associated to this symmetry, given by (see eq. (3.83))

$$J^\mu = \frac{\partial\mathcal{L}}{\partial\phi_{i,\mu}}\Delta\phi_i = \phi_i'^\mu\epsilon_{ij}\phi_j\theta. \tag{3.159}$$

It is convenient to factorize out the angle of the infinitesimal rotation and define a new current

$$j^\mu = \frac{1}{\theta}J^\mu = \phi_i'^\mu\epsilon_{ij}\phi_j = \phi_1'^\mu\phi_2 - \phi_2'^\mu\phi_1. \tag{3.160}$$

The conservation of the current follows from the equality of the masses of the two fields, as it can be verified directly from the equations of motion following from eq. (3.149):

$$\partial_\mu j^\mu = (\Box\phi_1)\phi_2 - (\Box\phi_2)\phi_1 = -(m_1^2 - m_2^2)\phi_1\phi_2. \tag{3.161}$$

The conserved charge associated to the current is

$$Q = \int d^3x\, j^0 = \int d^3x(\dot{\phi}_1\phi_2 - \dot{\phi}_2\phi_1) \tag{3.162}$$

and it is the generator of the infinitesimal transformations of the fields

$$[Q,\phi_1] = -i\phi_2, \qquad [Q,\phi_2] = i\phi_1. \tag{3.163}$$

Correspondingly a finite transformation is given by

$$U = e^{iQ\theta}. \tag{3.164}$$

In fact,

$$\begin{aligned}
e^{iQ\theta}\phi_1 e^{-iQ\theta} &= \phi_1 + i\theta[Q,\phi_1] + \frac{i^2}{2!}\theta^2[Q,[Q,\phi_1]] + \cdots \\
&= \phi_1 + \theta\phi_2 - \frac{1}{2}\phi_1\theta^2 + \cdots \\
&= \phi_1\cos\theta + \phi_2\sin\theta.
\end{aligned} \tag{3.165}$$

In analogous way one can derive the transformation properties of $\phi_2$. The invariance of $\mathcal{L}$ under rotations in the plane $(\phi_1,\phi_2)$ is referred to as the invariance under the group $O(2)$[3]. The real basis for the fields used so far is not the most convenient one. In fact, the charge $Q$ mixes the two fields. One can understand better the properties of the charge operator in a basis in which the fields are not mixed. This basis is a complex one and it is given by the combinations

$$\phi = \frac{1}{\sqrt{2}}(\phi_1 + i\phi_2), \qquad \phi^\dagger = \frac{1}{\sqrt{2}}(\phi_1 - i\phi_2), \tag{3.166}$$

(the factor $1/\sqrt{2}$ has been inserted for a correct normalization of the fields). It follows

$$[Q,\phi] = \frac{1}{\sqrt{2}}[Q,\phi_1 + i\phi_2] = \frac{1}{\sqrt{2}}(-i\phi_2 - \phi_1) = -\phi \tag{3.167}$$

and analogously

$$[Q,\phi^\dagger] = \phi^\dagger. \tag{3.168}$$

---

[3]The notation $O(N)$ is used to characterize the orthogonal transformations in $N$ dimensions.

Therefore the field $\phi$ lowers the charge of an eigenstate of $Q$ by one unit, whereas $\phi^\dagger$ increases the charge by the same amount. In fact, if $Q|q\rangle = q|q\rangle$

$$Q(\phi|q\rangle) = ([Q,\phi] + \phi Q)|q\rangle = (-1 + q)\phi|q\rangle \tag{3.169}$$

and

$$\phi|q\rangle \propto |q - 1\rangle. \tag{3.170}$$

In analogous way

$$\phi^\dagger|q\rangle \propto |q + 1\rangle. \tag{3.171}$$

Inverting the relations (3.166) we get

$$\phi_1 = \frac{1}{\sqrt{2}}(\phi + \phi^\dagger), \qquad \phi_2 = -\frac{i}{\sqrt{2}}(\phi - \phi^\dagger), \tag{3.172}$$

from which

$$\mathcal{L} = \partial_\mu \phi^\dagger \partial^\mu \phi - m^2 \phi^\dagger \phi \tag{3.173}$$

and

$$j_\mu = i\left[(\partial_\mu \phi)\phi^\dagger - (\partial_\mu \phi^\dagger)\phi\right]. \tag{3.174}$$

The charge operator is

$$Q = i \int d^3x \ \phi^\dagger \partial_t^{(-)} \phi. \tag{3.175}$$

The commutation relations in the new basis are given by

$$[\phi(\vec{x},t), \dot{\phi}^\dagger(\vec{y},t)] = i\delta^3(\vec{x} - \vec{y}) \tag{3.176}$$

and

$$[\phi(\vec{x},t), \phi(\vec{y},t)] = [\phi^\dagger(\vec{x},t), \phi^\dagger(\vec{y},t)] = 0,$$
$$[\dot{\phi}(\vec{x},t), \dot{\phi}(\vec{y},t)] = [\dot{\phi}^\dagger(\vec{x},t), \dot{\phi}^\dagger(\vec{y},t)] = 0. \tag{3.177}$$

Let us notice that these commutation relations could have also been obtained directly by canonical quantization of the Lagrangian (3.173), since

$$\Pi_\phi = \frac{\partial \mathcal{L}}{\partial \dot{\phi}} = \dot{\phi}^\dagger, \qquad \Pi_{\phi^\dagger} = \frac{\partial \mathcal{L}}{\partial \dot{\phi}^\dagger} = \dot{\phi}. \tag{3.178}$$

The original $O(2)$ symmetry becomes now an invariance of the Lagrangian (3.173) under a phase transformation of the fields. This follows from (3.173) but it can be seen also directly

$$\phi \rightarrow \frac{1}{\sqrt{2}}(\phi_1' + i\phi_2') = e^{-i\theta}\phi \tag{3.179}$$

and

$$\phi^\dagger \to e^{i\theta}\phi^\dagger. \tag{3.180}$$

In this basis we speak of invariance under the group $U(1)^4$.

Using the expansion for the real fields

$$\phi_i(x) = \int d^3k \left[ f_{\vec{k}}(x)a_i(\vec{k}) + f_{\vec{k}}^*(x)a^\dagger{}_i(\vec{k}) \right], \tag{3.181}$$

we get

$$\phi(x) = \int d^3k \left[ f_{\vec{k}}(x)\frac{1}{\sqrt{2}}(a_1(\vec{k}) + ia_2(\vec{k})) + f_{\vec{k}}^*(x)\frac{1}{\sqrt{2}}(a^\dagger{}_1(\vec{k}) + ia^\dagger{}_2(\vec{k})) \right] \tag{3.182}$$

and introducing the combinations

$$a(\vec{k}) = \frac{1}{\sqrt{2}}(a_1(\vec{k}) + ia_2(\vec{k})), \qquad b(\vec{k}) = \frac{1}{\sqrt{2}}(a_1(\vec{k}) - ia_2(\vec{k})), \tag{3.183}$$

it follows that

$$\phi(x) = \int d^3k \left[ f_{\vec{k}}(x)a(\vec{k}) + f_{\vec{k}}^*(x)b^\dagger(\vec{k}) \right]$$

$$\phi^\dagger(x) = \int d^3k \left[ f_{\vec{k}}(x)b(\vec{k}) + f_{\vec{k}}^*(x)a^\dagger(\vec{k}) \right]. \tag{3.184}$$

Using these relations we can evaluate the commutation relations for the creation and annihilation operators in the complex basis

$$[a(\vec{k}), a^\dagger(\vec{k}')] = [b(\vec{k}), b^\dagger(\vec{k}')] = \delta^3(\vec{k} - \vec{k}'), \tag{3.185}$$

$$[a(\vec{k}), b(\vec{k}')] = [a(\vec{k}), b^\dagger(\vec{k}')] = 0. \tag{3.186}$$

We get also

$$: P_\mu := \int d^3k\, k_\mu \sum_{i=1}^{2} a^\dagger{}_i(\vec{k})a_i(\vec{k}) = \int d^3k\, k_\mu \left[ a^\dagger(\vec{k})a(\vec{k}) + b^\dagger(\vec{k})b(\vec{k}) \right]. \tag{3.187}$$

Therefore both the operators $a^\dagger(\vec{k})$ and $b^\dagger(\vec{k})$ create particle states with momentum $k^\mu$, as the original operators $a^\dagger{}_i$. The charge $Q$ is given by

$$Q = i \int d^3x \phi^\dagger \partial_t^{(-)} \phi$$

$$= \int d^3k_1 d^3k_2 \delta^3(\vec{k}_1 - \vec{k}_2) \left[ a^\dagger(\vec{k}_1)a(\vec{k}_1) - b(\vec{k}_1)b^\dagger(\vec{k}_1) \right], \tag{3.188}$$

---

[4]$U(N)$ is the group of unitary transformations acting on complex vectors of dimension $N$.

where we have used the orthogonality relations (3.45). For the normal ordered charge operator we get

$$: Q := \int d^3k \left[ a^\dagger(\vec{k}) a(\vec{k}) - b^\dagger(\vec{k}) b(\vec{k}) \right], \qquad (3.189)$$

showing explicitly that $a^\dagger$ and $b^\dagger$ create particles of charge $+1$ and $-1$ respectively. The two types of particles, with the same mass and opposite charges are called particle and antiparticle. It is important to notice that the current density $j_0$ is local in the fields, and therefore

$$[j_0(x), j_0(y)] = 0, \quad \forall\, (x - y)^2 < 0. \qquad (3.190)$$

By following the discussion done in Section 3.5, we construct the operators

$$Q(V_i) = \int_{V_i} d^3x\, j_0(x), \qquad (3.191)$$

which, for any $V_i$ and $V_j$, commute at equal times. Therefore it is possible to localize a state of definite charge in an arbitrary spatial region. This agrees with the argument we have developed in the case of the number of particles, because the pair creation process does not change the charge of the state.

Finally we notice that the field operator is a linear combination of annihilation, $a(\vec{k})$, and creation, $b^\dagger(\vec{k})$, operators, meaning that a local theory deals in a symmetric way with the annihilation of a particle and the creation of an antiparticle. For instance, to annihilate a charge $+1$ is equivalent to the creation of a charge $-1$.

## 3.7   Exercises

(1) Show that the two different definitions of the number particle operator, $N$, given in eqs. (3.121) and (3.124) coincide.
(2) Show that the number of particles $N$ and the Hamiltonian for the Klein-Gordon case are commuting operators.
(3) Derive the commutators in eqs. (3.185) and (3.186) using the expressions (3.184) for the field and its hermitian conjugate.
(4) Derive the expressions (3.187) and (3.189) for the momentum and the charge operators for a charged field.
(5) The Hamiltonian of the Klein-Gordon field is given by eq. (3.94). Using the quantum equations of motion for the operators in the Heisenberg representation (Heisenberg equations),

$$\dot{A} = i[H, A], \qquad (3.192)$$

for the field and its conjugate momentum, show that the Klein-Gordon equation follows.

(6) Show that for a scalar field the angular momentum density has zero divergence

$$\partial_\mu \mathcal{M}^\mu_{\nu\rho} = 0, \tag{3.193}$$

if the energy-momentum tensor is symmetric

$$T_{\mu\nu} = T_{\nu\mu}. \tag{3.194}$$

(7) Given a classical Klein-Gordon field satisfying the boundary conditions

$$\phi(t = 0, \vec{x}) = 0, \quad \dot{\phi}(t = 0, \vec{x}) = c, \tag{3.195}$$

where $c$ is a constant, determine the form of $\phi$ at a generic time. (Hint: Use eq. (3.55) in order to find out the coefficients of $\phi$ in a plane wave expansion.)

(8) Prove the following equation for a Klein-Gordon field

$$\phi(x) = \int d^3y \left[ \phi(y) \partial_{x_0} \Delta(x - y) - \dot{\phi}(y) \Delta(y - x) \right]. \tag{3.196}$$

Here $i\Delta(x - y) = [\phi(x), \phi(y)]$. Give an argument to justify the fact that this equation holds for any value of $y_0$. The consequence is that through this formula we can solve the Cauchy problem for the field $\phi$ given the boundary conditions on $\phi$ and $\dot{\phi}$.

(9) The Lagrangian density for a charged scalar field is invariant under the substitution

$$\phi \to \eta_C \phi^\dagger, \tag{3.197}$$

where $\eta_C$ is a phase factor. It is possible to define a unitary operator $\mathcal{C}$ such that

$$\mathcal{C}\phi\mathcal{C}^{-1} = \eta_C \phi^\dagger. \tag{3.198}$$

Show that

$$\mathcal{C}^2 = 1, \quad \mathcal{C}a(k)\mathcal{C}^{-1} = \eta_C b(k), \quad \mathcal{C}b(k)\mathcal{C}^{-1} = \eta_C^* a(k). \tag{3.199}$$

Furthermore prove that the charge operator $Q$, changes sign under $\mathcal{C}$

$$\mathcal{C}Q\mathcal{C}^{-1} = -Q. \tag{3.200}$$

(10) Given a Klein-Gordon field $\phi$, introduce the 2-component vector

$$\psi = \begin{pmatrix} \theta \\ \chi \end{pmatrix} \qquad (3.201)$$

where

$$\theta = \frac{1}{2}\left(\phi + \frac{i}{m}\dot{\phi}\right), \quad \chi = \frac{1}{2}\left(\phi - \frac{i}{m}\dot{\phi}\right). \qquad (3.202)$$

Show that the Klein-Gordon equation can be re-written in the form

$$i\frac{\partial \psi}{\partial t} = H\psi \qquad (3.203)$$

with $H$ a $2 \times 2$ matrix. Find the form of $H$ and show that it is not hermitian in the positive definite metric defined as

$$\langle \psi | \psi \rangle = \int d^3\vec{x}\, \psi^\dagger \psi. \qquad (3.204)$$

On the other hand it is hermitian in the non positive definite metric defined as

$$\langle \psi || \psi \rangle = \int d^3\vec{x}\, \psi^\dagger \sigma_3 \psi \qquad (3.205)$$

where $\sigma_3$ is a Pauli matrix.

# Chapter 4

# The Dirac field

## 4.1 The Dirac equation

In 1928 Dirac tried to solve the problem of a non-positive probability density, present in the Klein-Gordon case, formulating a new wave equation. Dirac thought that in order to get a positive probability it was necessary to have a wave equation of first order in the time derivative (as it happens for the Schrödinger equation). Therefore Dirac looked for a way to reduce the Klein-Gordon equation (of second order in the time derivative) to a first-order differential equation. The Pauli formulation of the electron spin put Dirac on the right track. In fact, Pauli showed that in order to describe the spin, it was necessary to generalize the Schrödinger wave function (a complex number) to a two-component object

$$\psi \to \psi_\alpha = \begin{pmatrix} \psi_1 \\ \psi_2 \end{pmatrix}, \tag{4.1}$$

modifying also the wave equation to a matrix equation

$$i\frac{\partial \psi_\alpha}{\partial t} = \sum_{\beta=1}^{2} H_{\alpha\beta} \psi_\beta, \tag{4.2}$$

where the Hamiltonian $H$ is, in general, a $2 \times 2$ matrix. The electron spin is then described by a special set of $2 \times 2$ matrices, the Pauli matrices, $\vec{\sigma}$

$$\vec{S} = \frac{1}{2}\vec{\sigma}. \tag{4.3}$$

Dirac realized that it was possible to write the squared norm of a spatial vector as

$$|\vec{k}|^2 = (\vec{\sigma} \cdot \vec{k})^2, \tag{4.4}$$

as it follows from

$$[\sigma_i, \sigma_j]_+ = 2\delta_{ij}, \tag{4.5}$$

where $[A, B]_+ = AB + BA$ is the anticommutator of the operators $A$ and $B$.

Following this suggestion Dirac tried to write down a first order differential equation for a many-component wave function

$$i\frac{\partial \psi}{\partial t} = -i\vec{\alpha} \cdot \vec{\nabla}\psi + \beta m\psi \equiv H\psi, \tag{4.6}$$

where $\vec{\alpha}$ and $\beta$ are matrices. The requirements that this equation should satisfy are

- the wave function $\psi$, solution of the Dirac equation, should satisfy also the Klein-Gordon equation in order to get the correct dispersion relation between energy and momentum;
- the equation should admit a conserved current with the fourth component being positive definite;
- the equation should be covariant with respect to Lorentz transformations (see later).

In order to satisfy the first requirement, we iterate the Dirac equation and ask that the resulting second order differential equation coincides with the Klein-Gordon equation

$$-\frac{\partial^2 \psi}{\partial t^2} = (-i\vec{\alpha} \cdot \vec{\nabla} + \beta m)^2 \psi$$

$$= \left(-\alpha^i \alpha^j \frac{\partial^2}{\partial x^i \partial x^j} + \beta^2 m^2 - i(\beta\vec{\alpha} + \vec{\alpha}\beta) \cdot \vec{\nabla}\right)\psi$$

$$= \left(-\frac{1}{2}[\alpha^i, \alpha^j]_+ \frac{\partial^2}{\partial x^i \partial x^j} + \beta^2 m^2 - i(\beta\vec{\alpha} + \vec{\alpha}\beta) \cdot \vec{\nabla}\right)\psi. \tag{4.7}$$

We see that it is necessary to require the following matrix relations

$$[\alpha^i, \alpha^j]_+ = 2\delta_{ij}, \qquad [\alpha^i, \beta]_+ = 0, \qquad \beta^2 = 1. \tag{4.8}$$

We would like the Hamiltonian, $H$, to be hermitian. To this end we require $\vec{\alpha}$ and $\beta$ to be hermitian matrices. Since for any choice of the index $i$, $(\alpha^i)^2 = 1$, it follows that the eigenvalues of $\vec{\alpha}$ and $\beta$ must be $\pm 1$. We can also prove the following relations

$$Tr(\beta) = Tr(\alpha^i) = 0. \tag{4.9}$$

For instance, from $\alpha^i \beta = -\beta \alpha^i$, we get $\alpha^i = -\beta \alpha^i \beta$, and therefore

$$Tr(\alpha^i) = -Tr(\beta \alpha^i \beta) = -Tr(\alpha^i) = 0, \tag{4.10}$$

where we have made use of the cyclic property of the trace. The consequence is that the matrices $\alpha^i$ and $\beta$ can be realized only in a space of even dimensions. This is perhaps the biggest difficulty that Dirac had to cope with. In fact, the $\alpha^i$'s enjoy the same properties of the Pauli matrices. However, in a $2 \times 2$ matrix space, a further anticommuting matrix $\beta$ does not exist. It took some time before Dirac realized that the previous relations could have been satisfied by $4 \times 4$ matrices.

An explicit realization of the Dirac matrices is the following

$$\alpha^i = \begin{pmatrix} 0 & \sigma_i \\ \sigma_i & 0 \end{pmatrix}, \qquad \beta = \begin{pmatrix} 1 & 0 \\ 0 & -1 \end{pmatrix}, \tag{4.11}$$

as it can be easily checked.

Let us now show that the second requirement is also satisfied. We multiply the Dirac equation by $\psi^\dagger$ from the left, and then we consider the equation for $\psi^\dagger$

$$-i\frac{\partial \psi^\dagger}{\partial t} = i(\vec{\nabla}\psi^\dagger) \cdot \vec{\alpha} + m\psi^\dagger \beta, \tag{4.12}$$

multiplied from the right by $\psi$. Subtracting the resulting equations we get

$$i\psi^\dagger \frac{\partial \psi}{\partial t} + i\frac{\partial \psi^\dagger}{\partial t}\psi = \psi^\dagger(-i\vec{\alpha}\cdot\vec{\nabla} + \beta m)\psi - (i\vec{\nabla}\psi^\dagger \cdot \vec{\alpha} + \psi^\dagger \beta m)\psi = -i\vec{\nabla}\cdot(\psi^\dagger \vec{\alpha}\psi), \tag{4.13}$$

that is

$$i\frac{\partial}{\partial t}(\psi^\dagger \psi) + i\frac{\partial}{\partial x^j}(\psi^\dagger \alpha^j \psi) = 0. \tag{4.14}$$

We see that the current

$$j^\mu = (\psi^\dagger \psi, \psi^\dagger \alpha^i \psi) \tag{4.15}$$

is a conserved one:

$$\frac{\partial j^\mu}{\partial x^\mu} = 0. \tag{4.16}$$

Furthermore its fourth component $j^0 = \psi^\dagger \psi$ is positive definite. Of course we still have to prove that $j^\mu$ is a four-vector, implying that

$$\int d^3x \, \psi^\dagger \psi \tag{4.17}$$

is invariant under Lorentz transformations.

## 4.2   Covariance properties of the Dirac equation

To discuss the transformation properties of the Dirac equation under
Lorentz, it is convenient to write the equation in a slightly different form.
Let us multiply the equation by $\beta$

$$i\beta\frac{\partial\psi}{\partial t} = -i\beta\vec{\alpha}\cdot\vec{\nabla}\psi + m\psi \tag{4.18}$$

and define the following matrices

$$\gamma^0 = \beta = \begin{pmatrix} 1 & 0 \\ 0 & -1 \end{pmatrix}, \quad \gamma^i = \beta\alpha^i = \begin{pmatrix} 0 & \sigma_i \\ -\sigma_i & 0 \end{pmatrix}. \tag{4.19}$$

Then the equation becomes

$$\left(i\gamma^0\frac{\partial}{\partial x^0} + i\gamma^i\frac{\partial}{\partial x^i} - m\right)\psi = 0, \tag{4.20}$$

or, in a more compact way

$$(i\hat{\partial} - m)\psi = 0, \tag{4.21}$$

where

$$\hat{\partial} = \gamma^\mu\frac{\partial}{\partial x^\mu} = \gamma^\mu\partial_\mu. \tag{4.22}$$

The matrices $\gamma^\mu$ satisfy the following anticommutation relations

$$\left[\gamma^i, \gamma^j\right]_+ = \beta\alpha^i\beta\alpha^j + \beta\alpha^j\beta\alpha^i = -\left[\alpha^i, \alpha^j\right]_+ = -2\delta_{ij}, \tag{4.23}$$

$$\left[\gamma^0, \gamma^i\right]_+ = \left[\beta, \beta\alpha^i\right]_+ = \alpha^i + \beta\alpha^i\beta = 0, \tag{4.24}$$

or

$$\left[\gamma^\mu, \gamma^\nu\right]_+ = 2g^{\mu\nu}. \tag{4.25}$$

Notice that

$$(\gamma^i)^\dagger = (\beta\alpha^i)^\dagger = \alpha^i\beta = -\gamma^i \tag{4.26}$$

and

$$(\gamma^i)^2 = -1. \tag{4.27}$$

In order for the Dirac equation to be covariant, the following two conditions
have to be satisfied:

- given the Dirac wave function $\psi(x)$ in the Lorentz frame, $S$, an observer
  in a different frame, $S'$ should be able to evaluate, in terms of $\psi(x)$, the
  wave function $\psi'(x')$ describing the same physical state as $\psi(x)$ does in
  $S$;

- according to the relativity principle, $\psi'(x')$ must be a solution of an equation that in $S'$ has the same form as the Dirac equation in $S$. That is to say

$$\left(i\tilde{\gamma}^\mu \frac{\partial}{\partial x'^\mu} - m\right)\psi'(x') = 0. \tag{4.28}$$

The matrices $\tilde{\gamma}^\mu$ should satisfy the same algebra as the matrices $\gamma^\mu$, because in both cases the wave functions should satisfy the Klein-Gordon equation (which is invariant in form). Therefore, neglecting a possible unitary transformation, the two sets of matrices can be identified. As a consequence, the Dirac equation in $S'$ will be

$$\left(i\gamma^\mu \frac{\partial}{\partial x'^\mu} - m\right)\psi'(x') = 0. \tag{4.29}$$

Since both the Dirac equation and the Lorentz transformations are linear, we will require the wave functions in two different Lorentz frames to be linearly correlated

$$\psi'(x') = \psi'(\Lambda x) = S(\Lambda)\psi(x), \tag{4.30}$$

where $S(\Lambda)$ is a $4 \times 4$ matrix operating on the complex vector $\psi(x)$ and $\Lambda$ is the Lorentz transformation. On physical grounds, the matrix $S(\Lambda)$ should be invertible

$$\psi(x) = S^{-1}(\Lambda)\psi'(x'), \tag{4.31}$$

but using the relativity principle, since one goes from the frame $S'$ to the frame $S$ through the transformation $\Lambda^{-1}$, we must have

$$\psi(x) = S(\Lambda^{-1})\psi'(x'), \tag{4.32}$$

from which

$$S^{-1}(\Lambda) = S(\Lambda^{-1}). \tag{4.33}$$

Considering the Dirac equation in the frame $S$

$$\left(i\gamma^\mu \frac{\partial}{\partial x^\mu} - m\right)\psi(x) = 0, \tag{4.34}$$

we can write

$$\left(i\gamma^\mu \frac{\partial}{\partial x^\mu} - m\right)S^{-1}(\Lambda)\psi'(x') = 0. \tag{4.35}$$

Multiplying from the left by $S(\Lambda)$ and using

$$\frac{\partial}{\partial x^\mu} = \frac{\partial x'_\nu}{\partial x^\mu}\frac{\partial}{\partial x'_\nu} = \Lambda_{\nu\mu}\frac{\partial}{\partial x'_\nu}, \quad x'_\nu = \Lambda_{\nu\mu}x^\mu, \tag{4.36}$$

it follows

$$\left( iS(\Lambda)\gamma^\mu S^{-1}(\Lambda)\Lambda_{\nu\mu}\frac{\partial}{\partial x'_\nu} - m \right) \psi'(x') = 0. \tag{4.37}$$

Comparing with eq. (4.28), we get

$$S(\Lambda)\gamma^\mu S^{-1}(\Lambda)\Lambda_{\nu\mu} = \gamma_\nu \tag{4.38}$$

or

$$S^{-1}(\Lambda)\gamma_\nu S(\Lambda) = \Lambda_{\nu\mu}\gamma^\mu. \tag{4.39}$$

For an infinitesimal transformation we write

$$\Lambda_{\mu\nu} = g_{\mu\nu} + \epsilon_{\mu\nu}, \tag{4.40}$$

with $\epsilon_{\mu\nu} = -\epsilon_{\nu\mu}$ (see eq. (1.44)). By expanding $S(\Lambda)$ to the first order in $\epsilon_{\mu\nu}$

$$S(\Lambda) = 1 - \frac{i}{4}\sigma_{\mu\nu}\epsilon^{\mu\nu} \tag{4.41}$$

and using (4.39), we find the following condition on $\sigma_{\mu\nu}$

$$\left( 1 + \frac{i}{4}\sigma_{\rho\lambda}\epsilon^{\rho\lambda} \right)\gamma_\nu\left( 1 - \frac{i}{4}\sigma_{\alpha\beta}\epsilon^{\alpha\beta} \right) = (g_{\mu\nu} + \epsilon_{\nu\mu})\gamma^\mu, \tag{4.42}$$

from which

$$\frac{i}{4}\epsilon^{\rho\lambda}[\sigma_{\rho\lambda}, \gamma_\nu] = \epsilon_{\nu\mu}\gamma^\mu = \frac{1}{2}\epsilon^{\rho\lambda}(g_{\rho\nu}\gamma_\lambda - g_{\lambda\nu}\gamma_\rho) \tag{4.43}$$

and finally

$$[\sigma_{\rho\lambda}, \gamma_\nu] = -2i(g_{\rho\nu}\gamma_\lambda - g_{\lambda\nu}\gamma_\rho). \tag{4.44}$$

It is not difficult to show that the solution of this equation is given by

$$\sigma_{\rho\lambda} = \frac{i}{2}[\gamma_\rho, \gamma_\lambda]. \tag{4.45}$$

In fact,

$$\begin{aligned}
[\sigma_{\rho\lambda}, \gamma_\nu] &= \frac{i}{2}[\gamma_\rho\gamma_\lambda - \gamma_\lambda\gamma_\rho, \gamma_\nu] = \frac{i}{2}[\gamma_\rho\gamma_\lambda\gamma_\nu - \gamma_\nu\gamma_\rho\gamma_\lambda - \gamma_\lambda\gamma_\rho\gamma_\nu + \gamma_\nu\gamma_\lambda\gamma_\rho] \\
&= \frac{i}{2}[(2g_{\rho\lambda} - \gamma_\lambda\gamma_\rho)\gamma_\nu - \gamma_\nu(2g_{\rho\lambda} - \gamma_\lambda\gamma_\rho) - \gamma_\lambda\gamma_\rho\gamma_\nu + \gamma_\nu\gamma_\lambda\gamma_\rho] \\
&= -i[\gamma_\lambda\gamma_\rho\gamma_\nu - \gamma_\nu\gamma_\lambda\gamma_\rho] \\
&= -i[\gamma_\lambda(2g_{\rho\nu} - \gamma_\nu\gamma_\rho) - (2g_{\nu\lambda} - \gamma_\lambda\gamma_\nu)\gamma_\rho] \\
&= -2i[g_{\rho\nu}\gamma_\lambda - g_{\nu\lambda}\gamma_\rho]. \tag{4.46}
\end{aligned}$$

A finite Lorentz transformation is obtained by exponentiation

$$S(\Lambda) = e^{-\frac{i}{4}\sigma_{\mu\nu}\epsilon^{\mu\nu}}, \tag{4.47}$$

with

$$\sigma_{\mu\nu} = \frac{i}{2}[\gamma_\mu, \gamma_\nu]. \tag{4.48}$$

We can now verify that the current $j^\mu$, defined in eq. (4.15), transforms as a four-vector. To this end we introduce the following notation

$$\bar\psi(x) = \psi^\dagger(x)\beta = \psi^\dagger(x)\gamma_0. \tag{4.49}$$

It follows

$$j^0 = \psi^\dagger\psi = \bar\psi\gamma^0\psi, \quad j^i = \psi^\dagger\alpha^i\psi = \bar\psi\beta\alpha^i\psi = \bar\psi\gamma^i\psi, \tag{4.50}$$

or

$$j^\mu = \bar\psi\gamma^\mu\psi. \tag{4.51}$$

The transformation properties of $\bar\psi$ under Lorentz transformations are particularly simple. By noticing that

$$\gamma^0\gamma^{\mu\dagger}\gamma^0 = \gamma^\mu \tag{4.52}$$

and

$$\sigma_{\mu\nu}{}^\dagger = -\frac{i}{2}[\gamma_\mu, \gamma_\nu]^\dagger = \frac{i}{2}[\gamma_\mu{}^\dagger, \gamma_\nu{}^\dagger], \tag{4.53}$$

it follows

$$\gamma_0\sigma_{\mu\nu}{}^\dagger\gamma_0 = \sigma_{\mu\nu} \tag{4.54}$$

and therefore

$$\gamma_0 S^\dagger(\Lambda)\gamma_0 = S^{-1}(\Lambda), \tag{4.55}$$

from which

$$\bar\psi'(x') = \bar\psi(x)S^{-1}(\Lambda). \tag{4.56}$$

Finally we get

$$j'^\mu(x') = \bar\psi'(x')\gamma^\mu\psi'(x') = \bar\psi(x)S^{-1}(\Lambda)\gamma^\mu S(\Lambda)\psi(x) = \Lambda^\mu{}_\nu j^\nu(x). \tag{4.57}$$

Therefore $j^\mu$ has the desired transformation properties. The representation for the Lorentz generators, in the same basis used previously for the $\gamma_\mu$ matrices, is

$$\sigma_{0i} = \frac{i}{2}[\gamma_0, \gamma_i] = \frac{i}{2}(\beta^2\alpha_i - \beta\alpha_i\beta) = -i\alpha^i = -i\begin{pmatrix} 0 & \sigma_i \\ \sigma_i & 0 \end{pmatrix}, \tag{4.58}$$

$$\sigma_{ij} = \frac{i}{2}[\gamma_i, \gamma_j] = -\frac{i}{2}[\alpha_i, \alpha_j] = \epsilon_{ijk}\begin{pmatrix} \sigma_k & 0 \\ 0 & \sigma_k \end{pmatrix}. \tag{4.59}$$

The generators of the spatial rotations are nothing but the Pauli matrices, as one should expect for spin $1/2$ particles.

The behavior of the Dirac wave function under parity $\vec{x} \to -\vec{x}$ can be obtained in analogous way. In this case

$$\Lambda_P{}^\nu_\mu = \begin{pmatrix} 1 & & & \\ & -1 & & \\ & & -1 & \\ & & & -1 \end{pmatrix} \tag{4.60}$$

and therefore

$$S^{-1}(\Lambda_P)\gamma^\mu S(\Lambda_P) = \gamma_\mu. \tag{4.61}$$

This relation is satisfied by the choice

$$S(\Lambda_P) = \eta_P \gamma_0, \tag{4.62}$$

where $\eta_P$ is a non-observable arbitrary phase. Then

$$\psi(x) \to \psi'(x') = \eta_P \gamma_0 \psi(x), \qquad x'^\mu = (x^0, -\vec{x}). \tag{4.63}$$

It is useful to classify the bilinear expressions in the Dirac wave function under Lorentz transformations. Let us consider expressions of the type $\bar{\psi}A\psi$, where $A$ is an arbitrary $4 \times 4$ matrix. As a basis for the $4 \times 4$ matrices we can take the following set of 16 linearly independent matrices

$$\begin{aligned} \Gamma^S &= 1, \\ \Gamma^V_\mu &= \gamma_\mu, \\ \Gamma^A_\mu &= \gamma_5 \gamma_\mu, \\ \Gamma^T_{\mu\nu} &= \sigma_{\mu\nu}, \\ \Gamma^P &= \gamma_5, \end{aligned} \tag{4.64}$$

where the matrix $\gamma_5$ is defined as

$$\gamma_5 = \gamma^5 = i\gamma^0 \gamma^1 \gamma^2 \gamma^3. \tag{4.65}$$

This matrix has the following properties

$$\gamma_5^\dagger = i\gamma^3 \gamma^2 \gamma^1 \gamma^0 = i\gamma^0 \gamma^1 \gamma^2 \gamma^3 = \gamma_5, \tag{4.66}$$

$$\gamma_5^2 = 1, \qquad [\gamma_5, \gamma_\mu]_+ = 0, \tag{4.67}$$

and, in the previous representation of the $\gamma$-matrices:

$$\gamma_5 = \begin{pmatrix} 0 & 1 \\ 1 & 0 \end{pmatrix}. \tag{4.68}$$

One can easily verify that the bilinear expressions have the following behavior under Lorentz transformations

$$\bar{\psi}\psi \approx \text{scalar},$$
$$\bar{\psi}\gamma_\mu\psi \approx \text{four-vector},$$
$$\bar{\psi}\gamma_5\gamma_\mu\psi \approx \text{axial four-vector},$$
$$\bar{\psi}\sigma_{\mu\nu}\psi \approx \text{antisymmetric } 2^{\text{nd}} \text{ rank tensor},$$
$$\bar{\psi}\gamma_5\psi \approx \text{pseudoscalar}. \tag{4.69}$$

As an example, let us verify the last of these statements through a parity transformation:

$$\bar{\psi}(x)\gamma_5\psi(x) \to \bar{\psi}'(x')\gamma_5\psi'(x') = \eta_P^\star\eta_P\bar{\psi}(x)\gamma_0\gamma_5\gamma_0\psi(x) = -\bar{\psi}(x)\gamma_5\psi(x). \tag{4.70}$$

## 4.3   The Dirac equation and the Lorentz group

In this Section we will show how the Dirac equation follows naturally from the theory of the representations of the Lorentz group.

We have seen in Section 1.3 that parity interchanges the spinor representation $(s_1, s_2)$ with $(s_2, s_1)$. If one is interested in representations of the complete Lorentz group, that is the Lorentz group extended by parity $(L_+^\uparrow \oplus L_-^\uparrow)$, we have the following two possibilities

$$(s, s), \quad (s_1, s_2) \oplus (s_2, s_1). \tag{4.71}$$

Representations corresponding to the first possibilities are, for instance, $(0, 0)$, the scalar representation, or $(1/2, 1/2)$, the four-vector representation. The simplest example of the second possibility is

$$\left(\frac{1}{2}, 0\right) \oplus \left(0, \frac{1}{2}\right). \tag{4.72}$$

This case corresponds to the Dirac representation. To understand better this point, let us first consider the four-vector representation. We have said that this is characterized by the bispinor $(1/2, 1/2)$. More explicitly, if we are given a four-vector $V^\mu$, we can construct the bispinor

$$V \equiv V_{\dot{\alpha}\beta} = (\sigma_\mu)_{\dot{\alpha}\beta}V^\mu = \begin{pmatrix} V^0 - V^3 & -V^1 + iV^2 \\ -V^1 - iV^2 & V^0 + V^3 \end{pmatrix}. \tag{4.73}$$

This equation can simply be inverted

$$V^\mu = \frac{1}{2}\mathrm{Tr}\left(\tilde{\sigma}^\mu V\right). \tag{4.74}$$

Now, suppose we would like to construct a first-order wave equation for an undotted spinor. We can act upon the undotted spinor by the gradient operator in its bispinorial representation

$$i\partial^\mu(\sigma_\mu)_{\dot\alpha\beta}\zeta^\beta. \tag{4.75}$$

By covariance, this expression can be zero, or else, it must be a dotted spinor. That is

$$i\partial^\mu(\sigma_\mu)_{\dot\alpha\beta}\zeta^\beta = 0 \tag{4.76}$$

or

$$i\partial^\mu(\sigma_\mu)_{\dot\alpha\beta}\zeta^\beta = m\eta_{\dot\alpha}. \tag{4.77}$$

The first case corresponds to a massless fermion (see later) and was considered for the first time in [Weyl (1929b)] . However Pauli criticized this equation on the basis that it was not parity invariant. Weyl was vindicated only in 1956 after the discovery of the parity nonconservation in weak interactions. In the second case one is forced to introduce the dotted spinor $\eta_{\dot\alpha}$ which, in turn, must also satisfy a first order wave equation that can be written, without loss of generality, as[1]

$$i\partial^\mu(\tilde{\sigma}_\mu)^{\alpha\dot\beta}\eta_{\dot\beta} = m\zeta^\alpha. \tag{4.78}$$

Let us now introduce the four-component spinor

$$\psi = \begin{pmatrix} \zeta^\alpha \\ \eta_{\dot\alpha} \end{pmatrix}. \tag{4.79}$$

The two spinor equations can then be written as a single equation

$$(i\gamma^\mu\partial_\mu - m)\psi = 0, \tag{4.80}$$

where

$$\gamma^\mu = \begin{pmatrix} 0 & (\tilde{\sigma}^\mu)^{\alpha\dot\beta} \\ (\sigma^\mu)_{\dot\alpha\beta} & 0 \end{pmatrix}. \tag{4.81}$$

One can easily verify that these matrices satisfy the algebra (4.25), and furthermore that

$$\gamma_5 = i\gamma^0\gamma^1\gamma^2\gamma^3 = \begin{pmatrix} 1 & 0 \\ 0 & -1 \end{pmatrix}. \tag{4.82}$$

---

[1]In principle two different constants $m_1$ and $m_2$ could appear in the two spinor equations. However, by a convenient rescaling of the spinors, the two equations can be written in terms of a single constant.

In the basis that we have considered in the previous Section, $\gamma^0$ was diagonal, whereas in the present base $\gamma_5$ is the diagonal one. The two bases are called respectively the energy and the chiral basis. The first name corresponds to the fact that for a particle at rest, the Dirac operator is diagonal. The second denomination comes from the name chirality operator for $\gamma_5$ (see later). If we write the four-component field in the energy basis as

$$\begin{pmatrix} \phi \\ \chi \end{pmatrix}, \tag{4.83}$$

the relation with the chiral basis is

$$\phi = \frac{1}{\sqrt{2}}(\zeta + \eta), \quad \chi = \frac{1}{\sqrt{2}}(\zeta - \eta). \tag{4.84}$$

## 4.4 Free particle solutions of the Dirac equation

In this Section we will study the wave plane solutions of the Dirac equation. In the rest frame of the particle we look for solutions of the type

$$\psi(t) = u e^{-imt}, \tag{4.85}$$

where $u$ is a four-component spinor. This solution has positive energy. Substituting inside the Dirac equation we get

$$(i\hat{\partial} - m)\psi(t) = (m\gamma_0 - m)u e^{-imt} = 0, \tag{4.86}$$

that is

$$(\gamma_0 - 1)u = 0. \tag{4.87}$$

Since $\gamma_0$ has eigenvalues $\pm 1$ it follows that there also solutions of the type $e^{imt}$, corresponding to negative energy states. More generally we can look for solutions of the form

$$\begin{aligned} \psi^{(+)}(x) &= e^{-ikx}u(k), & \text{positive energy,} \\ \psi^{(-)}(x) &= e^{ikx}v(k), & \text{negative energy.} \end{aligned} \tag{4.88}$$

Inserting into the Dirac equation

$$\begin{aligned} (\hat{k} - m)u(k) &= 0, \\ (\hat{k} + m)v(k) &= 0. \end{aligned} \tag{4.89}$$

In the rest frame we get

$$\begin{aligned} (\gamma_0 - 1)u(m, \vec{0}) &= 0, \\ (\gamma_0 + 1)v(m, \vec{0}) &= 0. \end{aligned} \tag{4.90}$$

There are two independent spinors of type $u$ and two of type $v$ satisfying these equations. In the basis where $\gamma_0$ is a diagonal matrix we can choose the following solutions

$$u^{(1)}(m, \vec{0}) = \begin{pmatrix} 1 \\ 0 \\ 0 \\ 0 \end{pmatrix}, \qquad u^{(2)}(m, \vec{0}) = \begin{pmatrix} 0 \\ 1 \\ 0 \\ 0 \end{pmatrix},$$

$$v^{(1)}(m, \vec{0}) = \begin{pmatrix} 0 \\ 0 \\ 0 \\ 1 \\ 0 \end{pmatrix}, \qquad v^{(2)}(m, \vec{0}) = \begin{pmatrix} 0 \\ 0 \\ 0 \\ 1 \end{pmatrix}. \tag{4.91}$$

In a general Lorentz frame the solutions can be obtained by boosting the solutions in the rest frame. That is, applying a Lorentz transformation from the rest frame to a generic one. Alternatively, we can notice that the following expression

$$(\hat{k} - m)(\hat{k} + m) = k^2 - m^2 \tag{4.92}$$

vanishes for $k^2 = m^2$. Therefore we can solve our problem (except for a normalization constant), by putting

$$u^{(\alpha)}(k) = c_\alpha (\hat{k} + m) u^{(\alpha)}(m, \vec{0}),$$
$$v^{(\alpha)}(k) = d_\alpha (-\hat{k} + m) v^{(\alpha)}(m, \vec{0}), \tag{4.93}$$

with $k^2 = m^2$. In order to determine the normalization constants $c_\alpha$ and $d_\alpha$ we make use of the orthogonality conditions satisfied by the rest frame solutions (see eq. (4.91))

$$\bar{u}^{(\alpha)}(m, \vec{0}) u^{(\beta)}(m, \vec{0}) = \delta_{\alpha\beta},$$
$$\bar{v}^{(\alpha)}(m, \vec{0}) v^{(\beta)}(m, \vec{0}) = -\delta_{\alpha\beta},$$
$$\bar{u}^{(\alpha)}(m, \vec{0}) v^{(\beta)}(m, \vec{0}) = 0. \tag{4.94}$$

Since these relations involve Lorentz scalars, $\bar{\psi}\psi$, we can ask that they are satisfied also for $u^{(\alpha)}(k)$ and $v^{(\alpha)}(k)$. Let us start with the $u$ spinors:

$$\bar{u}^{(\alpha)}(k) u^{(\beta)}(k) = c_\alpha^* c_\beta \bar{u}^{(\alpha)}(m, \vec{0})(\hat{k} + m)^2 u^{(\beta)}(m, \vec{0})$$
$$= c_\alpha^* c_\beta \bar{u}^{(\alpha)}(m, \vec{0})(2m^2 + 2m\hat{k}) u^{(\beta)}(m, \vec{0}). \tag{4.95}$$

By taking into account that $u^{(\alpha)}(m, \vec{0})$ and $u^{\dagger(\alpha)}(m, \vec{0})$ are eigenstates of $\gamma_0$ with eigenvalue $+1$, we get

$$\bar{u}^{(\alpha)}(m, \vec{0}) \gamma^\mu u^{(\beta)}(m, \vec{0}) = u^{\dagger(\alpha)}(m, \vec{0}) \gamma_0 \gamma^\mu \gamma_0 u^{(\beta)}(m, \vec{0})$$
$$= u^{\dagger(\alpha)}(m, \vec{0}) \gamma_\mu u^{(\beta)}(m, \vec{0}), \tag{4.96}$$

from which

$$\bar{u}^{(\alpha)}(m,\vec{0})\gamma^\mu u^{(\beta)}(m,\vec{0}) = g^{\mu 0} u^{\dagger(\alpha)}(m,\vec{0})\gamma_0 u^{(\beta)}(m,\vec{0}) = g^{\mu 0}\delta_{\alpha\beta}, \quad (4.97)$$

that is

$$\bar{u}^{(\alpha)}(k)u^{(\beta)}(k) = |c_\alpha|^2(2m^2 + 2mE)\delta_{\alpha\beta}. \quad (4.98)$$

Then we choose

$$c_\alpha = \frac{1}{\sqrt{2m(m+E)}}, \quad E = \sqrt{|\vec{k}|^2 + m^2}. \quad (4.99)$$

In analogous way we have

$$\begin{aligned}
\bar{v}^{(\alpha)}(k)v^{(\beta)}(k) &= d_\alpha^* d_\beta \bar{v}^{(\alpha)}(m,\vec{0})(-\hat{k}+m)^2 v^{(\beta)}(m,\vec{0}) \\
&= d_\alpha^* d_\beta \bar{v}^{(\alpha)}(m,\vec{0})(2m^2 - 2m\hat{k})v^{(\beta)}(m,\vec{0}) \quad (4.100)
\end{aligned}$$

and using the fact that $v^{(\alpha)}(m,\vec{0})$ and $v^{\dagger(\alpha)}(m,\vec{0})$ are eigenstates of $\gamma_0$ with eigenvalue $-1$, we get

$$\begin{aligned}
\bar{v}^{(\alpha)}(m,\vec{0})\gamma^\mu v^{(\beta)}(m,\vec{0}) &= -v^{\dagger(\alpha)}(m,\vec{0})\gamma_0\gamma^\mu\gamma_0 v^{(\beta)}(m,\vec{0}) \\
&= \bar{v}^{(\alpha)}(m,\vec{0})\gamma_\mu v^{(\beta)}(m,\vec{0}), \quad (4.101)
\end{aligned}$$

implying

$$\bar{v}^{(\alpha)}(m,\vec{0})\gamma^\mu v^{(\beta)}(m,\vec{0}) = g^{\mu 0}\delta_{\alpha\beta}. \quad (4.102)$$

Therefore we obtain

$$\bar{v}^{(\alpha)}(k)v^{(\beta)}(k) = -|d_\alpha|^2(2m^2 + 2mE)\delta_{\alpha\beta} \quad (4.103)$$

and, finally

$$d_\alpha = c_\alpha = \frac{1}{\sqrt{2m(m+E)}}. \quad (4.104)$$

The normalized solutions we have obtained are

$$u^{(\alpha)}(k) = \frac{\hat{k}+m}{\sqrt{2m(m+E)}} u^{(\alpha)}(m,\vec{0}), \; v^{(\alpha)}(k) = \frac{-\hat{k}+m}{\sqrt{2m(m+E)}} v^{(\alpha)}(m,\vec{0}). \quad (4.105)$$

Notice that positive and negative energy spinors are orthogonal. In the following it will be useful to express our solutions in terms of two component spinors, $\phi^{(\alpha)}(m,\vec{0})$ and $\chi^{(\alpha)}(m,\vec{0})$

$$u^{(\alpha)}(m,\vec{0}) = \begin{pmatrix} \phi^{(\alpha)}(m,\vec{0}) \\ 0 \end{pmatrix}, \quad v^{(\alpha)}(m,\vec{0}) = \begin{pmatrix} 0 \\ \chi^{(\alpha)}(m,\vec{0}) \end{pmatrix}. \quad (4.106)$$

From the explicit representation (4.19) of the $\gamma^\mu$ matrices we get

$$\hat{k} = \begin{pmatrix} E & -\vec{k}\cdot\vec{\sigma} \\ \vec{k}\cdot\vec{\sigma} & -E \end{pmatrix}, \tag{4.107}$$

from which

$$u^{(\alpha)}(k) = \begin{pmatrix} \sqrt{\dfrac{m+E}{2m}}\phi^{(\alpha)}(m,\vec{0}) \\ \dfrac{\vec{k}\cdot\vec{\sigma}}{\sqrt{2m(m+E)}}\phi^{(\alpha)}(m,\vec{0}) \end{pmatrix}, \tag{4.108}$$

$$v^{(\alpha)}(k) = \begin{pmatrix} \dfrac{\vec{k}\cdot\sigma}{\sqrt{2m(m+E)}}\chi^{(\alpha)}(m,\vec{0}) \\ \sqrt{\dfrac{m+E}{2m}}\chi^{(\alpha)}(m,\vec{0}) \end{pmatrix}. \tag{4.109}$$

In the following we will need the explicit expression for the projectors of the positive and negative energy solutions. To this end, let us observe that

$$\sum_{\alpha=1}^{2} u^{(\alpha)}(m,\vec{0})\bar{u}^{(\alpha)}(m,\vec{0}) = \begin{pmatrix} 1 \\ 0 \\ 0 \\ 0 \end{pmatrix}(1\,0\,0\,0) + \begin{pmatrix} 0 \\ 1 \\ 0 \\ 0 \end{pmatrix}(0\,1\,0\,0) = \frac{1+\gamma_0}{2} \tag{4.110}$$

and analogously

$$\sum_{\alpha=1}^{2} v^{(\alpha)}(m,\vec{0})\bar{v}^{(\alpha)}(m,\vec{0}) = -\frac{1-\gamma_0}{2}. \tag{4.111}$$

Using $\gamma_0\gamma_\mu = 2g_{\mu 0} - \gamma_\mu\gamma_0$ and $k^2 = m^2$, we get

$$\begin{aligned}
(\hat{k}+m)\gamma_0(\hat{k}+m) &= (\hat{k}+m)(2E - \hat{k}\gamma_0 + m\gamma_0) \\
&= 2E(\hat{k}+m) + (\hat{k}+m)(-\hat{k}+m)\gamma_0 \\
&= 2E(\hat{k}+m).
\end{aligned} \tag{4.112}$$

Therefore the positive energy projector is given by

$$\begin{aligned}
\Lambda_+(k) &= \sum_{\alpha=1}^{2} u^{(\alpha)}(k)\bar{u}^{(\alpha)}(k) = \frac{\hat{k}+m}{\sqrt{2m(m+E)}}\frac{1+\gamma_0}{2}\frac{\hat{k}+m}{\sqrt{2m(m+E)}} \\
&= \frac{1}{2m(m+E)}\frac{(\hat{k}+m)^2 + 2E(\hat{k}+m)}{2} = \frac{\hat{k}+m}{2m}.
\end{aligned} \tag{4.113}$$

In similar way we get the expression for the negative energy projector

$$\Lambda_-(k) = -\sum_{\alpha=1}^{2} v^{(\alpha)}(k)\bar{v}^{(\alpha)}(k) = \frac{-\hat{k}+m}{2m}. \tag{4.114}$$

It is easy to verify that the matrices $\Lambda_\pm(k)$ verify all the properties of a complete set of projection operators

$$\Lambda_\pm^2 = \Lambda_\pm, \qquad \Lambda_+\Lambda_- = 0, \qquad \Lambda_+ + \Lambda_- = 1. \tag{4.115}$$

In this normalization the density $\psi^\dagger\psi$ has the correct Lorentz transformation properties

$$\psi_{(\alpha)}^{(+)\dagger}(x)\psi_{(\beta)}^{(+)}(x) = \bar{u}^{(\alpha)}(k)\gamma_0 u^{(\beta)}(k)$$

$$= \frac{1}{2m(m+E)}\bar{u}^{(\alpha)}(m,\vec{0})(\hat{k}+m)\gamma_0(\hat{k}+m)u^{(\beta)}(m,\vec{0})$$

$$= \frac{1}{2m(m+E)}2E\bar{u}^{(\alpha)}(m,\vec{0})(\hat{k}+m)u^{(\beta)}(m,\vec{0})$$

$$= \frac{2E(m+E)}{2m(m+E)}\delta_{\alpha\beta} = \frac{E}{m}\delta_{\alpha\beta}, \tag{4.116}$$

where we have used eq. (4.112). Therefore the density for positive energy solutions transforms as the fourth component of a four-vector. The same is true for the negative energy solutions

$$\psi_{(\alpha)}^{(-)\dagger}(x)\psi_{(\beta)}^{(-)}(x) = \frac{E}{m}\delta_{\alpha\beta}. \tag{4.117}$$

We find also

$$u^{\dagger(\alpha)}(k)v^{(\beta)}(\tilde{k}) = 0, \quad k^\mu = (E,\vec{k}), \quad \tilde{k}^\mu = (E,-\vec{k}). \tag{4.118}$$

In fact,

$$u^{\dagger(\alpha)}(k)v^{(\beta)}(\tilde{k}) = \bar{u}^{(\alpha)}(m,\vec{0})\frac{(\hat{k}+m)\gamma_0(-\hat{\tilde{k}}+m)}{2m(m+E)}v^{(\beta)}(m,\vec{0})$$

$$= \bar{u}^{(\alpha)}(m,\vec{0})\frac{(\hat{k}+m)(-\hat{k}+m)\gamma_0}{2m(m+E)}v^{(\beta)}(\tilde{m},\vec{0})$$

$$= 0. \tag{4.119}$$

It follows that solutions with opposite energy and with the same three momentum are orthogonal

$$\psi^{(+)} = e^{-i(Ex^0-\vec{k}\cdot\vec{x})}u(k), \quad k^\mu = (E,\vec{k}),$$
$$\psi^{(-)} = e^{+i(Ex^0+\vec{k}\cdot\vec{x})}v(\tilde{k}), \quad \tilde{k}^\mu = (E,-\vec{k}). \tag{4.120}$$

The positive and negative energy solutions are doubly degenerate. It is possible to remove the degeneration through the construction of projectors for states with definite polarization. Let us consider again the solutions in the rest frame. The generator of the rotations along the $z$-axis is given by

$$\sigma_{12} = \begin{pmatrix} \sigma_3 & 0 \\ 0 & \sigma_3 \end{pmatrix}. \tag{4.121}$$

Clearly $u^{(1)}(m, \vec{0})$ and $v^{(1)}(m, \vec{0})$ are eigenstates of this operator (and therefore of the third component of the spin operator) with eigenvalues $+1$, whereas $u^{(2)}(m, \vec{0})$ and $v^{(2)}(m, \vec{0})$ belong to the eigenvalue $-1$. The projector for the eigenstates with eigenvalues $+1$ can be written as

$$\frac{1 + \sigma_{12}}{2} = \frac{1 + \sigma_{12} n_R^3}{2}, \tag{4.122}$$

where $n_R^\mu = (0, 0, 0, 1)$ is a unit space-like four-vector. Also we have

$$\sigma_{12} = \frac{i}{2}[\gamma_1, \gamma_2] = i\gamma^1 \gamma^2 = -\gamma^0 \gamma_5 \gamma^3 = \gamma_5 \gamma_3 \gamma_0 \tag{4.123}$$

and

$$\sigma_{12} n_R^3 = \gamma_5 \hat{n}_R \gamma_0. \tag{4.124}$$

The presence of $\gamma_0$ forbids a simple extension of this expression to a generic Lorentz frame. We can avoid this, by changing the definition of the projection in the rest frame system. Let us put

$$\Sigma(\pm n_R) = \frac{1 \pm \sigma_{12} n_R^3 \gamma_0}{2} = \frac{1}{2} \begin{pmatrix} 1 \pm \sigma_3 & 0 \\ 0 & 1 \mp \sigma_3 \end{pmatrix}. \tag{4.125}$$

In this case $\Sigma(n_R)$ and $\Sigma(-n_R)$ project out $u^{(1)}(m, \vec{0})$, $v^{(2)}(m, \vec{0})$ and $u^{(2)}(m, \vec{0})$, $v^{(1)}(m, \vec{0})$, respectively. That is, $\Sigma(\pm n_R)$ projects out the positive energy solutions with spin $\pm 1/2$ and the negative energy solutions with spin $\mp 1/2$. Then, we have

$$\Sigma(\pm n_R) = \frac{1 \pm \gamma_5 \hat{n}_R}{2}. \tag{4.126}$$

In the rest frame $n_R^2 = -1$, $n_R \cdot k = 0$. We can go to a generic frame preserving these conditions

$$\Sigma(\pm n) = \frac{1 \pm \gamma_5 \hat{n}}{2}, \quad n^2 = -1, \quad n \cdot k = 0. \tag{4.127}$$

The projector $\Sigma(\pm n)$ projects out energy positive states that in the rest frame have a polarization given by $\vec{S} \cdot \vec{n} = \pm 1/2$, and the negative energy states with polarization $\vec{S} \cdot \vec{n} = \mp 1/2$.

In the following we will use the following notation

$$u(k_R, n_R) = u^{(1)}(m, \vec{0}),$$
$$u(k_R, -n_R) = u^{(2)}(m, \vec{0}),$$
$$v(k_R, -n_R) = v^{(1)}(m, \vec{0}),$$
$$v(k_R, n_R) = v^{(2)}(m, \vec{0}). \tag{4.128}$$

These spinors satisfy the following relations

$$\Sigma(\pm n_R)u(k_R, \pm n_R) = u(k_R, \pm n_R), \quad \Sigma(\pm n_R)v(k_R, \pm n_R) = v(k_R, \pm n_R) \tag{4.129}$$

and

$$\Sigma(\pm n_R)u(k_R, \mp n_R) = \Sigma(\pm n_R)v(k_R, \mp n_R) = 0. \tag{4.130}$$

All these relations generalize immediately to an arbitrary reference frame (always requiring $n^2 = -1$ and $n \cdot k = 0$)

$$\Sigma(\pm n)u(k, \pm n) = u(k, \pm n), \quad \Sigma(\pm n)v(k, \pm n) = v(k, \pm n), \tag{4.131}$$

$$\Sigma(\pm n)u(k, \mp n) = \Sigma(\pm n)v(k, \mp n) = 0. \tag{4.132}$$

The properties of the spin projectors are

$$\Sigma(n) + \Sigma(-n) = 1, \quad \Sigma(\pm n)^2 = \Sigma(\pm n), \quad \Sigma(n)\Sigma(-n) = 0. \tag{4.133}$$

Let us just verify the second equation

$$\left(\frac{1 + \gamma_5\hat{n}}{2}\right)^2 = \frac{1 + (\gamma_5\hat{n})^2 + 2\gamma_5\hat{n}}{4} = \frac{2 + 2\gamma_5\hat{n}}{4} = \Sigma(n), \tag{4.134}$$

where we have made use of $n^2 = -1$. In analogous way

$$\Sigma(n)\Sigma(-n) = \frac{1 + \gamma_5\hat{n}}{2}\frac{1 - \gamma_5\hat{n}}{2} = \frac{1 - (\gamma_5\hat{n})^2}{4} = 0. \tag{4.135}$$

## 4.5 Wave packets and negative energy solutions

As we have shown the Dirac equation leads to a positive probability density. This solves the problem present in the Klein-Gordon case. On the other hand the Dirac equation does not solve the problem of the negative energy solutions (and it should not, as we have seen their relevance for locality in the Klein-Gordon case). In fact, the completeness of the spinors involves all the solutions

$$\sum_{\alpha=1}^{2} \left[ u^{(\alpha)}(k)\bar{u}^{(\alpha)}(k) - v^{(\alpha)}(k)\bar{v}^{(\alpha)}(k) \right] = \Lambda_+(k) + \Lambda_-(k) = 1. \tag{4.136}$$

In the case of a non-interacting theory there are no possibilities of transitions among positive and negative energy states but, when an interaction is turned on, such a possibility cannot be excluded. In fact, if we try to localize a Dirac particle within distances of order $1/m$ the negative energy solutions cannot be ignored. To clarify this point let us consider the time evolution of a Gaussian wave packet, assigned at time $t = 0$,

$$\psi(\vec{x}, 0) = \frac{1}{(\pi d^2)^{3/4}} e^{-\frac{|\vec{x}|^2}{2d^2}} w, \tag{4.137}$$

where $w$ is a fixed spinor, $w = (\phi, 0)$, with $w^\dagger w = 1$. As one can check, the wave packet is correctly normalized to one

$$\int d^3 x \, \psi^\dagger \psi = \frac{1}{(\pi d^2)^{3/2}} \int d^3 x \, e^{-\frac{|\vec{x}|^2}{d^2}} = 1. \tag{4.138}$$

The solution of the Dirac equation with this boundary condition is obtained by expanding over all the plane wave solutions

$$\psi(\vec{x}, t) = \int d^3 k \frac{1}{\sqrt{(2\pi)^3}} \sqrt{\frac{m}{E}} \sum_{\alpha=1}^{2} \Big[ b(k, \alpha) u^{(\alpha)}(k) e^{-ikx}$$
$$+ d^\star(k, \alpha) v^{(\alpha)}(k) e^{ikx} \Big] \tag{4.139}$$

and evaluating the expansion coefficients $b(k, \alpha)$ and $d^\star(k, \alpha)$, by requiring that the solution coincides with eq. (4.137) at time $t = 0$, that is

$$\psi(\vec{x}, 0) = \int d^3 k \frac{1}{\sqrt{(2\pi)^3}} \sqrt{\frac{m}{E}} \sum_{\alpha=1}^{2} \Big[ b(k, \alpha) u^{(\alpha)}(k)$$
$$+ d^\star(\tilde{k}, \alpha) v^{(\alpha)}(\tilde{k}) \Big] e^{i\vec{k}\cdot\vec{x}} = \frac{1}{(\pi d^2)^{3/4}} e^{-\frac{|\vec{x}|^2}{2d^2}} w. \tag{4.140}$$

where $\tilde{k}^\mu = (E, -\vec{k})$. Fourier transforming both sides of this equation

$$\int \psi(\vec{x}, 0) e^{-i\vec{k}\cdot\vec{x}} d^3 \vec{x} = \sqrt{(2\pi)^3} \sqrt{\frac{m}{E}} \sum_{\alpha=1}^{2} \Big[ b(k, \alpha) u^{(\alpha)}(k) + d^\star(\tilde{k}, \alpha) v^{(\alpha)}(\tilde{k}) \Big]$$
$$= \frac{1}{(\pi d^2)^{3/4}} \int d^3 x \, e^{-\frac{|\vec{x}|^2}{2d^2}} e^{-i\vec{k}\cdot\vec{x}} w$$
$$= \frac{1}{(\pi d^2)^{3/4}} (2\pi d^2)^{3/2} e^{-\frac{|\vec{k}|^2 d^2}{2}} w. \tag{4.141}$$

From which

$$\sqrt{\frac{m}{E}} \sum_{\alpha=1}^{2} \Big[ b(k, \alpha) u^{(\alpha)}(k) + d^\star(\tilde{k}, \alpha) v^{(\alpha)}(\tilde{k}) \Big] = \left( \frac{d^2}{\pi} \right)^{3/4} e^{-\frac{|\vec{k}|^2 d^2}{2}} w. \tag{4.142}$$

Using the orthogonality relations for the spinors we find the amplitudes

$$b(k, \alpha) = \sqrt{\frac{m}{E}} \left(\frac{d^2}{\pi}\right)^{3/4} e^{-\frac{|\vec{k}|^2 d^2}{2}} u^{\dagger(\alpha)}(k)w, \qquad (4.143)$$

$$d^\star(\tilde{k}, \alpha) = \sqrt{\frac{m}{E}} \left(\frac{d^2}{\pi}\right)^{3/4} e^{-\frac{|\vec{k}|^2 d^2}{2}} v^{\dagger(\alpha)}(\tilde{k})w. \qquad (4.144)$$

Expressing $u$ and $v$ in terms of two-component spinors (see eqs. (4.108) and (4.109)) we get

$$b(k, \alpha) = \sqrt{\frac{m}{E}} \left(\frac{d^2}{\pi}\right)^{3/4} e^{-\frac{|\vec{k}|^2 d^2}{2}} \sqrt{\frac{m+E}{2m}} \phi^{(\alpha)\dagger}(m, \vec{0})\phi, \qquad (4.145)$$

$$d^\star(\tilde{k}, \alpha) = \sqrt{\frac{m}{E}} \left(\frac{d^2}{\pi}\right)^{3/4} e^{-\frac{|\vec{k}|^2 d^2}{2}} \frac{1}{\sqrt{2m(m+E)}} \chi^{(\alpha)\dagger}(m, \vec{0})\vec{k} \cdot \vec{\sigma}\phi, \qquad (4.146)$$

from which we can evaluate the ratio of the negative energy amplitudes to the positive energy ones

$$\frac{d^\star(\tilde{k}, \alpha)}{b(k, \alpha)} \approx \frac{|\vec{k}|}{m+E}. \qquad (4.147)$$

The amplitudes (for both signs of the energy) contribute only if $|\vec{k}| \ll 1/d$ (due to the Gaussian exponential). Suppose that we want to localize the particle over distances larger than $1/m$, that is we require $d \gg 1/m$. Since the negative energy state amplitudes are important only for $|\vec{k}| > m \gg 1/d$, their contribution is depressed by the Gaussian exponential. On the other hand, if we try to localize the particle over distances $d \approx 1/m$, the negative energy states contribution becomes important for values of $|\vec{k}|$ of order $m$, or of order $1/d$, that is in the momentum region in which the corresponding amplitudes are not negligible. We see that the negative energy solutions are necessary in order to be consistent with the uncertainty principle.

## 4.6 Electromagnetic interaction of a relativistic point-like particle

In order to understand better the physics behind the Dirac equation we will now introduce the interaction with the electromagnetic field. For the moment we will do it at the level of classical field theory. This will allow us to discuss the spin of the Dirac particle and also the properties of the antiparticles. To this end we will start considering the interaction of a point-like particle with the electromagnetic field in the relativistic formalism. From this discussion we will be able to derive a recipe for introducing the coupling of any charged particle to the electromagnetic field. Later on we will see that this prescription is equivalent to the requirement of gauge invariance of the theory (see Section 6.4).

Let us recall that the classical expression for the electromagnetic four-current is given by

$$j^\mu = (\rho, \rho\vec{v}), \tag{4.148}$$

where $\rho$ is the charge density, and $\vec{v}$ the velocity field. A point-like particle, following a given world line, can be described in a parametric form by four functions $x^\mu(\tau)$, with $\tau$ an arbitrary line parameter. The corresponding charge density at the time $t$ is localized at the position $\vec{x}(\tau)$, evaluated at the parameter value $\tau$ such that $t = x^0(\tau)$ (see Fig. 4.1). Therefore

$$\rho(\vec{y}, t) = e\delta^3(\vec{y} - \vec{x}(\tau))|_{t=x^0(\tau)}. \tag{4.149}$$

It follows

$$j^\mu(y) = e\frac{dx^\mu}{dx^0}\delta^3(\vec{y} - \vec{x}(\tau))|_{y^0=x^0(\tau)}. \tag{4.150}$$

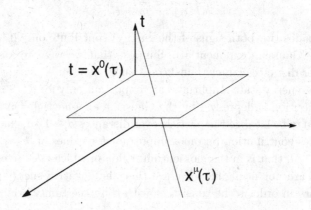

Fig. 4.1   The space-time trajectory of a point-like particle.

This expression can be put in a covariant form by using the following relation[2]

$$\int d\tau f(\tau)\delta(y^0 - x^0(\tau)) = \left(\frac{dx^0}{d\tau}\right)^{-1} f(\tau)\bigg|_{x^0(\tau)=y^0}. \tag{4.151}$$

---

[2]We assume $dx^0/d\tau > 0$ in order to have a correct parametrization of the trajectory.

We get

$$j^\mu(y) = e \int_{-\infty}^{+\infty} d\tau \frac{dx^\mu}{d\tau} \delta^4(y - x(\tau)). \qquad (4.152)$$

This four-current is conserved:

$$\partial_\mu j^\mu(y) = -e \int_{-\infty}^{+\infty} d\tau \frac{dx^\mu}{d\tau} \frac{\partial}{\partial x^\mu} \delta^4(y - x(\tau)) = 0. \qquad (4.153)$$

The only possible contributions could come from the end points $x(\pm\infty)$.

We recall also that the equations of motion for a free relativistic scalar particle can be derived by the following action

$$S = -m \int_{\tau_i}^{\tau_f} d\tau \sqrt{\dot{x}^2}, \qquad \dot{x}^\mu = \frac{dx^\mu}{d\tau}. \qquad (4.154)$$

We will be interested in deriving the Lagrangian describing the interaction between a charged particle and the electromagnetic field (we assume that the particle has charge $e$). We should be able to derive the following equations of motion

$$\frac{d}{dt} \frac{m\vec{v}}{\sqrt{1 - |\vec{v}|^2}} = e(\vec{E} + \vec{v} \wedge \vec{B}). \qquad (4.155)$$

We will show that the Lagrangian depends on the four-vector potential $A_\mu$ and not on the fields $\vec{E}$ and $\vec{B}$. In fact we will verify that the following action reproduces the previous equations of motion

$$S = -m \int_{\tau_i}^{\tau_f} d\tau \sqrt{\dot{x}^2} - \int d^4y A_\mu(y) j^\mu(y)$$

$$= -m \int_{\tau_i}^{\tau_f} d\tau \sqrt{\dot{x}^2} - e \int_{\tau_i}^{\tau_f} d\tau A_\mu(x(\tau)) \dot{x}^\mu(\tau). \qquad (4.156)$$

Using

$$\frac{\partial L}{\partial x^\mu} = -e \frac{\partial A_\nu}{\partial x^\mu} \dot{x}^\nu, \qquad \frac{\partial L}{\partial \dot{x}^\mu} = -m \frac{\dot{x}_\mu}{\sqrt{\dot{x}^2}} - eA_\mu \qquad (4.157)$$

and the Euler-Lagrangian equations

$$\frac{\partial L}{\partial x^\mu} - \frac{d}{d\tau} \frac{\partial L}{\partial \dot{x}^\mu} = 0, \qquad (4.158)$$

we get

$$-e \frac{\partial A_\nu}{\partial x^\mu} \dot{x}^\nu + m \frac{d}{d\tau} \frac{\dot{x}_\mu}{\sqrt{\dot{x}^2}} + e \frac{\partial A_\mu}{\partial x^\nu} \dot{x}^\nu = 0. \qquad (4.159)$$

Therefore

$$m \frac{d}{d\tau} \frac{\dot{x}_\mu}{\sqrt{\dot{x}^2}} = e(\partial_\mu A_\nu - \partial_\nu A_\mu) \dot{x}^\nu. \qquad (4.160)$$

Since $ds = d\tau\sqrt{\dot{x}^2}$, where $ds$ is the line element measured along the trajectory, we see that the four-velocity of the particle is

$$U^\mu = \frac{\dot{x}^\mu}{\sqrt{\dot{x}^2}}, \tag{4.161}$$

from which we get the equations of motion in a covariant form

$$m\frac{d}{ds}U_\mu = eF_{\mu\nu}U^\nu. \tag{4.162}$$

Here we have introduced the electromagnetic strength tensor

$$F_{\mu\nu} = \partial_\mu A_\nu - \partial_\nu A_\mu. \tag{4.163}$$

Using the relations between the potentials and the electric and magnetic fields

$$\vec{E} = -\vec{\nabla}A^0 - \frac{\partial\vec{A}}{\partial t}, \qquad \vec{B} = \vec{\nabla}\wedge\vec{A}, \tag{4.164}$$

we get

$$\vec{E} = (F^{10}, F^{20}, F^{30}), \qquad \vec{B} = (-F^{23}, -F^{31}, -F^{12}), \tag{4.165}$$

(we can also write $F^{ij} = -\epsilon_{ijk}B^k$). By choosing $\tau = x^0$ in eq. (4.160) we find

$$m\frac{d}{dt}\frac{-v^k}{\sqrt{1-|\vec{v}|^2}} = eF_{k0} + eF_{ki}\frac{dx^i}{dt} = -eE^k - \epsilon_{kij}B^j v^i, \tag{4.166}$$

reproducing eq. (4.155).

There are various ways to convince oneself about the necessity of the appearance of the four-potential inside the Lagrangian. For instance, consider the Maxwell equations

$$\partial^\mu F_{\mu\nu} = j_\nu, \qquad \partial^\mu \tilde{F}_{\mu\nu} = 0, \tag{4.167}$$

where $\tilde{F}_{\mu\nu} = \frac{1}{2}\epsilon_{\mu\nu\rho\sigma}F^{\rho\sigma}$ is the dual tensor. In Section 3.2 we have shown how to deduce the expression for the Lagrangian multiplying the field equations by an infinitesimal variation of the fields. In the actual case, to consider $F_{\mu\nu}$ as the fields to be varied, would create a problem because multiplying both sides of the Maxwell equations by $\delta F_{\mu\nu}$ we would not get a Lorentz scalar. This difficulty is avoided by taking $A_\mu$ as the independent degrees of freedom of the theory (we will show in the following that also this point of view has its own difficulties). In this case, due to the definition (4.163) of the electromagnetic tensor, the homogeneous Maxwell equations become identities (in fact, it is just solving these equations that one originally introduces the vector and the scalar potentials)

$$\tilde{F}_{\mu\nu} = \epsilon_{\mu\nu\rho\sigma}\partial^\rho A^\sigma \implies \partial^\mu \tilde{F}_{\mu\nu} = 0, \tag{4.168}$$

whereas the inhomogeneous ones give rise to

$$\partial^\mu(\partial_\mu A_\nu - \partial_\nu A_\mu) = \Box A_\nu - \partial_\nu \partial^\mu A_\mu = j_\nu. \tag{4.169}$$

The previous difficulty disappears because both sides of this equation multiplied by $A_\mu$ are Lorentz scalars. By regarding $j_\mu$ as a given external current, independent of $A_\mu$, we can now get easily the expression for the Lagrangian. By multiplying eq. (4.169) by $\delta A^\nu$ and integrating in $d^4x$ we get

$$
\begin{aligned}
0 &= \int_V d^4x \delta A^\nu (\Box A_\nu - \partial_\nu \partial^\mu A_\mu - j_\nu) \\
&= \int_V d^4x \Big[ -\delta(\partial_\mu A_\nu)\partial^\mu A^\nu + \partial_\mu(\delta A^\nu \partial^\mu A_\nu) \\
&\quad + \delta(\partial_\mu A_\nu)\partial^\nu A^\mu - \partial_\mu(\delta A^\nu \partial_\nu A^\mu) - \delta(A^\mu j_\mu) \Big] \\
&= \int_V d^4x \Big[ -\frac{1}{2}\delta(\partial_\mu A_\nu)\partial^\mu A^\nu - \frac{1}{2}\delta(\partial_\nu A_\mu)\partial^\nu A^\mu \\
&\quad + \frac{1}{2}\delta(\partial_\mu A_\nu)\partial^\nu A^\mu + \frac{1}{2}\delta(\partial_\nu A_\mu)\partial^\mu A^\nu - \delta(A^\mu j_\mu) \Big] + \text{surface terms} \\
&= -\frac{1}{2}\int_V d^4x (\delta F_{\mu\nu}) F^{\mu\nu} - \int_V d^4x \delta(A^\mu j_\mu) + \text{surface terms} \\
&= \delta \left[ \int_V d^4x \left( -\frac{1}{4} F_{\mu\nu} F^{\mu\nu} - A^\mu j_\mu \right) \right] + \text{surface terms}. \tag{4.170}
\end{aligned}
$$

We see that the action for an electromagnetic field interacting with an external current $j^\mu$ is given by (here $F_{\mu\nu}$ must be thought of as a function of $A_\mu$)

$$S = -\frac{1}{4}\int_V d^4x \, F_{\mu\nu} F^{\mu\nu} - \int_V d^4x \, j_\mu A^\mu. \tag{4.171}$$

Notice that the interacting term has the same structure we found for the point-like particle.

We stress again that the $A_\mu$'s are the canonical variables of the electrodynamics. In principle, one could reintroduce the fields by inverting the relations between fields and potentials. However, in this way, one would end up with a non-local action. From these considerations one can argue that the potentials play an important role in quantum mechanics, much more than in the classical case, where they are essentially a convenient trick. Recall also that the canonical variables satisfy local commutation relations (the commutator vanishes at space-like distances), implying that local observables should be local functions of the potentials. This is going to create some problem because the theory is invariant under gauge transformations, whereas the potentials are not

$$A_\mu(x) \to A_\mu(x) + \partial_\mu \Lambda(x), \tag{4.172}$$

(where $\Lambda(x)$ is an arbitrary function). Therefore, the observables of the theory should be gauge invariant implying that the potentials cannot be observed. An example of an observable which is both local in the potentials and gauge invariant is the electromagnetic tensor strength $F_{\mu\nu}$.

The pure electromagnetic part of the action (4.171) is naively gauge invariant, being a function of $F_{\mu\nu}$. As far as the interaction term is concerned, we have (assuming that the current is gauge invariant)

$$j_\mu A^\mu \to j_\mu A^\mu + j_\mu \partial^\mu \Lambda = j_\mu A^\mu + \partial^\mu(j_\mu \Lambda) - (\partial^\mu j_\mu)\Lambda. \tag{4.173}$$

Adding a four-divergence to the Lagrangian density does not change the equations of motion

$$\int_{t_1}^{t_2} dt \int d^3x \, \partial_\mu \chi^\mu = \int_{t_1}^{t_2} dt \frac{\partial}{\partial t} \int d^3 x \chi^0. \tag{4.174}$$

Therefore the invariance of the Lagrangian under gauge transformations (neglecting a four-divergence) is guaranteed, if the potentials are coupled to a gauge invariant and conserved current

$$\partial^\mu j_\mu = 0. \tag{4.175}$$

We have shown that this condition is indeed satisfied for the point-like particle.

In order to derive the general prescription to couple the electromagnetic potentials to a charged particle, let us go back to the action for the point-like particle. This prescription is known as the **minimal substitution**. By choosing $x^0 = \tau$ in the Lagrangian of eq. (4.156), we get

$$L = -m\sqrt{1 - |\vec{v}|^2} - e(A^0 - \vec{v} \cdot \vec{A}), \tag{4.176}$$

from which

$$\vec{p} = \frac{\partial L}{\partial \vec{v}} = \frac{m\vec{v}}{\sqrt{1 - |\vec{v}|^2}} + e\vec{A}. \tag{4.177}$$

The Hamiltonian is obtained through the usual Legendre transform

$$H = \vec{p} \cdot \vec{v} - L = m\frac{|\vec{v}|^2}{\sqrt{1 - |\vec{v}|^2}} + e\vec{v} \cdot \vec{A} + m\sqrt{1 - |\vec{v}|^2} + e(A^0 - \vec{v} \cdot \vec{A})$$

$$= \frac{m}{\sqrt{1 - |\vec{v}|^2}} + eA_0. \tag{4.178}$$

Therefore, we get the following relations

$$\vec{p} - e\vec{A} = m\frac{\vec{v}}{\sqrt{1 - |\vec{v}|^2}}, \qquad H - eA^0 = \frac{m}{\sqrt{1 - |\vec{v}|^2}}, \tag{4.179}$$

where the quantities in the right-hand sides of these two equations are the same as in the free case. It follows that we can go from the free case to the interacting one, through the simple substitution (minimal substitution)

$$p^\mu \to p^\mu - eA^\mu. \tag{4.180}$$

In the free case, inverting the relations between momenta and velocities

$$|\vec{v}|^2 = \frac{|\vec{p}|^2}{m^2 + |\vec{p}|^2}, \qquad 1 - |\vec{v}|^2 = \frac{m^2}{m^2 + |\vec{p}|^2}, \tag{4.181}$$

we get the Hamiltonian as a function of the canonical momenta

$$H_{\text{free}} = \sqrt{m^2 + |\vec{p}|^2}. \tag{4.182}$$

By performing the minimal substitution we get

$$H - eA^0 = \sqrt{m^2 + (\vec{p} - e\vec{A})^2}, \tag{4.183}$$

from which

$$H = eA^0 + \sqrt{m^2 + (\vec{p} - e\vec{A})^2}, \tag{4.184}$$

which is nothing but eq. (4.178), after using eq. (4.177). From the point of view of canonical quantization, the minimal substitution corresponds to the following substitution in the space-time derivatives

$$\partial^\mu \to \partial^\mu + ieA^\mu. \tag{4.185}$$

## 4.7   Nonrelativistic limit of the Dirac equation

In order to understand better the role of the spin in the Dirac equation we will now study the nonrelativistic limit in the presence of an electromagnetic field.

$$(i\hat{\partial} - m)\psi(x) = 0 \implies (i\hat{\partial} - e\hat{A} - m)\psi(x) = 0. \tag{4.186}$$

Notice that the Dirac équation is invariant under the transformation (4.172)

$$A_\mu(x) \to A_\mu(x) + \partial_\mu \alpha(x), \tag{4.187}$$

if, at the same time, we perform the following local phase transformation on the wave function

$$\psi(x) \to e^{-ie\alpha(x)} \psi(x). \tag{4.188}$$

The eq. (4.186) is also invariant under Lorentz transformations if, in going from the frame $S$ to the frame $S'$ ($x \to x' = \Lambda x$), the field $A_\mu$ transforms as

$$A_\mu(x) \to A'_\mu(x') = (\Lambda^{-1})^\nu_\mu A_\nu(x). \tag{4.189}$$

This shows that $A_\mu(x)$, under Lorentz transformations behaves as $\partial_\mu$:

$$\frac{\partial}{\partial x^\mu} \to \frac{\partial}{\partial x'^\mu} = \frac{\partial x^\nu}{\partial x'^\mu} \frac{\partial}{\partial x^\nu} = (\Lambda^{-1})^\nu_\mu \partial_\nu. \tag{4.190}$$

Equation (4.189) says simply that $A_\mu$ transforms as a four-vector under Lorentz transformations.

In order to study the non-relativistic limit, it is better to write $\psi(x)$ in the following form

$$\psi(x) = \begin{pmatrix} \tilde{\phi}(x) \\ \tilde{\chi}(x) \end{pmatrix}, \tag{4.191}$$

where $\tilde{\phi}(x)$ and $\tilde{\chi}(x)$ are two-component spinors. By defining

$$\vec{\pi} = \vec{p} - e\vec{A} \tag{4.192}$$

and using the representation in $2 \times 2$ blocks of the Dirac matrices given in eq. (4.11), we get, after multiplication by $\gamma_0$,

$$i\frac{\partial}{\partial t} \begin{pmatrix} \tilde{\phi}(x) \\ \tilde{\chi}(x) \end{pmatrix} = \begin{pmatrix} 0 & \vec{\sigma} \cdot \vec{\pi} \\ \vec{\sigma} \cdot \vec{\pi} & 0 \end{pmatrix} \begin{pmatrix} \tilde{\phi}(x) \\ \tilde{\chi}(x) \end{pmatrix} + \begin{pmatrix} m & 0 \\ 0 & -m \end{pmatrix} \begin{pmatrix} \tilde{\phi}(x) \\ \tilde{\chi}(x) \end{pmatrix} + eA^0 \begin{pmatrix} \tilde{\phi}(x) \\ \tilde{\chi}(x) \end{pmatrix}. \tag{4.193}$$

This gives rise to two coupled differential equations

$$i\frac{\partial \tilde{\phi}}{\partial t} = \vec{\sigma} \cdot \vec{\pi} \tilde{\chi} + (m + eA^0)\tilde{\phi},$$

$$i\frac{\partial \tilde{\chi}}{\partial t} = \vec{\sigma} \cdot \vec{\pi} \tilde{\phi} - (m - eA^0)\tilde{\chi}. \tag{4.194}$$

In the non-relativistic limit and, for weak fields, the mass term is the dominant one, and the energy positive solution will behave roughly as $e^{-imt}$. With this consideration in mind we put

$$\begin{pmatrix} \tilde{\phi}(x) \\ \tilde{\chi}(x) \end{pmatrix} = e^{-imt} \begin{pmatrix} \phi(x) \\ \chi(x) \end{pmatrix}, \tag{4.195}$$

and assume that $\phi$ and $\chi$ are functions slowly varying with time. In this way we obtain

$$i\frac{\partial\phi}{\partial t} = \vec{\sigma}\cdot\vec{\pi}\chi + eA^0\phi,$$

$$i\frac{\partial\chi}{\partial t} = \vec{\sigma}\cdot\vec{\pi}\phi - (2m - eA^0)\chi. \tag{4.196}$$

Assuming $eA^0 \ll 2m$, and $\partial\chi/\partial t \approx 0$ we have

$$\chi \approx \frac{\vec{\sigma}\cdot\vec{\pi}}{2m}\phi, \tag{4.197}$$

from which

$$i\frac{\partial\phi}{\partial t} = \left[\frac{(\vec{\sigma}\cdot\vec{\pi})^2}{2m} + eA^0\right]\phi. \tag{4.198}$$

One must be careful in evaluating $(\vec{\sigma}\cdot\vec{\pi})^2$, because the components of the vector $\vec{\pi}$ do not commute among themselves. In fact

$$[\pi^i, \pi^j] = [p^i - eA^i, p^j - eA^j] = ie\frac{\partial A^j}{\partial x^i} - ie\frac{\partial A^i}{\partial x^j}, \tag{4.199}$$

where we have made use of

$$[p^i, f(\vec{x})] = -i\frac{\partial f(\vec{x})}{\partial x^i}. \tag{4.200}$$

From $\vec{B} = \vec{\nabla}\wedge\vec{A}$ it follows that

$$[\pi^i, \pi^j] = ie\epsilon_{ijk}B^k \tag{4.201}$$

and

$$(\vec{\sigma}\cdot\vec{\pi})^2 = \sigma_i\sigma_j\pi^i\pi^j = \left(\frac{1}{2}[\sigma_i, \sigma_j] + \frac{1}{2}[\sigma_i, \sigma_j]_+\right)\pi^i\pi^j$$

$$= |\vec{\pi}|^2 + \frac{1}{4}[\sigma_i, \sigma_j][\pi^i, \pi^j] = |\vec{\pi}|^2 + \frac{i}{2}\epsilon_{ijk}\sigma_k(ie)\epsilon_{ijl}B^l, \tag{4.202}$$

that is

$$(\vec{\sigma}\cdot\vec{\pi})^2 = |\vec{\pi}|^2 - e\vec{\sigma}\cdot\vec{B}. \tag{4.203}$$

The equation for $\phi$ becomes

$$i\frac{\partial\phi}{\partial t} = \left[\frac{(\vec{p} - e\vec{A})^2}{2m} - \frac{e}{2m}\vec{\sigma}\cdot\vec{B} + eA^0\right]\phi. \tag{4.204}$$

This is nothing but the Pauli equation for an electron interacting with an electromagnetic field. In particular, the term proportional to the magnetic field represents the interaction with a magnetic dipole given by

$$\vec{\mu} = \frac{e}{2m}\vec{\sigma} = \frac{e}{m}\vec{S}, \tag{4.205}$$

where we have introduced the spin matrices $\vec{S} = \vec{\sigma}/2$. Recalling that the gyromagnetic ratio, $g$, for a particle of spin $\vec{S}$ and magnetic moment $\vec{\mu}$ is defined by the equation

$$\vec{\mu} = \frac{e}{2m} g \vec{S}, \qquad (4.206)$$

we see that the Dirac equation predicts a gyromagnetic ratio equal to two. We may see this also in a slightly different way, by considering the interaction with a weak uniform magnetic field. In this case the vector potential is given by

$$\vec{A} = \frac{1}{2} \vec{B} \wedge \vec{x}. \qquad (4.207)$$

Neglecting the quadratic term in the fields we have

$$(\vec{p} - e\vec{A})^2 \approx |\vec{p}|^2 - e(\vec{p} \cdot \vec{A} + \vec{A} \cdot \vec{p}). \qquad (4.208)$$

Using

$$\sum_i [p^i, A^i] = -i\vec{\nabla} \cdot \vec{A} = 0, \qquad (4.209)$$

it follows

$$\begin{aligned}
(\vec{p} - e\vec{A})^2 &\approx |\vec{p}|^2 - 2e\vec{p} \cdot \vec{A} = |\vec{p}|^2 - e\vec{p} \cdot (\vec{B} \wedge \vec{x}) \\
&= |\vec{p}|^2 - ep^i \epsilon_{ijk} B^j x^k = |\vec{p}|^2 - e\epsilon_{kij} x^k p^i B^j \\
&= |\vec{p}|^2 - e(\vec{x} \wedge \vec{p}) \cdot \vec{B} = |\vec{p}|^2 - e\vec{L} \cdot \vec{B}
\end{aligned} \qquad (4.210)$$

and finally

$$i\frac{\partial \phi}{\partial t} = \left[ \frac{|\vec{p}|^2}{2m} - \frac{e}{2m}(\vec{L} + 2\vec{S}) \cdot \vec{B} + eA^0 \right] \phi, \qquad (4.211)$$

which exhibits explicitly the value of the gyromagnetic ratio. Experimentally this is very close to two, and we shall see, in the following, that the difference is explained by quantum electrodynamics (QED). This is in fact one of the most important results of this theory. However, let us notice that, from the point of view of the Dirac equation, to find a value of the gyromagnetic ratio so close to the experimental value is not a real prediction. In fact, one could think of adding to the theory a further interaction term proportional to $F_{\mu\nu}\bar{\psi}\sigma^{\mu\nu}\psi$. This term is both Lorentz and gauge invariant. Such a term would give a further contribution to the magnetic moment of the electron, and therefore it would change the gyromagnetic ratio. We shall see that the requirement that QED is a renormalizable[3] forbids, in fact, the appearance of such a term.

---

[3]Renormalizable means that there exists a consistent procedure to eliminate the infinities arising when calculating higher order terms in perturbation theory (see later).

## 4.8 Charge conjugation, time reversal and PCT transformations

Dirac equation had a great success in explaining the fine structure of the hydrogen atom, but the problem of negative energy solutions that, in principle, makes the theory unstable, was still there. Dirac looked for a solution to this problem by taking advantage of the Pauli exclusion principle, which applies to half-integer spin particles. Dirac made the hypothesis that all the negative energy states were occupied by electrons. In such a situation, the Pauli principle forbids to any electron with positive energy to make a transition to a negative energy state. This solves the stability problem, but at the same time new phenomena are implied. For instance, an electron in a negative energy state could get enough energy (bigger than $2m$ which is the minimal energy gap between the negative and positive energy states) to make a transition to a state of positive energy. If we imagine that in the state of energy $-E$, $N$ electrons are present[4], and that one of these electrons undergoes the transition, the energy of the state changes as follows

$$\mathcal{E} - NE \rightarrow \mathcal{E} - (N-1)E = \mathcal{E} - NE + E, \qquad (4.212)$$

where $\mathcal{E}$ is the energy of all the other electrons (with energy different from $-E$) in the fundamental state. Notice that in the Dirac theory the fundamental state is the one with all the negative energy states occupied and zero electrons in the energy positive states. In a sense this is the physical explanation of the infinite energy of the vacuum that we found in the case of the Klein-Gordon field, and that we will find also in the Dirac case (see later). In a complete analogous way, also the charge of the vacuum is infinite and its variation in the previous transition is given by

$$Q + Ne \rightarrow Q + (N-1)e = Q + Ne - e, \qquad (4.213)$$

where $e$ is the charge of the electron ($e < 0$). We see that the vacuum energy and the electric charge increase respectively by $E$ and $-e$ in the transition. We can interpret this by saying that the hole left in the vacuum by the electron has charge $-e$ and energy $E$. In other words, we can think of the hole being a particle of positive energy and positive charge. This is the way in which the idea of antiparticles came around: the hole is

---

[4]We are simplifying things, because due to the momentum degeneracy there is actually an infinite number of electrons.

thought as the antiparticle of the electron. The transition of an electron of negative energy to a state of positive energy is then seen as the creation of a particle-antiparticle (the hole) pair. Of course this may happen only if a convenient amount of energy (for instance electromagnetic energy) is absorbed by the energy negative state. In the same way, once we have a hole in the vacuum, it may happen that a positive energy electron makes a transition to that particular hole state. In this case both the electron and the hole disappear. This is the pair annihilation phenomenon happening with a release of energy.

The hole theory is nowadays reinterpreted in terms of antiparticles, but this way of thinking has been extremely fruitful in many fields, as in the study of electrons in metals, in nuclear physics and so on.

Assuming seriously the hole theory means that the Dirac equation should admit, beyond the positive energy solutions corresponding to an electron, other positive energy solutions with the same mass of the electron, but with opposite charge. To see this point in a formal way, we look for a transformation of the electron wave function, $\psi(x)$, to the antielectron (positron) wave function $\psi^C(x)$, such that, if $\psi$ satisfies

$$(i\hat{\partial} - e\hat{A} - m)\psi(x) = 0, \tag{4.214}$$

then $\psi^C$ satisfies

$$(i\hat{\partial} + e\hat{A} - m)\psi^C(x) = 0. \tag{4.215}$$

$\psi^C$ is said the charge conjugated of $\psi$ and the operation relating the two wave functions is called **charge conjugation**. We will require the transformation to be local and such that the transformed of the antiparticle wave function gives back, except for a possible phase factor, the electron wave function. To build up $\psi^C$ we will start by taking the complex conjugate of $\psi$. This is clearly the only way of changing a negative energy solution, described by $e^{iEt}$, in a positive energy solution, described by $e^{-iEt}$. By taking the hermitian conjugate, multiplying by $\gamma_0$ (from the right) and transposing, we get

$$(i\hat{\partial} - e\hat{A} - m)\psi(x) = 0 \rightarrow -i\partial^\mu \bar{\psi}\gamma_\mu - e\bar{\psi}\hat{A} - m\bar{\psi} = 0$$
$$\rightarrow [\gamma^{\mu T}(-i\partial_\mu - eA_\mu) - m]\bar{\psi}^T = 0, \tag{4.216}$$

where

$$\bar{\psi}^T = \gamma_0{}^T \psi^\star. \tag{4.217}$$

If there is a matrix, $C$, such that

$$C\gamma_\mu{}^T C^{-1} = -\gamma_\mu, \tag{4.218}$$

multiplying eq. (4.216) by $C$, we get the wanted result

$$(i\hat{\partial} + e\hat{A} - m)C\bar{\psi}^T = 0. \tag{4.219}$$

This describes a particle with charge $-e$. Therefore, apart from a phase factor $\eta_C$, we can identify $\psi^C$ with $C\bar{\psi}^T$:

$$\psi^C = \eta_C C\bar{\psi}^T. \tag{4.220}$$

In the representation where $\gamma_0$ is diagonal we have

$$\gamma_0{}^T = \gamma_0, \quad \gamma_1{}^T = -\gamma_1, \quad \gamma_2{}^T = \gamma_2, \quad \gamma_3{}^T = -\gamma_3. \tag{4.221}$$

It is enough to choose $C$ commuting with $\gamma_1$ and $\gamma_3$ and anticommuting with $\gamma_0$ and $\gamma_2$. It follows that $C$ must be proportional to $\gamma_2\gamma_0$. Let us choose

$$C = i\gamma^2\gamma^0 = \begin{pmatrix} 0 & -i\sigma_2 \\ -i\sigma_2 & 0 \end{pmatrix}. \tag{4.222}$$

In this way, $C$ satisfies

$$-C = C^{-1} = C^T = C^\dagger. \tag{4.223}$$

To understand how the transformation works, let us consider, in the rest frame, a negative energy solution with spin down

$$\psi^{(-)}_{\text{down}} = e^{imt} \begin{pmatrix} 0 \\ 0 \\ 0 \\ 1 \end{pmatrix}, \tag{4.224}$$

$$\psi^{(-)C}_{\text{down}} = \eta_C C\bar{\psi}^T_{\text{down}} = \eta_C C\gamma_0 \psi^{(-)}_{\text{down}}{}^* = \eta_C i\gamma^2 \psi^{(-)}_{\text{down}}{}^*$$

$$= \eta_C e^{-imt} \begin{pmatrix} 0 & 0 & 0 & 1 \\ 0 & 0 & -1 & 0 \\ 0 & -1 & 0 & 0 \\ 1 & 0 & 0 & 0 \end{pmatrix} \begin{pmatrix} 0 \\ 0 \\ 0 \\ 1 \end{pmatrix} \tag{4.225}$$

and

$$\psi^{(-)C}_{\text{down}} = \eta_C e^{-imt} \begin{pmatrix} 1 \\ 0 \\ 0 \\ 0 \end{pmatrix} = \eta_C \psi^{(+)}_{\text{up}}. \tag{4.226}$$

That is, given a negative energy wave function describing an electron with spin down, its charge conjugated is a positive energy wave function describing a positron with spin up. For an arbitrary solution with definite energy and spin, by using the projectors of Section 4.4, we write

$$\psi = \frac{\epsilon\hat{p} + m}{2m} \frac{1 + \gamma_5\hat{n}}{2} \psi, \tag{4.227}$$

where $p_0 > 0$ and $\epsilon = \pm 1$ selects the energy sign. Since $C$ commutes with $\gamma_5$, $\gamma_5{}^\star = \gamma_5$, and

$$\gamma^0 \gamma^{\mu\star} \gamma^0 = \gamma^{\mu T}, \tag{4.228}$$

as it follows from

$$\gamma^0 \gamma^{\mu\dagger} \gamma^0 = \gamma^\mu, \qquad \gamma_0^T = \gamma_0, \tag{4.229}$$

we obtain

$$\psi^C = \eta_C C \gamma_0 \frac{\epsilon \hat{p}^\star + m}{2m} \frac{1 + \gamma_5 \hat{n}^\star}{2} \psi^\star = \eta_C C \frac{\epsilon \hat{p}^T + m}{2m} \frac{1 - \gamma_5 \hat{n}^T}{2} \gamma^0 \psi^\star$$

$$= \frac{-\epsilon \hat{p} + m}{2m} \frac{1 + \gamma_5 \hat{n}}{2} \psi^C. \tag{4.230}$$

We see that $\psi^C$ is described by the four-vectors $p^\mu$ and $n^\mu$ appearing in $\psi$, but with opposite sign of the energy. Then

$$u(p, n) = \eta_C v^C(p, n), \qquad v(p, n) = \eta_C u^C(p, n). \tag{4.231}$$

Since the spin projector selects the states of spin $\pm 1/2$ along $\vec{n}$ according to the sign of the energy, it follows that the charge conjugation inverts the spin projection of the particle. Notice also that, being $\psi^C$ a solution of the Dirac equation with $e \to -e$, the transformation

$$\psi \to \psi^C, \qquad A_\mu \to -A_\mu \tag{4.232}$$

is a symmetry of the Dirac equation. Since in this transformation we change sign to the four-potential, we say that the photon has charge conjugation $-1$.

Another discrete transformation we will consider here is time reversal. The physical meaning of this transformation can be illustrated in terms of a movie where we record all the observations made on the state described by the wave function $\psi(x)$. If we run the movie backward and we make a series of observations which are physically consistent, we say that the theory is invariant under time reversal. From a mathematical point of view this is a symmetry if, sending $t \to t' = -t$, it is possible to transform the wave function in such a way that it satisfies the original Dirac equation. If this happens, the transformed wave function describes an electron propagating backward in time. To build up explicitly the time reversal transformation, let us consider the electron in interaction with the electromagnetic field. It is convenient to write the Dirac equation in Hamiltonian form (see eq. (4.6))

$$i\frac{\partial \psi(\vec{x}, t)}{\partial t} = H\psi(\vec{x}, t), \tag{4.233}$$

with

$$H = eA^0 + \gamma^0\vec{\gamma} \cdot (-i\vec{\nabla} - e\vec{A}) + \gamma^0 m. \tag{4.234}$$

Let us define our transformation through the following equation

$$\psi'(\vec{x}, t') = \eta_T K\psi(\vec{x}, t), \quad t' = -t, \tag{4.235}$$

where $\eta_T$ is a phase factor, From eq. (4.233), omitting the spatial argument

$$i\frac{\partial}{\partial t}K^{-1}\psi'(t') = HK^{-1}\psi'(t'). \tag{4.236}$$

Multiplying this equation by $K$ we get

$$\frac{\partial}{\partial t'}K(-i)K^{-1}\psi'(t') = KHK^{-1}\psi'(t'). \tag{4.237}$$

The invariance can be realized in two ways

$$K(-i)K^{-1} = i; \qquad KHK^{-1} = H, \tag{4.238}$$

or

$$K(-i)K^{-1} = -i; \qquad KHK^{-1} = -H. \tag{4.239}$$

The second possibility can be excluded immediately, since under time reversal we have

$$\vec{\nabla} \to \vec{\nabla}, \qquad \vec{A} \to -\vec{A}, \qquad A^0 \to A^0. \tag{4.240}$$

As it follows recalling that the vector potential is generated by a distribution of currents (changing sign under time reversal), whereas the scalar potential is generated by a distribution of charges. Then, let us define $K$ as a $4 \times 4$ matrix $T$ times the operation of complex conjugation

$$K = T \times (\text{complex conjugation}). \tag{4.241}$$

From eq. (4.237) we get

$$i\frac{\partial}{\partial t'}\psi'(t') = TH^\star T^{-1}\psi'(t'). \tag{4.242}$$

The theory is invariant if

$$TH^\star T^{-1} = H. \tag{4.243}$$

By taking into account the transformation properties of the potentials we get

$$TH^\star T^{-1} = T(eA'_0 + (\gamma_0\vec{\gamma})^\star \cdot (i\vec{\nabla} + e\vec{A}') + \gamma_0{}^\star m)T^{-1}. \tag{4.244}$$

Since we want to reproduce $H$, we need a matrix $T$ such that

$$T\gamma_0(\vec{\gamma})^\star T^{-1} = -\gamma_0\vec{\gamma}, \qquad T\gamma_0 T^{-1} = \gamma_0, \qquad (4.245)$$

where we have used the reality properties of $\gamma_0$. In conclusion, $T$ must commute with $\gamma_0$ and satisfy

$$T\vec{\gamma}^\star T^{-1} = -\vec{\gamma}. \qquad (4.246)$$

In our representation, the matrices $\gamma^1$ and $\gamma^3$ are real, whereas $\gamma^2$ is pure imaginary. Therefore

$$T\gamma^0 T^{-1} = \gamma^0, \quad T\gamma^1 T^{-1} = -\gamma^1, \quad T\gamma^2 T^{-1} = \gamma^2, \quad T\gamma^3 T^{-1} = -\gamma^3. \qquad (4.247)$$

By choosing the phase, we put

$$T = i\gamma^1\gamma^3. \qquad (4.248)$$

With this choice $T$ satisfies

$$T^\dagger = T, \qquad T^2 = 1. \qquad (4.249)$$

To understand the correspondence with the classical results, where momentum and angular momentum change sign under time reversal, let us study how a positive energy solution transforms:

$$K\left[\frac{\hat{p}+m}{2m}\frac{1+\gamma_5\hat{n}}{2}\psi(t)\right] = T\left[\frac{\hat{p}^\star+m}{2m}\frac{1+\gamma_5\hat{n}^\star}{2}\right]\psi^\star(t)$$

$$= T\left[\frac{\hat{p}^\star+m}{2m}\right]T^{-1}T\left[\frac{1+\gamma_5\hat{n}^\star}{2}\right]T^{-1}T\psi^\star(t)$$

$$= \frac{\hat{\tilde{p}}+m}{2m}\frac{1+\gamma_5\hat{\tilde{n}}}{2}\psi'(t'), \qquad (4.250)$$

where, again $t' = -t$, and

$$\tilde{p} = (p^0, -\vec{p}), \qquad \tilde{n} = (n^0, -\vec{n}). \qquad (4.251)$$

The three discrete symmetry operations described so far, parity, $P$, charge conjugation, $C$, and time reversal, $(T)$, can be combined together into a symmetry transformation called $PCT$. Omitting all the phases, and recalling that the matrices $P$, $C$ and $T$ are defined respectively in eqs. (4.61), (4.222) and (4.241)

$$\psi_{PCT}(-x) = PC\overline{[K\psi(x)]}^T = PC\gamma_0(K\psi(x))^\star = i\gamma^0\gamma^2(-i\gamma^{1\star}\gamma^{3\star})\psi(x)$$

$$= i\gamma_5\psi(x) \qquad (4.252)$$

and it suggests a simple correspondence between the wave function of a positron moving backward in time ($\psi_{PCT}(-x)$), and the electron wave function. For a free particle of negative energy we have

$$\psi_{PCT}(-x) = i\gamma_5 \frac{-\hat{p} + m}{2m} \frac{1 + \gamma_5 \hat{n}}{2} \psi(x)$$

$$= \frac{\hat{p} + m}{2m} \frac{1 - \gamma_5 \hat{n}}{2} (i\gamma_5 \psi(x))$$

$$= \frac{\hat{p} + m}{2m} \frac{1 - \gamma_5 \hat{n}}{2} \psi_{PCT}(-x). \qquad (4.253)$$

Comparison with eq. (4.230), giving the charge conjugated of an energy negative state

$$\psi^C = \frac{\hat{p} + m}{2m} \frac{1 + \gamma_5 \hat{n}}{2} \psi^C, \qquad (4.254)$$

we see that the two expressions differ only in the spin direction. Similar conclusion can be reached by starting from the Dirac equation multiplied by $i\gamma_5$. We get ($x' = -x$, and $A'(x') = A(x)$)

$$i\gamma_5(i\hat{\partial}_x - e\hat{A}(x) - m)\psi(x) = (-i\hat{\partial}_x + e\hat{A}(x) - m)\psi_{PCT}(x')$$

$$= (i\hat{\partial}_{x'} + e\hat{A}(x') - m)\psi_{PCT}(x'), \quad (4.255)$$

showing that a positron moving backward in time satisfies the same equation as an electron moving forward. Equation (4.255) tells us that the *PCT* transformation on $\psi$, combined with the *PCT* transformation on the four-vector potential, that is $A_\mu(x) \to -A_\mu(-x)$, is a symmetry of the theory.

The interpretation of the positrons as negative energy electrons moving backward in time is the basis of the positron theory by [Stueckelberg (1942); Feynman (1948c, 1949b,c)]. In this approach it is possible to formulate the scattering theory without using field theory. In fact, the pair creation and pair annihilation processes can be reinterpreted in terms of scattering processes among electrons moving forward and backward in time.

## 4.9 Dirac field quantization

In this Section we abandon the study of the Dirac wave equation thought as a generalization of the Schrödinger equation, due to its difficulties to cope with many particle states. We will adopt here the point of view of quantum field theory. That is the relativistic equation will be reinterpreted as an equation for an operator-valued field. In the Klein-Gordon case we have shown that after quantization, we get a many-particle system satisfying

Bose-Einstein statistics. On the other hand we have also seen that the Dirac equation describes spin 1/2 particles, which should satisfy the Fermi-Dirac statistics. As a consequence we expect to run into troubles if we would insist in quantizing the Dirac field as we did for the Klein-Gordon case. To show how these troubles come about we will follow the canonical way of quantization, showing that this leads to problems with the positivity of the energy. Looking for a solution we will find also the way of solving the wrong statistics problem.

We will begin our study by looking for the action giving rise to the Dirac equation. We will take the quantities $\psi$ and $\bar{\psi}$ as independent ones. Following the usual procedure, we multiply the Dirac equation by $\delta\bar{\psi}$ (in such a way to form a Lorentz scalar) and integrate over the space-time volume $V$

$$0 = \int_V d^4x \; \delta\bar{\psi}(i\hat{\partial} - m)\psi = \delta \int_V d^4x \; \bar{\psi}(i\hat{\partial} - m)\psi, \qquad (4.256)$$

where we have assumed that $\psi$ and $\bar{\psi}$ are independent variables. We will then assume the following action

$$S = \int_V d^4x \; \bar{\psi}(i\hat{\partial} - m)\psi. \qquad (4.257)$$

It is simply verified that this action gives rise to the correct equation of motion for $\bar{\psi}$. In fact,

$$\frac{\partial\mathcal{L}}{\partial\psi} = -m\bar{\psi}, \qquad \frac{\partial\mathcal{L}}{\partial\psi_{,\mu}} = \bar{\psi}i\gamma^\mu, \qquad (4.258)$$

from which

$$-m\bar{\psi} - i\partial_\mu\bar{\psi}\gamma^\mu = 0. \qquad (4.259)$$

The canonical momenta result to be

$$\Pi_\psi = \frac{\partial\mathcal{L}}{\partial\dot{\psi}} = i\psi^\dagger, \qquad \Pi_{\psi^\dagger} = \frac{\partial\mathcal{L}}{\partial\dot{\psi}^\dagger} = 0. \qquad (4.260)$$

The canonical momenta do not depend on the velocities. In principle, this creates a problem for the Hamiltonian formalism. In fact a rigorous treatment requires an extension of the classical Hamiltonian treatment, which was performed by Dirac himself [Dirac (2001)]. In this particular case, the result one gets is the same as proceeding in a naive way. For this reason we will avoid to describe this extension, and we will proceed as in the standard case. Then the Hamiltonian density turns out to be

$$\mathcal{H} = \Pi_\psi\dot{\psi} - \mathcal{L} = i\psi^\dagger\dot{\psi} - \bar{\psi}(i\gamma^0\partial_0 + i\gamma^k\partial_k - m)\psi = \psi^\dagger(-i\vec{\alpha}\cdot\vec{\nabla} + \beta m)\psi. \qquad (4.261)$$

If one makes use of the Dirac equation, it is possible to write the Hamiltonian density as

$$\mathcal{H} = \psi^\dagger i \frac{\partial \psi}{\partial t}. \tag{4.262}$$

Contrarily to the Klein-Gordon case (see eq. (3.92)), the Hamiltonian density is not positive definite. Let us recall the general expression for the energy momentum tensor (see eq. (3.81))

$$T^\mu_\nu = \frac{\partial \mathcal{L}}{\partial \phi^i_{,\mu}} \phi^i_{,\nu} - g^\mu_\nu \mathcal{L}. \tag{4.263}$$

In our case we get

$$T^\mu_\nu = i\bar{\psi}\gamma^\mu \psi_{,\nu} - g^\mu_\nu (\bar{\psi}(i\hat{\partial} - m)\psi) \tag{4.264}$$

and using the Dirac equation

$$T^\mu_\nu = i\bar{\psi}\gamma^\mu \psi_{,\nu}, \tag{4.265}$$

we can immediately verify that this expression has vanishing four-divergence. Also

$$T^0_k = i\psi^\dagger \partial_k \psi, \tag{4.266}$$

from which we get the momentum of the field

$$P^k = \int d^3x \, T^{0k} \Longrightarrow \vec{P} = -i \int d^3x \, \psi^\dagger \vec{\nabla} \psi. \tag{4.267}$$

Now, let us consider the angular momentum density (see eq. (3.88))

$$\mathcal{M}^\mu_{\rho\nu} = x_\rho T^\mu_\nu - x_\nu T^\mu_\rho - \frac{\partial \mathcal{L}}{\partial \phi^i_{,\mu}} \Sigma^{ij}_{\rho\nu} \phi^j. \tag{4.268}$$

The matrices $\Sigma^{ij}_{\mu\nu}$ are defined in terms of the transformation properties of the field (see eq. (3.86))

$$\Delta \phi^i = -\frac{1}{2} \Sigma^{ij}_{\mu\nu} \epsilon^{\mu\nu} \phi^j. \tag{4.269}$$

From eqs. (4.30) and eq. (4.41) for an infinitesimal Lorentz transformation, we get

$$\Delta \psi(x) = \psi'(x') - \psi(x) = [S(\Lambda) - 1]\psi(x) \approx -\frac{i}{4}\sigma_{\mu\nu}\epsilon^{\mu\nu}\psi(x), \tag{4.270}$$

giving

$$\Sigma_{\mu\nu} = \frac{i}{2}\sigma_{\mu\nu} = -\frac{1}{4}[\gamma_\mu, \gamma_\nu]. \tag{4.271}$$

Our result is then

$$\mathcal{M}^\mu_{\rho\nu} = i\bar{\psi}\gamma^\mu \left( x_\rho \partial_\nu - x_\nu \partial_\rho - \frac{i}{2}\sigma_{\rho\nu} \right) \psi$$

$$= i\bar{\psi}\gamma^\mu \left( x_\rho \partial_\nu - x_\nu \partial_\rho + \frac{1}{4}[\gamma_\rho, \gamma_\nu] \right) \psi. \qquad (4.272)$$

By taking the spatial components we obtain

$$\vec{J} = (M^{23}, M^{31}, M^{12}) = \int d^3x \; \psi^\dagger \left( -i\vec{x} \wedge \vec{\nabla} + \frac{1}{2}\vec{\sigma} \otimes 1_2 \right) \psi, \qquad (4.273)$$

where $1_2$ is the identity matrix in 2 dimensions, and using eq. (4.59) we have defined

$$\vec{\sigma} \otimes 1_2 = \begin{pmatrix} \vec{\sigma} & 0 \\ 0 & \vec{\sigma} \end{pmatrix}. \qquad (4.274)$$

The expression of $\vec{J}$ shows the decomposition of the total angular momentum in the orbital and in the spin part. The theory has a further conserved quantity, the current $\bar{\psi}\gamma^\mu\psi$.

We will need the decomposition of the Dirac field in plane waves. To this end we will make use of the spinors $u(p, \pm n)$ and $v(p, \pm n)$ that we have defined at the end of Section 4.4. The expansion is similar to the one used in eq. (4.139), but with operator valued coefficients, $b$ and $d$.

$$\psi(x) = \sum_{\pm n} \int \frac{d^3 p}{\sqrt{(2\pi)^3}} \sqrt{\frac{m}{E_p}} \left[ b(p, n)u(p, n)e^{-ipx} + d^\dagger(p, n)v(p, n)e^{ipx} \right], \qquad (4.275)$$

$$\psi^\dagger(x) = \sum_{\pm n} \int \frac{d^3 p}{\sqrt{(2\pi)^3}} \sqrt{\frac{m}{E_p}} [d(p, n)\bar{v}(p, n)e^{-ipx}$$

$$+ b^\dagger(p, n)\bar{u}(p, n)e^{ipx}]\gamma_0, \qquad (4.276)$$

where $E_p = \sqrt{|\vec{p}|^2 + m^2}$. For convenience we will collect here the main properties of the spinors:

- Dirac equation

$$(\hat{p} - m)u(p, n) = \bar{u}(p, n)(\hat{p} - m) = 0,$$

$$(\hat{p} + m)v(p, n) = \bar{v}(p, n)(\hat{p} + m) = 0. \qquad (4.277)$$

- Orthogonality

$$\bar{u}(p, n)u(p, n') = -\bar{v}(p, n)v(p, n') = \delta_{nn'},$$

$$u^\dagger(p, n)u(p, n') = v^\dagger(p, n)v(p, n') = \frac{E_p}{m}\delta_{nn'},$$

$$\bar{v}(p, n)u(p, n') = v^\dagger(p, n)u(\tilde{p}, n') = 0, \qquad (4.278)$$

where, if $p^\mu = (E_p, \vec{p})$, then $\tilde{p}^\mu = (E_p, -\vec{p})$.

- Completeness

$$\sum_{\pm n} u(p,n)\bar{u}(p,n) = \frac{\hat{p}+m}{2m},$$

$$\sum_{\pm n} v(p,n)\bar{v}(p,n) = \frac{\hat{p}-m}{2m}. \tag{4.279}$$

We are now in the position to express the Hamiltonian in terms of the operators $b(p,n)$ and $d(p,n)$. Using eq. (4.262) and integrating over the space coordinates, we find

$$H = \sum_{\pm n, \pm n'} \int d^3p \, d^3p' \, \frac{E_{p'}m}{\sqrt{E_p E_{p'}}}$$

$$\times \Big[ d(p,n)b(\tilde{p},n')e^{-i(E_p+E_{p'})t}v^\dagger(p,n)u(\tilde{p},n')\delta^3(\vec{p}+\vec{p}')$$

$$+ b^\dagger(p,n)b(p,n')e^{+i(E_p-E_{p'})t}u^\dagger(p,n)u(p,n')\delta^3(\vec{p}-\vec{p}')$$

$$- d(p,n)d^\dagger(p,n')e^{-i(E_p-E_{p'})t}v^\dagger(p,n)v(p,n')\delta^3(\vec{p}-\vec{p}')$$

$$- b^\dagger(p,n)d(\tilde{p},n')e^{+i(E_p+E_{p'})t}$$

$$\times u^\dagger(p,n)v(\tilde{p},n')\delta^3(\vec{p}+\vec{p}')\Big]. \tag{4.280}$$

Performing one of the momentum integrations and using the orthogonality relations, we get

$$H = \sum_{\pm n} \int d^3p \, E_p[b^\dagger(p,n)b(p,n) - d(p,n)d^\dagger(p,n)]. \tag{4.281}$$

In analogous way we find

$$\vec{P} = \sum_{\pm n} \int d^3p \, \vec{p}[b^\dagger(p,n)b(p,n) - d(p,n)d^\dagger(p,n)]. \tag{4.282}$$

If we try to interpret these expressions as we did in the Klein-Gordon case, we would assume that the operator $d(p,n)$ creates from the vacuum a state of energy $-E_p$ and momentum $-\vec{p}$. Dirac tried to solve the problem assuming that the vacuum was filled up by the negative energy solutions. Due to the Pauli principle this would make impossible for any other negative energy state to be created. Let us called the vacuum filled up by the negative energy solutions the **Dirac vacuum**. Then the operator $d(p,n)$ should give zero when acting upon this state. We then define as the true vacuum of the theory the Dirac vacuum and require

$$d(p,n)|0\rangle_{\text{Dirac}} = 0. \tag{4.283}$$

That is, in the Dirac vacuum the operator $d(p, n)$ behaves as an annihilation operator (as we have anticipated in writing). Since the Dirac vacuum is obtained by applying to the original vacuum a bunch of operators $d(p, n)$, with all possible $p_\mu$ and $n_\mu$, the previous equation can be satisfied assuming the algebraic identity

$$(d(p, n))^2 = 0. \tag{4.284}$$

We can satisfy this relation in a uniform algebraic way by requiring that the operators $d(p, n)$ anticommute among themselves

$$[d(p, n), d(p', n')]_+ = 0. \tag{4.285}$$

This led [Jordan and Wigner (1928)] to the idea of quantizing the Dirac field in terms of anticommutators

$$\left[b(p, n), b^\dagger(p', n')\right]_+ = \left[d(p, n), d^\dagger(p', n')\right]_+ = \delta_{nn'}\delta^3(\vec{p} - \vec{p'}). \tag{4.286}$$

The problem of positivity is then solved automatically, since the four-momentum operator can be written as

$$P^\mu = \sum_{\pm n} \int d^3p \, p^\mu \left[b^\dagger(p, n)b(p, n) + d^\dagger(p, n)d(p, n) - \left[d(p, n), d^\dagger(p, n)\right]_+\right]. \tag{4.287}$$

Due to the anticommutation relations, the last term is an infinite negative constant which, physically, can be associated to the energy of the infinite electrons filling up the Dirac vacuum (called also the Dirac sea). If we ignore this constant (as we did in the Klein-Gordon case, and with the same warning), the energy operator is positive definite. The use of the anticommutators solves also the problem of the wrong statistics. In fact, the wave functions are now antisymmetric for the exchange of two Dirac particles (from now on we will put $|0\rangle_{\text{Dirac}} = |0\rangle$):

$$b^\dagger(p_1, n_1)b^\dagger(p_2, n_2)|0\rangle = -b^\dagger(p_2, n_2)b^\dagger(p_1, n_1)|0\rangle. \tag{4.288}$$

Therefore, the quanta of the Dirac field satisfy the Fermi-Dirac statistics. Once we have realized all that, we can safely forget about the hole theory and related stuff. In fact, looking at the four-momentum operator, we can simply say that $d^\dagger(p, n)$ creates and $d(p, n)$ annihilates a positron state. Then we think of the vacuum as a state with no electrons and/or positrons (that is without electrons and holes).

In the Klein-Gordon case we interpreted the conserved current as the electromagnetic current. We shall show now that in the Dirac case the expression $\bar{\psi}\gamma_\mu\psi$, has the same interpretation. Let us start evaluating the spatial integral of the density

$$\int d^3x \; \psi^\dagger\psi, \qquad (4.289)$$

in terms of the creation and annihilation operators

$$\int d^3x \; \psi^\dagger\psi = \sum_{\pm n} \int d^3p \left[b^\dagger(p,n)b(p,n) + d(p,n)d^\dagger(p,n)\right]. \qquad (4.290)$$

As we know, this expression is formally positive definite. However, if we couple the Dirac field to the electromagnetism through the minimal substitution we find that the free action (4.257) becomes

$$S = \int_V d^4x\bar{\psi}(i\hat{\partial} - e\hat{A} - m)\psi. \qquad (4.291)$$

Therefore the electromagnetic field is coupled to the conserved current

$$j^\mu = e\bar{\psi}\gamma^\mu\psi. \qquad (4.292)$$

This forces us to say that the integral of the fourth component of the current should be the charge operator, and as such it should not be positive definite. In fact, we find

$$Q = e \int d^3x \; \psi^\dagger\psi = \sum_{\pm n} \int d^3p \; e[b^\dagger(p,n)b(p,n) - d^\dagger(p,n)d(p,n)$$

$$+ \left[d(p,n), d^\dagger(p,n)\right]_+]. \qquad (4.293)$$

The subtraction of the infinite charge associated to the Dirac sea leaves us with an operator which is not anymore positive definite. We see also that the operators $b^\dagger$ create particles of charge $e$ (electrons) whereas $d^\dagger$ create particles of charge $-e$ (positrons). Notice that the interpretation of $Q$ as the charge operator would not have worked by using commutation relations.

A further potential problem is connected with causality. In the Klein-Gordon case we have shown that the causality properties are guaranteed, for local observables, by the canonical commutation relations for the fields. But this is just the property we have given up in the Dirac case. In order to discuss this point, let us start evaluating the equal time anticommutator for the Dirac field

$$[\psi(\vec{x},t),\psi^\dagger(\vec{y},t)]_+ = \int \frac{d^3p}{(2\pi)^3} \frac{m}{E_p}$$
$$\times \left[\left(\frac{\hat{p}+m}{2m}\right)e^{i\vec{p}\cdot(\vec{x}-\vec{y})} + \left(\frac{\hat{p}-m}{2m}\right)e^{-i\vec{p}\cdot(\vec{x}-\vec{y})}\right]\gamma_0$$
$$= \int \frac{d^3p}{(2\pi)^3}\frac{m}{E_p}\left[\left(\frac{\hat{p}+m}{2m}\right) + \left(\frac{\hat{\tilde{p}}-m}{2m}\right)\right]\gamma_0 e^{i\vec{p}\cdot(\vec{x}-\vec{y})}$$
$$= \int \frac{d^3p}{(2\pi)^3}\frac{m}{E_p}\frac{2E_p}{2m}e^{-i\vec{p}\cdot(\vec{x}-\vec{y})} = \delta^3(\vec{x}-\vec{y}), \tag{4.294}$$

where the four-vector $\tilde{p}^\mu$ is defined as in eq. (4.278), that is $\tilde{p}^\mu = (p^0, -\vec{p})$. In the same fashion we get

$$[\psi(\vec{x},t),\psi(\vec{y},t)]_+ = \left[\psi^\dagger(\vec{x},t),\psi^\dagger(\vec{y},t)\right]_+ = 0. \tag{4.295}$$

By using eq. (4.260), the anticommutator between $\psi$ and $\psi^\dagger$ can be written as

$$[\Pi_\psi(\vec{x},t),\psi(\vec{y},t)]_+ = i\delta^3(\vec{x}-\vec{y}). \tag{4.296}$$

This shows that also in the Dirac case one can use the canonical formalism of quantization, but replacing commutators with anticommutators. Evaluating the anticommutator for arbitrary space-time separations we obtain

$$[\psi(x),\psi^\dagger(y)]_+ = \int \frac{d^3p}{(2\pi)^3}\frac{m}{E_p}$$
$$\times \left[\left(\frac{\hat{p}+m}{2m}\right)e^{-ip(x-y)} + \left(\frac{\hat{p}-m}{2m}\right)e^{ip(x-y)}\right]\gamma_0$$
$$= \left[\left(i\hat{\partial}+m\right)_x\gamma_0\right]\int\frac{d^3p}{(2\pi)^3}\frac{1}{2E_p}\left[e^{-ip(x-y)}-e^{ip(x-y)}\right]$$
$$= \left[\left(i\hat{\partial}+m\right)_x\gamma_0\right]i\Delta(x-y), \tag{4.297}$$

where $\Delta(x)$ is the invariant function obtained in eq. (3.139) in the evaluation of the commutator for the Klein-Gordon field. From the properties of $\Delta(x)$, it follows that the anticommutator of the Dirac fields vanishes at space-like distances. Furthermore by evaluating the commutator of the Dirac field one can show that this does not vanish for a space-like separation. It follows that the Dirac field cannot be an observable quantity. This observation by itself would put in a serious trouble the idea of quantizing the Dirac field via anticommutation relations. So, we have to solve the causality problem. The crucial observation lies in the following identity

$$[AB,C] = A[B,C] + [A,C]B = A[B,C]_+ - [A,C]_+B, \tag{4.298}$$

which holds for arbitrary operators. The identity shows that $AB$ commutes with $C$ if $A$ and $B$ both commute or anticommute with $C$. An immediate consequence is that a local quantity containing an even number of Dirac fields commutes with itself at space-like distances. So, in order to reconcile the causality with the quantization of the Dirac field we have to give up the possibility that this field is an observable. However, all the important physical quantities, as energy-momentum tensor and electromagnetic current are bilinear in the Fermi fields, and therefore they are observable quantities.

What we have shown here is that, in order to give a sense to the quantization of the Dirac field, we have to use anticommutation relations which, in turn, imply that the corresponding quanta obey the Fermi-Dirac statistics. This is nothing but an example of the celebrated **spin statistics theorem** that was proved by [Pauli (1940)]. This theorem says that in a Lorentz invariant local field theory, integer and half-integer particles must satisfy Bose-Einstein and Fermi-Dirac statistics respectively.

## 4.10 Massless spin 1/2 particles

Let us consider the Dirac equation for a massless particle, $m = 0$,

$$i\hat{\partial}\psi(x) = 0. \tag{4.299}$$

By taking a positive energy solution

$$\psi(x) = e^{-ikx}\psi(k), \tag{4.300}$$

it follows from (4.299), $k^2 = 0$. Therefore, for positive energy solutions, $E = k_0 = |\vec{k}|$, we can write

$$E\gamma_0\,\psi(k) = \vec{k}\cdot\vec{\gamma}\,\psi(k) \to E\,\psi(k) = \vec{k}\cdot\vec{\alpha}\,\psi(k). \tag{4.301}$$

Recalling from Section 4.2 that

$$\alpha^i = \begin{pmatrix} 0 & \sigma_i \\ \sigma_i & 0 \end{pmatrix}, \quad \beta = \begin{pmatrix} 1 & 0 \\ 0 & -1 \end{pmatrix}, \quad \gamma_5 = \begin{pmatrix} 0 & 1 \\ 1 & 0 \end{pmatrix}, \tag{4.302}$$

we see that multiplying by $\gamma_5$ the second equation in (4.301) we get

$$\gamma_5\psi(k) = \frac{\vec{k}\cdot\vec{\Sigma}}{|\vec{k}|}\psi(k), \tag{4.303}$$

where

$$\vec{\Sigma} = \gamma_5\vec{\alpha} = \begin{pmatrix} \vec{\sigma} & 0 \\ 0 & \vec{\sigma} \end{pmatrix}. \tag{4.304}$$

The operator $\vec{\Sigma} \cdot \vec{k}/|\vec{k}|$ is called "helicity operator" since it tells us the direction of the spin with respect to the momentum. In fact, its eigenvalues are $\pm 1$, as it follows from

$$\left(\frac{\vec{\Sigma} \cdot \vec{k}}{|\vec{k}|}\right)^2 = 1. \tag{4.305}$$

The operator $\gamma_5$ is also called **chirality**. Therefore eq. (4.303) says that, in the massless case, for positive energy solutions chirality and helicity coincide. For the negative energy solutions they differ in sign since $E = -|\vec{k}|$. Notice that in the massless case there is no point in using $4 \times 4$ matrices. In fact the Dirac equation can be written as

$$i\frac{\partial \psi(x)}{\partial t} = -i\vec{\alpha} \cdot \vec{\nabla}\psi(x). \tag{4.306}$$

Since the only matrices appearing here are the $\alpha_i$'s, with the algebra

$$[\alpha_i, \alpha_j]_+ = 2\delta_{ij}, \tag{4.307}$$

it is clear that they can be represented in terms of $2 \times 2$ Pauli matrices, $\pm\sigma_i$'s. Here the choice of the sign is related to positive or negative chirality. This can be seen also starting by the $4 \times 4$ representation but choosing a different basis for the algebra of the $\gamma$ matrices, the **chiral basis**, where $\gamma_5$ is taken diagonal. Then the $\gamma$-matrices are given by (see Section 4.3)

$$\alpha^i = \begin{pmatrix} \sigma_i & 0 \\ 0 & -\sigma_i \end{pmatrix}, \quad \gamma_0 = \begin{pmatrix} 0 & -1 \\ -1 & 0 \end{pmatrix}, \quad \gamma_5 = \begin{pmatrix} 1 & 0 \\ 0 & -1 \end{pmatrix}. \tag{4.308}$$

In this representation the fields with positive and negative chiralities are given respectively by

$$\psi_+ = \begin{pmatrix} \phi \\ 0 \end{pmatrix}, \quad \psi_- = \begin{pmatrix} 0 \\ \chi \end{pmatrix}, \tag{4.309}$$

with the two component fields satisfying

$$(-k^0 + \vec{k} \cdot \vec{\sigma})\phi = 0, \quad (k^0 + \vec{k} \cdot \vec{\sigma})\chi = 0. \tag{4.310}$$

The relativistic description of a massless particle in terms of two-component spinors was first given by [Weyl (1929b)]. However this equation was rejected since it does not preserve parity. In fact in this representation the parity operator $\gamma_0$ is not diagonal and exchanges the two chirality solutions. Also the charge conjugation operator exchanges the two solutions. In fact the charge conjugation matrix $C$ (see eq. (4.222)) is given by

$$C = -i\alpha^2 = \begin{pmatrix} \sigma_2 & 0 \\ 0 & -\sigma_2 \end{pmatrix}. \tag{4.311}$$

This matrix is diagonal but in the definition of the charge conjugation there is also a conjugation and a $\gamma_0$ matrix (see eq. (4.220)). As a result the charge conjugation exchanges the chiralities and send positive energy solutions into negative energy ones. However, the combined operation $CP$ leaves the Weyl equation invariant. Today we know that parity is violated and in fact, the fermions in the standard models are described by massless Dirac equations and only after symmetry breaking they become massive. As a consequence the fermions can be described by Weyl equations that after the breaking pair together in such a way to generate massive Dirac fields.

The Weyl equations can be easily obtained from the considerations made in Section 4.3. In the massless case (and forgetting about a representation with definite parity) we do not need to introduce both representations $(1/2, 0)$ and $(0, 1/2)$. In fact, eq. (4.76)

$$i\partial^\mu(\sigma_\mu)_{\dot\alpha\beta}\zeta^\beta = 0, \tag{4.312}$$

for spinors belonging to the $(1/2, 0)$ representation is consistent by itself and we do not need to introduced a dotted spinor. By comparison with the results found in Section 4.3, we see that the two component spinors defined in eq. (4.309), $\phi$ and $\chi$ coincide respectively with $\zeta^\alpha$ and $\eta_{\dot\alpha}$.

We will not study in detail the solutions of the massless Dirac equation (4.299). However the normalization condition should be modified in this case. A convenient choice is

$$u^\dagger(p, n)u^\beta(p, n') = v^\dagger(p, n)v^\beta(pn') = 2E\delta_{nn'}. \tag{4.313}$$

Alternatively, when evaluating physical observable as a cross-section, we can assume $m \neq 0$ at the beginning, and take the limit $m \to 0$ at the end of the calculations. We will see an example of this in Section 9.2.

## 4.11 Exercises

(1) Verify the relation

$$\gamma_5\sigma^{\mu\nu} = \frac{i}{2}\epsilon^{\mu\nu\rho\lambda}\sigma_{\rho\lambda}. \tag{4.314}$$

(2) Verify the transformation properties of the Dirac bilinears in eq. (4.69).

(3) The 16 matrices, $\Gamma^\alpha = (1, \gamma^\mu, \gamma^\mu\gamma_5, \sigma^{\mu\nu})$, form a complete basis in the space of the $4 \times 4$ matrices. Show that any $4 \times 4$ matrix can be written

as

$$A = \sum_\alpha b_\alpha \Gamma^\alpha, \tag{4.315}$$

with

$$b_\alpha = \frac{1}{4}\text{Tr}[A\Gamma_\alpha]. \tag{4.316}$$

(4) Show that the complex conjugate of the matrix element of a 4×4 matrix between Dirac spinors can be written as

$$(\bar{u}_f O u_i)^* = \bar{u}_i \bar{O} u_f \tag{4.317}$$

where

$$\bar{O} = \gamma_0 O^\dagger \gamma_0. \tag{4.318}$$

Then, evaluate the expression $\bar{\Gamma}_\alpha$ for the 16 matrices $\Gamma^\alpha$.

(5) Verify the following identity

$$\bar{u}(p')\sigma_{\mu\nu}(p+p')^\nu u(p) = i\bar{u}(p')(p'-p)_\mu u(p). \tag{4.319}$$

(6) The Foldy-Wouthuysen transformation [Foldy and Wouthuysen (1950)]

$$H' = UHU^\dagger, \tag{4.320}$$

where $H$ is the Dirac Hamiltonian and

$$U = e^{i\vec{\alpha}\cdot\vec{p}\theta(p)} \tag{4.321}$$

is such to make $H'$ proportional to the matrix $\beta$ for a convenient choice of the function $\theta(p)$. Determine this function.

(7) Derive eqs. (4.282) and (4.293) for the three momentum $\vec{P}$ and for the charge operators, using the expressions (4.275) and (4.276) for the fields $\psi$ and $\psi^\dagger$.

(8) In the form given by eq. (4.257), the Lagrangian density of the Dirac field is not hermitian. Show that it is hermitian up to a four-divergence and devise a way to write it in an explicit hermitian way.

(9) Evaluate the following equal time commutators

$$[\mathcal{M}^\mu_{\rho\nu}(x), \psi(y)]_{x^0=y^0}, \quad [T^{\mu\nu}(x), \psi(y)]_{x^0=y^0}, \quad [j^\mu(x), \psi(y)]_{x^0=y^0}. \tag{4.322}$$

(10) Given an operator $A$ of the form

$$A = \sum_j \alpha_j a_j^\dagger a_j, \tag{4.323}$$

with

$$[a_j, a_k]_+ = [a_j^\dagger, a_k^\dagger]_+ = 0, \quad [a_j, a_k^\dagger]_+ = \delta_{ik}, \tag{4.324}$$

and an eigenstate of $A$, $|\rho\rangle$, corresponding to the eigenvalue $\rho$, show that $a_j^\dagger|\rho\rangle$ is an eigenstate of $A$ with eigenvalue $\rho + \alpha_j$ and that $a_j|\rho\rangle$ corresponds to the eigenvalue $\rho - \alpha_j$.

# Chapter 5

# Vector fields

## 5.1 The electromagnetic field

In Section 4.6 we have shown that the action for the electromagnetic field must be expressed in terms of the four-vector potential $A_\mu$. We recall also that the Lagrangian density for the free case is given by (see eq. (4.171))

$$\mathcal{L} = -\frac{1}{4} F_{\mu\nu} F^{\mu\nu}, \tag{5.1}$$

where

$$F_{\mu\nu} = \partial_\mu A_\nu - \partial_\nu A_\mu. \tag{5.2}$$

The resulting equations of motion are:

$$\Box A_\mu - \partial_\mu (\partial^\nu A_\nu) = 0. \tag{5.3}$$

We recall also that the potentials are defined up to a gauge transformation

$$A_\mu(x) \to A'_\mu(x) = A_\mu(x) + \partial_\mu \Lambda(x). \tag{5.4}$$

In fact, $A_\mu$ and $A'_\mu$ satisfy the same equations of motion (and give rise to the same electromagnetic field). In other words, the action for the electromagnetic field is gauge invariant. Up to now, we have considered only symmetries depending on a finite number of parameters. For instance, in the case of the $O(2)$ symmetry for the charged scalar field, the transformation depends on a single parameter, the rotation angle. In the case of the gauge symmetry, one deals with a continuous number of parameters, given by the function $\Lambda(x)$. In fact, in each space-time point, we can change the definition of $A_\mu$ by adding the gradient of $\Lambda$ evaluated at that point. The main consequence of this type of invariance is to reduce the effective degrees of freedom of the theory from 4 to 2. In order to prove this statement, let us start from the classical theory. We recall that it is possible to use the gauge invariance to require some particular condition on the field $A_\mu$ (gauge fixing). For instance, we can perform a gauge transformation in such a way that the transformed field satisfies

$$\partial_\mu A^\mu = 0. \tag{5.5}$$

In fact, given an arbitrary $A_\mu$, we can gauge transform it by choosing $\Lambda(x)$ such that

$$\Box\Lambda + \partial^\mu A_\mu = 0. \tag{5.6}$$

Then, the transformed field $A'_\mu = A_\mu + \partial_\mu\Lambda$ has vanishing four-divergence. When $A_\mu$ satisfies the condition $\partial_\mu A^\mu = 0$, we say that the potential is in the **Lorenz gauge**[1]. However $A_\mu$ is not yet completely determined. In fact, we might perform a further gauge transformation

$$A_\mu \to A'_\mu = A_\mu + \partial_\mu\Lambda', \tag{5.7}$$

with

$$\Box\Lambda' = 0. \tag{5.8}$$

Then the new field $A'_\mu$ satisfies $\partial^\mu A'_\mu = 0$ and we are still in the Lorenz gauge. In this gauge the equations of motion simplify and reduce to the wave equation, which is nothing but the Klein-Gordon equation with $m = 0$. Since the Lorenz gauge condition is relativistically invariant, the theory formulated in this gauge is explicitly covariant, making this choice of gauge particularly convenient. On the other hand, the counting of the effective degrees of freedom is not so evident. From this point of view a better choice is the **Coulomb gauge**, which is defined (in the non interacting case) as the gauge where the scalar potential and the spatial divergence of the vector potential vanish. To see that such a gauge exists, let us perform the following gauge transformation

$$A'_\mu(x) = A_\mu(x) - \partial_\mu \int_0^t A_0(\vec{x}, t')dt'. \tag{5.9}$$

Clearly

$$A'_0 = 0. \tag{5.10}$$

Then we perform a second gauge transformation

$$A''_\mu = A'_\mu - \partial_\mu\Lambda, \tag{5.11}$$

---

[1]This gauge was introduced by Ludvig Valentin Lorenz (1829-1891), a Danish mathematician and physicist, not to be confused with Hendrich Antoon Lorentz (1853-1928), the author of the Lorentz transformations.

in such a way that $\vec{\nabla} \cdot \vec{A}'' = 0$. To this end we choose $\Lambda(x)$ such that

$$\vec{\nabla} \cdot \vec{A}'' = \vec{\nabla} \cdot \vec{A}' + \vec{\nabla}^2 \Lambda = 0. \tag{5.12}$$

This equation can be solved by recalling that

$$\vec{\nabla}^2 \frac{1}{|\vec{x}|} = -4\pi \delta^3(\vec{x}). \tag{5.13}$$

Then

$$\Lambda(\vec{x}, t) = \frac{1}{4\pi} \int \frac{d^3 x'}{|\vec{x} - \vec{x}'|} \vec{\nabla} \cdot \vec{A}'(\vec{x}', t) \tag{5.14}$$

and

$$\frac{\partial \Lambda(\vec{x}, t)}{\partial t} = \frac{1}{4\pi} \int \frac{d^3 x'}{|\vec{x} - \vec{x}'|} \vec{\nabla} \cdot \frac{\partial}{\partial t} \vec{A}'(\vec{x}', t). \tag{5.15}$$

From the Gauss equation for the electric field it follows

$$\vec{\nabla} \cdot \vec{E} = -\vec{\nabla}^2 A_0 - \vec{\nabla} \cdot \frac{\partial}{\partial t} \vec{A} = 0 \tag{5.16}$$

and, in terms of $A'_\mu$

$$\vec{\nabla} \cdot \vec{E} = -\vec{\nabla} \cdot \frac{\partial}{\partial t} \vec{A}' = 0. \tag{5.17}$$

Therefore

$$\frac{\partial \Lambda(\vec{x}, t)}{\partial t} = 0, \tag{5.18}$$

and $A_0'' = A_0' = 0$. This shows that the second gauge transformation does not destroy the vanishing of the scalar potential. In conclusion, we have shown that it is possible to choose a gauge such that

$$A_0 = \vec{\nabla} \cdot \vec{A} = 0. \tag{5.19}$$

It follows that the electromagnetic field has only two independent degrees of freedom. Another way of showing that $A_\mu$ has two degrees of freedom is through the equations of motion. Let us consider the four-dimensional Fourier transform of $A_\mu(x)$

$$A_\mu(x) = \int d^4 k \, e^{ikx} A_\mu(k). \tag{5.20}$$

Substituting this expression inside the equations of motion we get

$$-k^2 A_\mu(k) + k_\mu(k^\nu A_\nu(k)) = 0. \tag{5.21}$$

Let us now decompose $A_\mu(k)$ in terms of four independent four-vectors, which can be chosen as $k^\mu = (E, \vec{k})$, $\tilde{k}^\mu = (E, -\vec{k})$, and two further four-vectors $e_\mu^\lambda(k)$, $\lambda = 1, 2$, orthogonal to $k^\mu$

$$k^\mu e_\mu^\lambda = 0, \qquad \lambda = 1, 2. \tag{5.22}$$

The decomposition of $A_\mu(k)$ reads

$$A_\mu(k) = a_\lambda(k)e_\mu^\lambda + b(k)k_\mu + c(k)\tilde{k}_\mu. \tag{5.23}$$

From the equations of motion we get

$$-k^2(a_\lambda e_\mu^\lambda + bk_\mu + c\tilde{k}_\mu) + k_\mu(bk^2 + c(k \cdot \tilde{k})) = 0. \tag{5.24}$$

The term in $b(k)$ cancels, therefore it is left undetermined by the equations of motion. For the other quantities we have

$$k^2 a_\lambda(k) = c(k) = 0. \tag{5.25}$$

The arbitrariness of $b(k)$ is a consequence of the gauge invariance. In fact if we gauge transform $A_\mu(x)$

$$A_\mu(x) \to A_\mu(x) + \partial_\mu \Lambda(x), \tag{5.26}$$

then

$$A_\mu(k) \to A_\mu(k) + ik_\mu \Lambda(k), \tag{5.27}$$

where

$$\Lambda(x) = \int d^4k \, e^{ikx} \Lambda(k). \tag{5.28}$$

Since the gauge transformation amounts to the translation $b(k) \to b(k) + i\Lambda(k)$, we can always choose $b(k) = 0$. Therefore we are left with the two degrees of freedom described by the amplitudes $a_\lambda(k)$, $\lambda = 1, 2$. Furthermore these amplitudes are different from zero only if the dispersion relation $k^2 = 0$ is satisfied. This shows that the corresponding quanta have zero mass. With the choice $b(k) = 0$, the field $A_\mu(k)$ becomes

$$A_\mu(k) = a_\lambda(k)e_\mu^\lambda(k), \tag{5.29}$$

proving that $k^\mu A_\mu(k) = 0$. Therefore the choice $b(k) = 0$ is equivalent to fix the Lorenz gauge.

## 5.2  Quantization of the electromagnetic field

Let us now consider the quantization of the electromagnetic field. If we could ignore the constraints on $A_\mu$, which imply that the gauge field has only two independent components, we could require non-trivial commutation relations for all the components of the field. That is

$$[A_\mu(\vec{x}, t), \Pi^\nu(\vec{y}, t)] = ig_\mu^\nu \delta^3(\vec{x} - \vec{y}), \tag{5.30}$$

$$[A_\mu(\vec{x}, t), A_\nu(\vec{y}, t)] = [\Pi^\mu(\vec{x}, t), \Pi^\nu(\vec{y}, t)] = 0, \tag{5.31}$$

with

$$\Pi^\mu = \frac{\partial \mathcal{L}}{\partial \dot{A}_\mu}. \tag{5.32}$$

To evaluate the conjugated momenta, it is better to write the Lagrangian density (see eq. (5.1)) in the following form

$$\mathcal{L} = -\frac{1}{4}[A_{\mu,\nu} - A_{\nu,\mu}][A^{\mu,\nu} - A^{\nu,\mu}] = -\frac{1}{2}A_{\mu,\nu}A^{\mu,\nu} + \frac{1}{2}A_{\mu,\nu}A^{\nu,\mu}. \tag{5.33}$$

Therefore

$$\frac{\partial \mathcal{L}}{\partial A_{\mu,\nu}} = -A^{\mu,\nu} + A^{\nu,\mu} = F^{\mu\nu}, \tag{5.34}$$

implying

$$\Pi^\mu = \frac{\partial \mathcal{L}}{\partial \dot{A}_\mu} = F^{\mu 0}. \tag{5.35}$$

It follows

$$\Pi^0 = \frac{\partial \mathcal{L}}{\partial \dot{A}_0} = 0. \tag{5.36}$$

We see that it is impossible to satisfy the condition

$$[A_0(\vec{x}, t), \Pi^0(\vec{y}, t)] = i\delta^3(\vec{x} - \vec{y}). \tag{5.37}$$

We can try to find a solution to this problem modifying the Lagrangian density in such a way that $\Pi^0 \neq 0$. But doing so we will not recover the Maxwell equations. However, we can take advantage of the gauge invariance, modifying the Lagrangian density in such a way to recover the equations of motion in a particular gauge. For instance, in the Lorenz gauge we have the equations of motion

$$\Box A_\mu(x) = 0, \tag{5.38}$$

which can be obtained from the Lagrangian density

$$\mathcal{L} = -\frac{1}{2} A_{\mu,\nu} A^{\mu,\nu}, \tag{5.39}$$

(just remember the Klein-Gordon Lagrangian). We will see that the minus sign is necessary to recover a positive Hamiltonian density. We now express this Lagrangian density in terms of the gauge invariant one, given in eq. (5.1). To this end we observe that the difference between the two Lagrangian densities is nothing but the second term of eq. (5.33)

$$\frac{1}{2} A_{\mu,\nu} A^{\nu,\mu} = \partial^\mu \left[ \frac{1}{2} A_{\mu,\nu} A^\nu \right] - \frac{1}{2} (\partial^\mu A_{\mu,\nu}) A^\nu$$

$$= \partial^\mu \left[ \frac{1}{2} A_{\mu,\nu} A^\nu \right] - \partial^\nu \left[ \frac{1}{2} (\partial^\mu A_\mu) A_\nu \right] + \frac{1}{2} (\partial^\mu A_\mu)^2. \tag{5.40}$$

Then, up to a four-divergence, we can write the new Lagrangian density in the form

$$\mathcal{L} = -\frac{1}{4} F_{\mu\nu} F^{\mu\nu} - \frac{1}{2} (\partial^\mu A_\mu)^2. \tag{5.41}$$

One can check that this form gives the correct equations of motion. In fact from

$$\frac{\partial \mathcal{L}}{\partial A_{\mu,\nu}} = -A^{\mu,\nu} + A^{\nu,\mu} - g^{\mu\nu} (\partial^\lambda A_\lambda), \qquad \frac{\partial \mathcal{L}}{\partial A_\mu} = 0, \tag{5.42}$$

we get

$$0 = -\Box A^\mu + \partial^\mu (\partial^\nu A_\nu) - \partial^\mu (\partial^\lambda A_\lambda) = -\Box A^\mu. \tag{5.43}$$

The term

$$-\frac{1}{2} (\partial^\mu A_\mu)^2, \tag{5.44}$$

which is not gauge invariant, is called the **gauge fixing term**. More generally, we could add to the original Lagrangian density a term of the form

$$-\frac{\lambda}{2} (\partial^\mu A_\mu)^2. \tag{5.45}$$

The corresponding equations of motion would be

$$\Box A_\mu - (1 - \lambda) \partial^\mu (\partial^\lambda A_\lambda) = 0. \tag{5.46}$$

For $\lambda = 1$, these equations coincide with the Maxwell equations in the Lorenz gauge. Therefore, in the following we will use $\lambda = 1$. From eq. (5.42) we see that

$$\Pi^0 = \frac{\partial \mathcal{L}}{\partial \dot{A}_0} = -\partial^\mu A_\mu. \tag{5.47}$$

In the Lorenz gauge we find again $\Pi^0 = 0$. To avoid the corresponding problem we can require $\partial^\mu A_\mu = 0$ not to hold as an operator equation, but rather as a condition upon the physical states

$$\langle \text{phys} | \partial^\mu A_\mu | \text{phys} \rangle = 0. \tag{5.48}$$

Correspondingly the price to pay in order to quantize the theory in a covariant way is to work in a Hilbert space bigger than the physical one. In fact, the Hilbert space will be given by the subspace of physical states, the ones satisfying eq. (5.48), plus the ones that do not satisfy the constraint (5.48). A bonus of this procedure is that it allows us to make use of local commutation relations. On the contrary, in the Coulomb gauge, one needs to introduce non-local commutation relations for the canonical variables. We will come back later to the condition (5.48).

Since we don't have to worry any more about the equation $\Pi^0 = 0$, we can proceed with our program of canonical quantization. The canonical momentum densities are

$$\Pi^\mu = \frac{\partial \mathcal{L}}{\partial \dot{A}_\mu} = F^{\mu 0} - g^{\mu 0}(\partial^\lambda A_\lambda), \tag{5.49}$$

or, explicitly

$$\Pi^0 = -\partial^\lambda A_\lambda = -\dot{A}_0 - \vec{\nabla} \cdot \vec{A},$$
$$\Pi^i = \partial^i A^0 - \partial^0 A^i = -\dot{A}^i + \partial^i A^0. \tag{5.50}$$

Since the spatial gradient of the field commutes with the field itself at equal time, the canonical commutator (5.30) gives rise to

$$[A_\mu(\vec{x}, t), \dot{A}_\nu(\vec{y}, t)] = -i g_{\mu\nu} \delta^3(\vec{x} - \vec{y}). \tag{5.51}$$

To get the quanta of the field, we look for plane wave solutions of the wave equation. We need four independent four-vectors in order to expand the solutions in the momentum space. In a given frame, let us consider the unit four-vector defining the time axis. This must be a time-like vector, $n^2 = 1$, and we will choose $n^0 > 0$. For instance, $n^\mu = (1, 0, 0, 0)$. Then we take two four-vectors $\epsilon_\mu^{(\lambda)}$, $\lambda = 1, 2$, in the plane orthogonal to $n^\mu$ and $k^\mu$:

$$k^\mu \epsilon_\mu^{(\lambda)} = n^\mu \epsilon_\mu^{(\lambda)} = 0, \qquad \lambda = 1, 2. \tag{5.52}$$

Notice that now $k^2 = 0$, since we are considering solutions of the wave equation. The four-vectors $\epsilon_\mu^{(\lambda)}$, being orthogonal to $n^\mu$ are space-like, then they will be chosen orthogonal and normalized in the following way

$$\epsilon_\mu^{(\lambda)} \epsilon^{(\lambda')\mu} = -\delta_{\lambda\lambda'}. \tag{5.53}$$

Next, we define a unit space-like four-vector, orthogonal to $n^\mu$ and lying in the plane $(k, n)$

$$n_\mu \epsilon^{(3)\mu} = 0, \qquad (5.54)$$

with

$$\epsilon_\mu^{(3)} \epsilon^{(3)\mu} = -1. \qquad (5.55)$$

By construction $\epsilon_\mu^{(3)}$ is orthogonal to $\epsilon_\mu^{(\lambda)}$. This four-vector is completely fixed by the previous conditions, and we get

$$\epsilon_\mu^{(3)} = \frac{k_\mu - (n \cdot k) n_\mu}{(n \cdot k)}. \qquad (5.56)$$

As a last unit four-vector we choose $n^\mu$

$$\epsilon_\mu^{(0)} = n_\mu. \qquad (5.57)$$

These four-vectors are orthonormal and we can write

$$\epsilon_\mu^{(\lambda)} \epsilon^{(\lambda')\mu} = g^{\lambda\lambda'} \qquad (5.58)$$

and being linearly independent, they satisfy the completeness relation

$$\epsilon_\mu^{(\lambda)} \epsilon_\nu^{(\lambda')} g_{\lambda\lambda'} = g_{\mu\nu}. \qquad (5.59)$$

Using the explicit expressions for $\epsilon_\mu^{(3)}$ and $\epsilon_\mu^{(0)}$, together with the completeness we find the following result for the transverse physical polarization vectors

$$\sum_{\lambda=1,2} \epsilon_\mu^{(\lambda)} \epsilon_\nu^{(\lambda)} = -g_{\mu\nu} - \frac{k_\mu k_\nu - (n \cdot k)(n_\mu k_\nu + n_\nu k_\mu)}{(n \cdot k)^2}. \qquad (5.60)$$

In the frame where $n^\mu = (1, \vec{0})$ and $k^\mu = (k, 0, 0, k)$, we have

$$\epsilon^{(1)\mu} = (0, 1, 0, 0), \quad \epsilon^{(2)\mu} = (0, 0, 1, 0), \quad \epsilon^{(3)\mu} = (0, 0, 0, 1). \qquad (5.61)$$

The plane wave expansion of $A_\mu$ is

$$A_\mu(x) = \int \frac{d^3 k}{\sqrt{2\omega_k (2\pi)^3}} \sum_{\lambda=0}^{3} \epsilon_\mu^{(\lambda)}(k) \left[ a_\lambda(k) e^{-ikx} + a_\lambda^\dagger(k) e^{ikx} \right], \qquad (5.62)$$

where we have imposed the hermiticity condition for $A_\mu(x)$ (classically it is a real field). For any fixed $\mu$, this expansion is the same as the one that we wrote for the Klein-Gordon field, with the substitution $\epsilon_\mu^{(\lambda)} a_\lambda(k) \to a(k)$. Then, from eq. (3.55)

$$\epsilon_\mu^{(\lambda)}(k) a_\lambda(k) = i \int d^3 x \, f_{\vec{k}}^*(x) \partial_t^{(-)} A_\mu(x), \qquad (5.63)$$

with the functions $f_{\vec{k}}(x)$ defined as in Section 3.2. Using the orthogonality of the $\epsilon^{(\lambda)\mu}$'s we find

$$a_\lambda(k) = ig_{\lambda\lambda'} \int d^3x \; \epsilon^{(\lambda')\mu}(k) f_{\vec{k}}^*(x) \partial_t^{(-)} A_\mu(x) \tag{5.64}$$

and analogously

$$a_\lambda^\dagger(k) = ig_{\lambda\lambda'} \int d^3x \; \epsilon^{(\lambda')\mu}(k) A_\mu(x) \partial_t^{(-)} f_{\vec{k}}(x). \tag{5.65}$$

Comparison with the calculation done in eq. (2.129) shows that

$$[a_\lambda(k), a_{\lambda'}^\dagger(k')] = -\int d^3x \; f_{\vec{k}}^*(x) i\partial_t^{(-)} f_{\vec{k}'}(x) g_{\lambda\lambda'} \tag{5.66}$$

and using the orthogonality relations given in eq. (3.45)

$$[a_\lambda(k), a_{\lambda'}^\dagger(k')] = -g_{\lambda\lambda'}\delta^3(\vec{k} - \vec{k}'). \tag{5.67}$$

Analogously

$$[a_\lambda(k), a_{\lambda'}(k')] = [a_\lambda^\dagger(k), a_{\lambda'}^\dagger(k')] = 0. \tag{5.68}$$

Again, comparison with the Klein-Gordon commutators implies

$$[A_\mu(x), A_\nu(y)] = -ig_{\mu\nu}\Delta(x - y), \tag{5.69}$$

with the invariant function $\Delta(x)$ defined as in eq. (3.139), but with $m = 0$. The commutation rules we have derived for the operators $a_\lambda(k)$ create some problem. Let us consider a one-particle state

$$|1, \lambda\rangle = \int d^3k \; f(k) a_\lambda^\dagger(k)|0\rangle, \tag{5.70}$$

its norm is given by

$$\begin{aligned}
\langle 1, \lambda | 1, \lambda \rangle &= \int d^3k \; d^3k' \; f^*(k) f(k') \langle 0|a_\lambda(k) a_\lambda^\dagger(k')|0\rangle \\
&= \int d^3k \; d^3k' \; f^*(k) f(k') \langle 0|[a_\lambda(k), a_\lambda^\dagger(k')]|0\rangle \\
&= -g_{\lambda\lambda} \int d^3k \; |f(k)|^2.
\end{aligned} \tag{5.71}$$

Therefore the states with $\lambda = 0$ have negative norm. This problem is not completely unexpected. In fact, physically we know that the only physical states are the transverse ones ($\lambda = 1, 2$).

To clarify the situation let us go back to the gauge fixing condition $\langle\text{phys}|\partial^\mu A_\mu|\text{phys}\rangle = 0$. Remember that its aim is to select the physical part of the total Hilbert space. Therefore the relevant question is if the physical states satisfying this condition have positive norm. To discuss this point notice that the gauge fixing condition is bilinear in the states, therefore it could destroy the linearity of the Hilbert space. So we will try to formulate this condition in a linear way. A possibility would be to write

$$\partial^\mu A_\mu |\text{phys}\rangle = 0. \tag{5.72}$$

But this would be a too strong requirement. Not even the vacuum state satisfies it. However, if we consider the positive and negative frequency parts of the field

$$A_\mu^{(+)}(x) = \int \frac{d^3 k}{\sqrt{2\omega_k (2\pi)^3}} \sum_{\lambda=0}^{3} \epsilon_\mu^{(\lambda)}(k) a_\lambda(k) e^{-ikx}, \quad A_\mu^{(-)}(x) = (A^{(+)}(x))^\dagger, \tag{5.73}$$

it is possible to weaken the condition requiring

$$\partial^\mu A_\mu^{(+)}(x) |\text{phys}\rangle = 0. \tag{5.74}$$

In this way the original requirement is automatically satisfied

$$\langle \text{phys.}|(\partial^\mu A_\mu^{(+)} + \partial^\mu A_\mu^{(-)})|\text{phys}\rangle = 0. \tag{5.75}$$

To make this condition more explicit let us evaluate the four-divergence of $A_\mu^{(+)}$

$$i\partial^\mu A_\mu^{(+)}(x) = \int \frac{d^3 k}{\sqrt{2\omega_k (2\pi)^3}} e^{-ikx} \sum_{\lambda=0,3} k^\mu \epsilon_\mu^{(\lambda)}(k) a_\lambda(k). \tag{5.76}$$

Using eq. (5.56), we get

$$k^\mu \epsilon_\mu^{(3)} = -(n \cdot k), \qquad k^\mu \epsilon_\mu^{(0)} = (n \cdot k), \tag{5.77}$$

from which

$$[a_0(k) - a_3(k)]|\text{phys}\rangle = 0. \tag{5.78}$$

Notice that

$$[a_0(k) - a_3(k), a_0^\dagger(k') - a_3^\dagger(k')] = -\delta^3(\vec{k} - \vec{k}') + \delta^3(\vec{k} - \vec{k}') = 0. \tag{5.79}$$

Let us denote by $|\Phi_{\vec{k}}(n_0, n_3)\rangle$ the state with $n_0$ scalar photons (that is with polarization $\epsilon_\mu^{(0)}(k)$), and with $n_3$ longitudinal photons (that is with polarization $\epsilon_\mu^{(3)}(k)$). Then the following states satisfy the condition (5.74)

$$|\Phi_{\vec{k}}^{(m)}\rangle = \frac{1}{m!}(a_0^\dagger(k) - a_3^\dagger(k))^m |\Phi_{\vec{k}}(0,0)\rangle. \tag{5.80}$$

These states have vanishing norm

$$|||\Phi_{\vec{k}}^{(m)}\rangle||^2 = 0. \tag{5.81}$$

More generally we can make the following observation. Let us consider the number operator for scalar and longitudinal photons

$$N = \int d^3k \; (a_3^\dagger(k)a_3(k) - a_0^\dagger(k)a_0(k)). \tag{5.82}$$

Notice the minus sign arising from the commutation relations, and which ensures that $N$ has positive eigenvalues. For instance

$$Na_0^\dagger(k)|0\rangle = -\int d^3k' \; a_0^\dagger(k')[a_0(k'), a_0^\dagger(k)]|0\rangle = a_0^\dagger(k)|0\rangle. \tag{5.83}$$

Let us consider a physical state $|\varphi_n\rangle$ with a total number, $n$, of scalar and longitudinal photons. Then

$$\langle\varphi_n|N|\varphi_n\rangle = 0, \tag{5.84}$$

as it follows from eq. (5.78). Therefore

$$n\langle\varphi_n|\varphi_n\rangle = 0. \tag{5.85}$$

We see that all the physical states with a definite number of scalar and longitudinal photons have zero norm, except for the vacuum state $(n = 0)$, that is

$$\langle\varphi_n|\varphi_n\rangle = \delta_{n,0}. \tag{5.86}$$

A generic physical state with zero transverse photons is a linear superposition of the states $|\varphi_n\rangle$

$$|\Psi\rangle = c_0|\varphi_0\rangle + \sum_{n\neq0} c_n|\varphi_n\rangle. \tag{5.87}$$

This state has a positive definite norm

$$\langle\Psi|\Psi\rangle = |c_0|^2 \geq 0. \tag{5.88}$$

The proof that a physical state has a positive norm can be extended to states with transverse photons. Of course, the coefficients $c_n$, appearing in the expression of a physical state, are completely arbitrary, but this is not going to modify the values of the observables. For instance, consider the Hamiltonian. We have

$$H = \int d^3x \; : [\Pi^\mu \dot{A}_\mu - \mathcal{L}] :$$

$$= \int d^3x \; : \left[ F^{\mu 0} \dot{A}_\mu - (\partial^\lambda A_\lambda)\dot{A}_0 \right.$$

$$\left. + \frac{1}{4}F_{\mu\nu}F^{\mu\nu} + \frac{1}{2}(\partial^\lambda A_\lambda)^2 \right] : . \tag{5.89}$$

One can easily show that the Hamiltonian is given by the sum of all the degrees of freedom appearing in $A_\mu$ (see the Klein-Gordon case, eq. (3.94))

$$H = \frac{1}{2} \int d^3x \; : \left[ \sum_{i=1}^{3} \left( \dot{A}_i^2 + (\vec{\nabla} A_i)^2 \right) - \dot{A}_0^2 - \vec{\nabla} A_0^2 \right] :$$

$$= \int d^3k \; \omega_k \; : \left[ \sum_{\lambda=1}^{3} a_\lambda^\dagger(k) a_\lambda(k) - a_0^\dagger(k) a_0(k) \right] : . \qquad (5.90)$$

Since $a_0$ and $a_3$ act in the same way on the physical states, we get

$$\langle \text{phys}|H|\text{phys} \rangle = \langle \text{phys}| \int d^3k \; \omega_k \sum_{\lambda=1}^{2} a_\lambda^\dagger(k) a_\lambda(k) |\text{phys} \rangle. \qquad (5.91)$$

The generic physical state is a linear combination of terms of the form $|\varphi_T\rangle \otimes |\Psi\rangle$, with $|\Psi\rangle$ defined as in eq. (5.87). Since only $|\varphi_T\rangle$ contributes to the evaluation of an observable quantity, we can always choose $|\Psi\rangle$ proportional to $|\varphi_0\rangle$. However, this does not mean that we are always working in the restricted physical space, because in a sum over the intermediate states we need to include all the degrees of freedom. This is crucial for the explicit covariance and locality of the theory.

The arbitrariness in defining the state $|\Psi\rangle$ has a very simple interpretation. It corresponds to add to $A_\mu$ a four-gradient, that is to perform a gauge transformation. Consider the following matrix element

$$\langle \Psi|A_\mu(x)|\Psi \rangle = \sum_{n,m} c_n^\star c_m \langle \varphi_n|A_\mu(x)|\varphi_m \rangle. \qquad (5.92)$$

Since $A_\mu$ change the occupation number by one unit and all the states $|\varphi_n\rangle$ have zero norm (except for the state with $n = 0$), the only non-vanishing contributions come from $n = 0$, $m = 1$ and $n = 1$, $m = 0$

$$\langle \Psi|A_\mu(x)|\Psi \rangle = c_0^\star c_1 \langle 0| \int \frac{d^3k}{\sqrt{2\omega_k(2\pi)^3}} \; e^{-ikx} [\epsilon_\mu^{(3)}(k) a_3(k)$$

$$+ \epsilon_\mu^{(0)}(k) a_0(k)] |\varphi_1\rangle + \text{c.c.} \qquad (5.93)$$

In order to satisfy the gauge condition, the state $|\varphi_1\rangle$ must be of the form

$$|\varphi_1\rangle = \int d^3q \; f(\vec{q}) [a_3^\dagger(q) - a_0^\dagger(q)] |0\rangle \qquad (5.94)$$

and therefore

$$\langle \Psi|A_\mu(x)|\Psi \rangle = \int \frac{d^3k}{\sqrt{2\omega_k(2\pi)^3}} [\epsilon_\mu^{(3)}(k) + \epsilon_\mu^{(\lambda)}(k)][c_0^\star c_1 e^{-ikx} f(\vec{k}) + \text{c.c.}].$$

$$(5.95)$$

From eqs. (5.56) and (5.57) we have

$$\epsilon_\mu^{(3)} + \epsilon_\mu^{(0)} = \frac{k_\mu}{(k \cdot n)}, \tag{5.96}$$

implying

$$\langle \Psi | A_\mu(x) | \Psi \rangle = \partial_\mu \Lambda(x), \tag{5.97}$$

with

$$\Lambda(x) = \int \frac{d^3 k}{\sqrt{2\omega_k (2\pi)^3}} \frac{1}{n \cdot k} (ic_0^* c_1 e^{-ikx} f(\vec{k}) + \text{c.c.}). \tag{5.98}$$

It is important to notice that this gauge transformation leaves $A_\mu$ in the Lorenz gauge. In fact

$$\Box \Lambda = 0, \tag{5.99}$$

since the momentum $k$ inside the integral satisfies the mass-shell condition $k^2 = 0$.

## 5.3 Massive vector fields

For completeness we will comment briefly about the massive vector fields, also called Proca fields. We will not discuss them any further in this book, but it will be important to understand some of their properties as, for instance, the fact that they have 3 independent degrees of freedom[2]. In particular, this point will be relevant when we discuss the Higgs mechanism (see the next Chapter).

Massive vector bosons are described by a vector field $V_\mu$ satisfying an equation of motion which generalizes the Maxwell equation

$$(\Box + m^2)V_\mu - \partial_\mu(\partial^\nu V_\nu) = 0. \tag{5.100}$$

The Lagrangian density giving rise to this equation is

$$\mathcal{L} = -\frac{1}{4}F_{\mu\nu}F^{\mu\nu} + \frac{1}{2}m^2 V^2, \tag{5.101}$$

where

$$F_{\mu\nu} = \partial_\mu V_\nu - \partial_\nu V_\mu. \tag{5.102}$$

Notice the change of sign with respect to the Klein-Gordon Lagrangian density due to the choice of the metric $g_{\mu\nu}$, implying a minus sign for the spatial components. If we take the four-divergence of eq. (5.100) we get

$$m^2 \partial_\mu V^\mu = 0. \tag{5.103}$$

---

[2] A more detailed treatment can be found in [Itzykson and Zuber (1980)].

Therefore $V_\mu$ has zero four-divergence as a consequence of the equations of motion. Let us now count the number of independent degrees of freedom of $V_\mu$. To this end let us consider the Fourier transform of $V_\mu(x)$

$$V_\mu(x) = \int d^4 k e^{ikx} V_\mu(k). \tag{5.104}$$

Let us introduce four independent four-vectors[3],

$$k^\mu = (E, \vec{k}), \quad \epsilon^{\mu(i)} = (0, \vec{n}_i), \quad i = 1, 2, \quad \epsilon^{\mu(3)} = \frac{1}{m}\left(|\vec{k}|, \frac{E\vec{k}}{|\vec{k}|}\right), \tag{5.105}$$

with $\vec{k} \cdot \vec{n}_i = 0$, $\vec{n}_1 \cdot \vec{n}_2 = 0$ and $|\vec{n}_i|^2 = 1$. Then we can write $V_\mu(k)$ as

$$V_\mu(k) = \epsilon_\mu^{(\lambda)} a_\lambda(k) + k_\mu b(k). \tag{5.106}$$

By inserting this expression inside eqs. (5.100)

$$(k^2 - m^2)(\epsilon_\mu^{(\lambda)} a_\lambda(k) + k_\mu b(k)) - k_\mu k^2 b(k) = 0, \tag{5.107}$$

that is

$$(k^2 - m^2)a_\lambda = 0, \qquad m^2 b = 0. \tag{5.108}$$

Remember that in the electromagnetic case the field $b$ was not determined due to the gauge invariance, whereas in the massive case it is fixed to be zero by the equations of motion. Therefore the field has three degrees of freedom corresponding to two transverse and one longitudinal polarizations. All these degrees of freedom satisfy a Klein-Gordon equation with mass $m$. It is important to stress that the massive vector field has one extra-degree of freedom (the longitudinal), with respect to the massless case. We see also that the three independent fields described by $a_\lambda(k)$ satisfy a Klein-Gordon equation with mass $m$. Correspondingly the field $A_\mu$ can be expressed as[4]

$$A_\mu(x) = \sum_{\lambda=1,2,3} \int d^3 k [f_{\vec{k}} a_\lambda(\vec{k}) \epsilon_\mu^{(\lambda)} + f_{\vec{k}}^* a_\lambda^\dagger(\vec{k}) \epsilon_\mu^{(\lambda)*}], \tag{5.109}$$

with the non-vanishing commutators between creation and annihilation operators given by

$$[a_\lambda(\vec{k}), a_{\lambda'}^\dagger(\vec{k})] = \delta_{\lambda\lambda'} \delta^3(\vec{k} - \vec{k}'). \tag{5.110}$$

As a final remark, let us notice that introducing a fourth normalized four-vector

$$\epsilon_\mu^{(0)} = \frac{k_\mu}{m}, \tag{5.111}$$

---

[3]The following discussion is very similar to the one done for the electromagnetic field. However, notice the differences arising from the fact that in the present case $m \neq 0$.
[4]Compare with eq. (3.57).

we get a basis in the four-dimensional Minkowski space, satisfying the completeness relation

$$\sum_{\lambda,\lambda'=0}^{3} \epsilon_\mu^{(\lambda)} \epsilon_\nu^{(\lambda')} g_{\lambda\lambda'} = g_{\mu\nu}. \tag{5.112}$$

Therefore the sum over the physical polarizations is given by

$$\sum_{\lambda=1}^{3} \epsilon_\mu^{(\lambda)} \epsilon_\nu^{(\lambda)} = -g_{\mu\nu} + \frac{k_\mu k_\nu}{m^2}. \tag{5.113}$$

## 5.4 Exercises

(1) Derive the relation (5.56).
(2) Derive the expressions (5.64) and (5.65) for the annihilation and creation operators of the electromagnetic field.
(3) Derive expression (5.90) for the Hamiltonian operator of the electromagnetic field.
(4) Show that, for $k^2 = 0$, the following quantities

$$P_{\mu\nu} = g_{\mu\nu} - \frac{k_\mu \bar{k}_\nu + k_\nu \bar{k}_\mu}{k \cdot \bar{k}}, \quad P_{\mu\nu}^\perp = \frac{k_\mu \bar{k}_\nu + k_\nu \bar{k}_\mu}{k \cdot \bar{k}}, \tag{5.114}$$

where $\bar{k}^\mu = (k^0, -\vec{k})$, are orthogonal projectors that is

$$P_{\mu\nu}P^{\nu\rho} = \delta_\mu^\rho, \quad P_{\mu\nu}^\perp P^{\perp\nu\rho} = \delta_\mu^\rho, \quad P_{\mu\nu}^\perp P^{\nu\rho} = 0,$$
$$P_{\mu\nu}^\perp + P_{\mu\nu} = g_{\mu\nu}. \tag{5.115}$$

# Chapter 6

# Symmetries in field theories

## 6.1   The linear $\sigma$-model

In this Section we will study, from a classical point of view, some field theory with particular properties of symmetry. We will start considering the linear $\sigma$-model. This is a model for $N$ scalar fields, with a symmetry $O(N)$. The Lagrangian density is given by

$$\mathcal{L} = \frac{1}{2} \sum_{i=1}^{N} \partial_\mu \phi_i \partial^\mu \phi_i - \frac{1}{2} \mu^2 \sum_{i=1}^{N} \phi_i \phi_i - \frac{\lambda}{4} \left( \sum_{i=1}^{N} \phi_i \phi_i \right)^2 . \qquad (6.1)$$

$\mathcal{L}$ is invariant under linear transformations acting upon the vector $\vec{\phi} = (\phi_1, \cdots, \phi_N)$ and leaving invariant its norm

$$|\vec{\phi}|^2 = \sum_{i=1}^{N} \phi_i \phi_i. \qquad (6.2)$$

Consider an infinitesimal transformation (from now on we will omit the index of sum over repeated indices)

$$\delta \phi_i = \epsilon_{ij} \phi_j. \qquad (6.3)$$

The condition for the norm being invariant is

$$|\vec{\phi} + \delta \vec{\phi}|^2 = |\vec{\phi}|^2, \qquad (6.4)$$

from which

$$\vec{\phi} \cdot \delta \vec{\phi} = 0, \qquad (6.5)$$

or, in components,

$$\phi_i \epsilon_{ij} \phi_j = 0. \qquad (6.6)$$

131

This is satisfied by

$$\epsilon_{ij} = -\epsilon_{ji}, \tag{6.7}$$

showing that the rotations in $N$ dimensions depend on $N(N-1)/2$ parameters. For a finite transformation we have

$$|\vec{\phi}'|^2 = |\vec{\phi}|^2, \tag{6.8}$$

with

$$\phi_i' = S_{ij}\phi_j, \tag{6.9}$$

implying

$$SS^T = 1. \tag{6.10}$$

In fact, by exponentiating the infinitesimal transformation one gets

$$S = e^\epsilon, \tag{6.11}$$

with

$$\epsilon^T = -\epsilon. \tag{6.12}$$

From this expression it follows that $S$ is an orthogonal transformation. The matrices $S$ form the rotation group in $N$ dimensions, $O(N)$.

Noether's theorem says that there is a conserved current for any continuous symmetry of the theory. In this case we will get $N(N-1)/2$ conserved quantities. It is useful, for further generalizations, to write the infinitesimal transformations in the form

$$\delta\phi_i = \epsilon_{ij}\phi_j = -\frac{i}{2}\epsilon_{AB}T_{ij}^{AB}\phi_j, \qquad i,j = 1,\cdots,N, \quad A,B = 1,\cdots,N, \tag{6.13}$$

which is similar to what we did in Section 3.3 when we discussed the Lorentz transformations. By comparison we see that the matrices $T^{AB}$ are given by

$$T_{ij}^{AB} = i(\delta_i^A \delta_j^B - \delta_j^A \delta_i^B). \tag{6.14}$$

It is not difficult to show that these matrices satisfy the algebra

$$[T^{AB}, T^{CD}] = -i\delta^{AC}T^{BD} + i\delta^{AD}T^{BC} - i\delta^{BD}T^{AC} + i\delta^{BC}T^{AD}. \tag{6.15}$$

This is nothing but the Lie algebra of the group $O(N)$, and the $T^{AB}$ are the infinitesimal generators of the group. Applying now the Noether theorem we find the conserved current

$$j_\mu = \frac{\partial \mathcal{L}}{\partial(\partial^\mu \phi_i)}\delta\phi_i = -\frac{i}{2}\phi_{i,\mu}\epsilon_{AB}T_{ij}^{AB}\phi_j, \tag{6.16}$$

since the $N(N-1)/2$ parameters $\epsilon_{AB}$ are linearly independent, we find the $N(N-1)/2$ conserved currents

$$J_\mu^{AB} = -i\phi_{i,\mu} T_{ij}^{AB} \phi_j. \tag{6.17}$$

In the case of $N=2$ the symmetry is the same as in Section 3.6, and the only conserved current is given by

$$J_\mu^{12} = -J_\mu^{21} = \phi_{1,\mu}\phi_2 - \phi_{2,\mu}\phi_1, \tag{6.18}$$

in agreement with (3.160). One can easily check that the charges associated to the conserved currents close the same Lie algebra as the generators $T^{AB}$. More generally, if we have conserved currents given by

$$j_\mu^A = -i\phi_{i,\mu} T_{ij}^A \phi_j, \tag{6.19}$$

with

$$[T^A, T^B] = if^{ABC} T^C, \tag{6.20}$$

using the canonical commutation relations, we get the commutators for the conserved charges

$$Q^A = \int d^3x\, j_0^A(x) = -i \int d^3x\, \phi_i T_{ij}^A \phi_j, \tag{6.21}$$

that is

$$[Q^A, Q^B] = if^{ABC} Q^C. \tag{6.22}$$

An important example is the $\sigma$-model for $N=4$. In this case we parametrize our fields in the form

$$\vec{\phi} = (\pi_1, \pi_2, \pi_3, \sigma) = (\vec{\pi}, \sigma). \tag{6.23}$$

These fields can be arranged into a $2 \times 2$ matrix

$$M = \sigma + i\vec{\tau} \cdot \vec{\pi}, \tag{6.24}$$

where $\vec{\tau}$ are the Pauli matrices. Recalling that $\tau_2$ is pure imaginary, $\tau_1$ and $\tau_3$ real, and that $\tau_2$ anticommutes with $\tau_1$ and $\tau_3$, we get

$$M = \tau_2 M^* \tau_2. \tag{6.25}$$

Furthermore the following relation is satisfied

$$|\vec{\phi}|^2 = \sigma^2 + |\vec{\pi}|^2 = \frac{1}{2} Tr(M^\dagger M). \tag{6.26}$$

Then we can write the Lagrangian (6.1) in the form

$$\mathcal{L} = \frac{1}{4} Tr(\partial_\mu M^\dagger \partial^\mu M) - \frac{1}{4}\mu^2 Tr(M^\dagger M) - \frac{1}{16}\lambda \left(Tr(M^\dagger M)\right)^2. \tag{6.27}$$

This Lagrangian is invariant under the following transformation of the matrix $M$

$$M \to LMR^\dagger, \tag{6.28}$$

where $L$ and $R$ are two special (with determinant equal to 1) unitary matrices. That is $L, R \in SU(2)$. The reason to restrict these matrices to be special is that only in this way the transformed matrix satisfies the condition (6.25). In fact, if $A$ is a $2 \times 2$ matrix with $\det A = 1$, then

$$\tau_2 A^T \tau_2 = A^{-1}. \tag{6.29}$$

Therefore, for $M' = LMR^\dagger$ we get

$$\tau_2 M'^* \tau_2 = \tau_2 L^* M^* R^T \tau_2 = \tau_2 L^* \tau_2 (\tau_2 M^* \tau_2) \tau_2 R^T \tau_2 \tag{6.30}$$

and from (6.29)

$$\tau_2 L^* \tau_2 = \tau_2 L^{\dagger^T} \tau_2 = L^{\dagger^{-1}} = L,$$
$$\tau_2 R^T \tau_2 = R^{-1} = R^\dagger, \tag{6.31}$$

since $L$ and $R$ are independent transformations, the invariance group in this basis is $SU(2)_L \otimes SU(2)_R$. In fact, this group and $O(4)$ are related by the following observation: the transformation $M \to LMR^\dagger$ is a linear transformation on the matrix elements of $M$, but from the relation (6.26) we see that $M \to LMR^\dagger$ leaves the norm of the vector $\vec{\phi} = (\vec{\pi}, \sigma)$. Since the same conditions are satisfied by the orthogonal transformation acting upon the field $\vec{\phi} = (\sigma, \vec{\pi})$, the transformation $M \to LMR^\dagger$ must belong to $O(4)$. This shows that the two groups $SU(2) \otimes SU(2)$ and $O(4)$ are homomorphic[1](actually there is a 2 to 1 relationship, since $-L$ and $-R$ define the same $S$ as $L$ and $R$).

We can evaluate the effect of an infinitesimal transformation. To this end we will consider separately left and right transformations. We parametrize the transformations as follows

$$L \approx 1 - \frac{i}{2}\vec{\theta}_L \cdot \vec{\tau}, \quad R \approx 1 - \frac{i}{2}\vec{\theta}_R \cdot \vec{\tau}, \tag{6.32}$$

then we get

$$\delta_L M = \left(-\frac{i}{2}\vec{\theta}_L \cdot \vec{\tau}\right)M = \left(-\frac{i}{2}\vec{\theta}_L \cdot \vec{\tau}\right)(\sigma + i\vec{\pi} \cdot \vec{\tau}) = \frac{1}{2}\vec{\theta}_L \cdot \vec{\pi} + \frac{i}{2}(\vec{\theta}_L \wedge \vec{\pi} - \vec{\theta}_L \sigma) \cdot \vec{\tau}, \tag{6.33}$$

---

[1]The relation between $O(4)$ and $SU(2)_L \otimes SU(2)_R$ is similar to the one existing between the Lorentz group $O(3,1)$ and $SL(2,C)$ (see Section 1.3).

where we have used

$$\tau_i \tau_j = \delta_{ij} + i\epsilon_{ijk}\tau_k. \tag{6.34}$$

Since

$$\delta_L M = \delta_L \sigma + i\delta_L \vec{\pi} \cdot \vec{\tau}, \tag{6.35}$$

we get

$$\delta_L \sigma = \frac{1}{2}\vec{\theta}_L \cdot \vec{\pi}, \qquad \delta_L \vec{\pi} = \frac{1}{2}(\vec{\theta}_L \wedge \vec{\pi} - \vec{\theta}_L \sigma). \tag{6.36}$$

Analogously we obtain

$$\delta_R \sigma = -\frac{1}{2}\vec{\theta}_R \cdot \vec{\pi}, \qquad \delta_R \vec{\pi} = \frac{1}{2}(\vec{\theta}_R \wedge \vec{\pi} + \vec{\theta}_R \sigma). \tag{6.37}$$

The combined transformation is given by

$$\delta\sigma = \frac{1}{2}(\vec{\theta}_L - \vec{\theta}_R) \cdot \vec{\pi}, \qquad \delta\vec{\pi} = \frac{1}{2}[(\vec{\theta}_L + \vec{\theta}_R) \wedge \vec{\pi} - (\vec{\theta}_L - \vec{\theta}_R)\sigma] \tag{6.38}$$

and we can check immediately that

$$\sigma\delta\sigma + \vec{\pi} \cdot \delta\vec{\pi} = 0, \tag{6.39}$$

as it must be for a transformation leaving the form $\sigma^2 + |\vec{\pi}|^2$ invariant. Of particular interest are the transformations with $\vec{\theta}_L = \vec{\theta}_R \equiv \theta$. In this case we have $L = R$ and

$$M \to LML^\dagger. \tag{6.40}$$

These transformations span a subgroup $SU(2)$ of $SU(2)_L \otimes SU(2)_R$ called the diagonal subgroup, and the corresponding infinitesimal transformations are

$$\delta\sigma = 0, \qquad \delta\vec{\pi} = \vec{\theta} \wedge \vec{\pi}. \tag{6.41}$$

We see that the transformations corresponding to the diagonal $SU(2)$ are rotations in the 3-dimensional space spanned by $\vec{\pi}$. These rotations define a subgroup $O(3)$ of the original symmetry group $O(4)$. From Noether's theorem we get the conserved currents

$$j_\mu^L = \frac{1}{2}\sigma_{,\mu}\vec{\theta}_L \cdot \vec{\pi} + \frac{1}{2}\vec{\pi}_{,\mu} \cdot (\vec{\theta}_L \wedge \vec{\pi} - \vec{\theta}_L \sigma) \tag{6.42}$$

and extracting the coefficients of $\vec{\theta}_L$

$$\vec{J}_\mu^L = \frac{1}{2}(\sigma_{,\mu}\vec{\pi} - \vec{\pi}_{,\mu}\sigma - \vec{\pi}_{,\mu} \wedge \vec{\pi}). \tag{6.43}$$

Analogously

$$\vec{J}_\mu^R = \frac{1}{2}(-\sigma_{,\mu}\vec{\pi} + \vec{\pi}_{,\mu}\sigma - \vec{\pi}_{,\mu} \wedge \vec{\pi}). \tag{6.44}$$

Using the canonical commutation relations one can verify that the corresponding charges satisfy the Lie algebra of $SU(2)_L \otimes SU(2)_R$

$$[Q_i^L, Q_j^L] = i\epsilon_{ijk} Q_k^L, \quad [Q_i^R, Q_j^R] = i\epsilon_{ijk} Q_k^R, \quad [Q_i^L, Q_j^R] = 0. \quad (6.45)$$

By taking the following combinations of currents

$$\vec{J}_\mu^V = \vec{J}_\mu^L + \vec{J}_\mu^R, \quad \vec{J}_\mu^A = \vec{J}_\mu^L - \vec{J}_\mu^R, \quad (6.46)$$

one has

$$\vec{J}_\mu^V = \vec{\pi} \wedge \vec{\pi},_\mu \quad (6.47)$$

and

$$\vec{J}_\mu^A = \sigma,_\mu \vec{\pi} - \vec{\pi},_\mu \sigma. \quad (6.48)$$

The corresponding algebra of conserved charges is

$$[Q_i^V, Q_j^V] = i\epsilon_{ijk} Q_k^V, \quad [Q_i^V, Q_j^A] = i\epsilon_{ijk} Q_k^A, \quad [Q_i^A, Q_j^A] = i\epsilon_{ijk} Q_k^V. \quad (6.49)$$

These equations show that $Q_i^V$ are the infinitesimal generators of the diagonal subgroup $SU(2)$ of $SU(2)_L \otimes SU(2)_R$. This can also be seen from

$$[Q_i^V, \pi_j] = i\epsilon_{ijk} \pi_k, \quad [Q_i^V, \sigma] = 0. \quad (6.50)$$

In the following we will be interested in treating the interacting field theories by using perturbation theory. As in the quantum mechanical case, this is a well defined procedure only when we are considering the theory close to a minimum of the energy of the system. In fact, if we expand around a maximum the oscillations of the system can become very large leading outside of the domain of perturbation theory. In the case of the linear $\sigma$-model the energy is given by

$$H = \int d^3x \, \mathcal{H} = \int d^3x \left[ \sum_{i=1}^N \frac{\partial \mathcal{L}}{\partial \dot{\phi}_i} \dot{\phi}_i - \mathcal{L} \right]$$

$$= \int d^3x \left[ \frac{1}{2} \sum_{i=1}^N (\dot{\phi}_i^2 + |\vec{\nabla}\phi_i|^2) + V(|\vec{\phi}|^2) \right]. \quad (6.51)$$

Since in the last equation the first two terms are positive definite, it follows that the absolute minimum is obtained for constant field configurations, that is for

$$\frac{\partial V(|\vec{\phi}|^2)}{\partial \phi_i} = 0. \quad (6.52)$$

Let us call by $v_i$ the generic solution of this equation (in general it could happen that the absolute minimum is degenerate). Then the condition for getting a minimum is that the eigenvalues of the matrix of the second derivatives of the potential at the stationary point are definite positive. In this case we define new fields by shifting the original ones by $v_i$

$$\phi_i \to \phi_i' = \phi_i - v_i. \tag{6.53}$$

The Lagrangian density becomes

$$\mathcal{L} = \frac{1}{2}\partial_\mu\phi_i'\partial^\mu\phi_i' - V(|\vec{\phi}' + \vec{v}|^2). \tag{6.54}$$

Expanding $V$ in series of $\phi_i'$ we get

$$V = V(|\vec{v}|^2) + \frac{1}{2}\frac{\partial^2 V}{\partial\phi_i\partial\phi_j}\Big|_{\vec{\phi}=\vec{v}}\phi_i'\phi_j' + \cdots. \tag{6.55}$$

This equation shows that the particle masses are given by the eigenvalues of the second derivative of the potential at the minimum. In the case of the linear $\sigma$-model we have

$$V = \frac{1}{2}\mu^2|\vec{\phi}|^2 + \frac{\lambda}{4}(|\vec{\phi}|^2)^2. \tag{6.56}$$

Therefore

$$\frac{\partial V}{\partial\phi_i} = \mu^2\phi_i + \lambda\phi_i|\vec{\phi}|^2. \tag{6.57}$$

In order to have a solution to the stationary condition we must have $\phi_i = 0$, or

$$|\vec{\phi}|^2 = -\frac{\mu^2}{\lambda}. \tag{6.58}$$

This equation has real solutions only if $\mu^2/\lambda < 0$. However, in order to have a potential bounded from below one has to require $\lambda > 0$, therefore we may have non-zero solutions to the minimum condition only if $\mu^2 < 0$. But notice that in this case $\mu^2$ cannot be identified with a physical mass since the masses are given by the eigenvalues of the matrix of the second derivatives of the potential at the minimum and they are positive definite by definition. We will study this case in the following Sections. When $\mu^2 > 0$ the minimum is at $\phi_i = 0$ and one can study the theory by taking the term $\lambda(|\vec{\phi}|^2)^2$ as a small perturbation (that is requiring both $\lambda$ and the values of $\phi_i$, the fluctuations around the minimum to be small). The free theory is given by the quadratic terms in the Lagrangian density, and describes $N$ particles of common mass $m$. Furthermore, both the free and the interacting theories are $O(N)$ symmetric.

## 6.2   Spontaneous symmetry breaking

In this Section we will show that the linear $\sigma$-model with $\mu^2 < 0$ is an example of a general phenomenon named **spontaneous symmetry breaking**. This phenomenon lies at the basis of the modern description of phase transitions and it has acquired a capital relevance in the last years in all fields of physics. In particular, in the case of the theory of elementary particles, it is the basis for the description of the weak interactions.

The idea is very simple and consists in the observation that a theory with Hamiltonian invariant under a symmetry group may not show explicitly the symmetry at the level of the solutions. As we shall see this may happen when the following conditions are realized:

- The theory is invariant under a symmetry group $G$.
- The fundamental state of the theory is degenerate and transforms in a non-trivial way under the symmetry group.

Just as an example consider a scalar field described by a Lagrangian invariant under parity

$$P: \quad \phi \to -\phi. \tag{6.59}$$

The Lagrangian density will be of the type

$$\mathcal{L} = \frac{1}{2}\partial_\mu\phi\partial^\mu\phi - V(\phi^2). \tag{6.60}$$

If the vacuum state is non-degenerate, barring a phase factor, we must have

$$P|0\rangle = |0\rangle, \tag{6.61}$$

since $P$ commutes with the Hamiltonian. It follows

$$\langle 0|\phi|0\rangle = \langle 0|P^{-1}P\phi P^{-1}P|0\rangle = \langle 0|P\phi P^{-1}|0\rangle = -\langle 0|\phi|0\rangle, \tag{6.62}$$

from which

$$\langle 0|\phi|0\rangle = 0. \tag{6.63}$$

The case is different if the fundamental state is degenerate. This would be the case in the example (6.60), if

$$V(\phi^2) = \frac{\mu^2}{2}\phi^2 + \frac{\lambda}{4}\phi^4, \tag{6.64}$$

with $\mu^2 < 0$. In fact, this potential has two minima located at

$$\phi = \pm v, \quad v = \sqrt{-\frac{\mu^2}{\lambda}}. \tag{6.65}$$

By denoting with $|R\rangle$ and $|L\rangle$ the two states corresponding to the classical configurations $\phi = \pm v$, we have

$$P|R\rangle = |L\rangle \neq |R\rangle. \tag{6.66}$$

Therefore

$$\langle R|\phi|R\rangle = \langle R|P^{-1}P\phi P^{-1}P|R\rangle = -\langle L|\phi|L\rangle, \tag{6.67}$$

which, contrarily to the previous case, does not imply that the expectation value of the field vanishes. In the following we will be rather interested in the case of continuous symmetries. So let us consider two scalar fields, and a Lagrangian density with symmetry $O(2)$

$$\mathcal{L} = \frac{1}{2}\partial_\mu \vec{\phi} \cdot \partial^\mu \vec{\phi} - \frac{1}{2}\mu^2 \vec{\phi} \cdot \vec{\phi} - \frac{\lambda}{4}(\vec{\phi} \cdot \vec{\phi})^2, \tag{6.68}$$

where

$$\vec{\phi} \cdot \vec{\phi} = \phi_1^2 + \phi_2^2. \tag{6.69}$$

For $\mu^2 > 0$ there is a unique ground state (minimum of the potential) $\vec{\phi} = 0$, whereas for $\mu^2 < 0$ there are infinite degenerate states corresponding to the points belonging to the manifold

$$|\vec{\phi}|^2 = \phi_1^2 + \phi_2^2 = v^2, \tag{6.70}$$

with $v$ defined as in (6.65). By denoting with $R(\theta)$ the operator rotating the fields in the plane $(\phi_1, \phi_2)$ we have, for the non-degenerate case,

$$R(\theta)|0\rangle = |0\rangle \tag{6.71}$$

and

$$\langle 0|\phi|0\rangle = \langle 0|R^{-1}R\phi R^{-1}R|0\rangle = \langle 0|\phi^\theta|0\rangle = 0, \tag{6.72}$$

since $\phi^\theta \neq \phi$. In the case $\mu^2 < 0$ (degenerate case), we have

$$R(\theta)|0\rangle = |\theta\rangle, \tag{6.73}$$

where $|\theta\rangle$ is one of the infinitely many degenerate fundamental states lying on the circle $|\vec{\phi}|^2 = v^2$. Then

$$\langle 0|\phi_i|0\rangle = \langle 0|R^{-1}(\theta)R(\theta)\phi_i R^{-1}(\theta)R(\theta)|0\rangle = \langle \theta|\phi_i^\theta|\theta\rangle, \tag{6.74}$$

with

$$\phi_i^\theta = R(\theta)\phi_i R^{-1}(\theta) \neq \phi_i. \tag{6.75}$$

Again, the expectation value of the field (contrarily to the non-degenerate case) does not need to vanish. The situation can be described qualitatively saying that the existence of a degenerate fundamental state forces the system to choose one of the equivalent states, and consequently to break the symmetry. But the breaking is only at the level of the solutions, the Lagrangian and the equations of motion preserve the symmetry. One can easily construct classical systems exhibiting spontaneous symmetry breaking. For instance, a classical particle in a double-well potential. This system is parity invariant, $x \to -x$, where $x$ is the particle position. The equilibrium positions are at the points, $\pm x_0$, corresponding to the minima of the potential. If we put a particle close to $x_0$, it will perform oscillations around that point and the original symmetry is lost. A further example is given by a ferromagnet extending in all 3 dimensions. The corresponding Hamiltonian is invariant under rotations in the 3-dimensional space. However, below the Curie temperature, the ferromagnet exhibits spontaneous magnetization, breaking the rotational symmetry in space to the symmetry of rotations around the direction of the magnetization. These situations are typical for the so-called second order phase transitions. One can describe them through the Landau free-energy, which depends on two different kind of parameters:

- **Control parameters**, as $\mu^2$ for the scalar field, or as the temperature for the ferromagnet.
- **Order parameters**, as the expectation value of the scalar field or as the magnetization.

The system goes from one phase to another varying the control parameters, and the phase transition is characterized by the order parameter which assumes different values in different phases. In the previous examples, the order parameter was zero in the symmetric phase and different from zero in the broken phase.

The situation looks more involved at the quantum level, since spontaneous symmetry breaking cannot happen in finite systems. This follows from the existence of the tunnel effect. Let us consider again a particle in a double-well potential, and recall that we have defined the ground states through the correspondence with the classical minima

$$x = x_0 \to |R\rangle,$$
$$x = -x_0 \to |L\rangle. \tag{6.76}$$

But the tunnel effect gives rise to a transition between these two states and as a consequence it removes the degeneracy. In fact, due to the nonvanishing transition probability between the two degenerate states, the Hamiltonian acquires a non-zero matrix element between the states $|R\rangle$ and $|L\rangle$. By denoting with $\underline{H}$ the matrix of the Hamiltonian between these two states, we get

$$\underline{H} = \begin{pmatrix} \epsilon_0 & \epsilon_1 \\ \epsilon_1 & \epsilon_0 \end{pmatrix}. \tag{6.77}$$

The eigenvalues of $\underline{H}$ are

$$(\epsilon_0 + \epsilon_1, \epsilon_0 - \epsilon_1). \tag{6.78}$$

We have no more degeneracy and the eigenstates are

$$|S\rangle = \frac{1}{\sqrt{2}}(|R\rangle + |L\rangle), \tag{6.79}$$

with eigenvalue $E_S = \epsilon_0 + \epsilon_1$, and

$$|A\rangle = \frac{1}{\sqrt{2}}(|R\rangle - |L\rangle), \tag{6.80}$$

with eigenvalue $E_A = \epsilon_0 - \epsilon_1$. One can show that $\epsilon_1 < 0$ and therefore the ground state is the symmetric one, $|S\rangle$. This situation gives rise to the so-called quantum oscillations. We can express the states $|R\rangle$ and $|L\rangle$ in terms of the energy eigenstates

$$|R\rangle = \frac{1}{\sqrt{2}}(|S\rangle + |A\rangle),$$
$$|L\rangle = \frac{1}{\sqrt{2}}(|S\rangle - |A\rangle). \tag{6.81}$$

Then, let us prepare a state, at $t = 0$, by putting the particle in the right minimum. This is not an energy eigenstate and its time evolution is given by

$$|R, t\rangle = \frac{1}{\sqrt{2}}\left(e^{-iE_S t}|S\rangle + e^{-iE_A t}|A\rangle\right)$$
$$= \frac{1}{\sqrt{2}}e^{-iE_S t}\left(|S\rangle + e^{-it\Delta E}|A\rangle\right), \tag{6.82}$$

with $\Delta E = E_A - E_S$. Therefore, for $t = \pi/\Delta E$ the state $|R\rangle$ transforms into the state $|L\rangle$ and comes back to $|R\rangle$ for $t = 2\pi/\Delta E$. As a consequence the state oscillates between states described by the kets $|R\rangle$ and $|L\rangle$, with a period

$$T = \frac{2\pi}{\Delta E}. \tag{6.83}$$

In nature there are finite systems as sugar molecules, which seem to exhibit spontaneous symmetry breaking. In fact, we observe both right-handed and left-handed sugar molecules. The explanation is simply that the energy difference $\Delta E$ is so small that the oscillation period is of the order of $10^4 - 10^6$ years.

The splitting of the fundamental states decreases with the height of the potential between two minima. Therefore, for infinite systems, the previous mechanism does not work, and we may have spontaneous symmetry breaking. In fact, coming back to the scalar field example, its expectation value on the vacuum must be a constant, as it follows from the translational invariance of the vacuum

$$\langle 0|\phi(x)|0\rangle = \langle 0|e^{iPx}\phi(0)e^{-iPx}|0\rangle = \langle 0|\phi(0)|0\rangle = v \qquad (6.84)$$

and the energy difference between the maximum at $\phi = 0$, and the minimum at $\phi = v$, becomes infinite in the limit of infinite volume

$$H(\phi = 0) - H(\phi = v) = -\int_V d^3x \left[\frac{\mu^2}{2}v^2 + \frac{\lambda}{4}v^4\right] = \frac{\mu^4}{4\lambda}\int_V d^3x = \frac{\mu^4}{4\lambda}V.$$

$$(6.85)$$

## 6.3   The Goldstone theorem

One of the most interesting consequences of spontaneous symmetry breaking is the Goldstone theorem. This theorem says that for any spontaneously broken continuous symmetry, there exists a massless particle[2] (the Goldstone boson). The theorem holds rigorously in a relativistic local field theory, under the following hypotheses:

- The spontaneous broken symmetry must be a continuous one.
- The theory must be manifestly covariant.
- The Hilbert space of the theory must have a definite positive norm.

We will limit ourselves to analyze the theorem in the case of a classical scalar field theory. Let us start considering the Lagrangian for the linear $\sigma$-model with invariance $O(N)$

$$\mathcal{L} = \frac{1}{2}\partial_\mu\vec{\phi}\cdot\partial^\mu\vec{\phi} - \frac{\mu^2}{2}\vec{\phi}\cdot\vec{\phi} - \frac{\lambda}{4}(\vec{\phi}\cdot\vec{\phi})^2, \qquad (6.86)$$

where, according to our previous discussion, we assume $\lambda > 0$. The conditions for $V$ to be stationary is

$$\frac{\partial V}{\partial \phi_l} = \mu^2\phi_l + \lambda\phi_l|\vec{\phi}|^2 = 0, \qquad (6.87)$$

---

[2]In the non-relativistic case the counting of the massless states is different and it depends on the dispersion relation for the Goldstone particles [Nielsen and Chada (1976)].

with solutions

$$\phi_l = 0, \qquad |\vec{\phi}|^2 = v^2, \qquad v = \sqrt{\frac{-\mu^2}{\lambda}}. \tag{6.88}$$

The character of the stationary points can be studied by evaluating the second derivatives

$$\frac{\partial^2 V}{\partial \phi_l \partial \phi_m} = \delta_{lm}(\mu^2 + \lambda|\vec{\phi}|^2) + 2\lambda\phi_l\phi_m. \tag{6.89}$$

We have two possibilities:

- $\mu^2 > 0$, we have only one real solution given by $\vec{\phi} = 0$, which is a minimum, since

$$\frac{\partial^2 V}{\partial \phi_l \partial \phi_m} = \delta_{lm}\mu^2 > 0. \tag{6.90}$$

- $\mu^2 < 0$, there are infinite solutions, among which $\vec{\phi} = 0$ is a maximum. The points of the sphere $|\vec{\phi}|^2 = v^2$ are degenerate minima. In fact, by choosing $\phi_l = v\delta_{lN}$ as a representative point, we get

$$\frac{\partial^2 V}{\partial \phi_l \partial \phi_m} = 2\lambda v^2 \delta_{lN}\delta_{mN} > 0. \tag{6.91}$$

Expanding the potential around this minimum we get

$$V(\vec{\phi}) \approx V\Big|_{\text{min}} + \frac{1}{2}\frac{\partial^2 V}{\partial \phi_l \partial \phi_m}\Big|_{\text{min}} (\phi_l - v\delta_{lN})(\phi_m - v\delta_{mN}). \tag{6.92}$$

If we are interested in the perturbative expansion, this must be made around a minimum and the right fields to be used are $\phi_l - v\delta_{lN}$. The mass matrix is given by the coefficients of the quadratic term

$$M_{lm}^2 = \frac{\partial^2 V}{\partial \phi_l \partial \phi_m}\Big|_{\text{min}} = -2\mu^2 \delta_{lN}\delta_{mN} = \begin{pmatrix} 0 & 0 & \cdot & 0 \\ 0 & 0 & \cdot & 0 \\ \cdot & \cdot & \cdot & \cdot \\ 0 & 0 & \cdot & -2\mu^2 \end{pmatrix}. \tag{6.93}$$

Therefore the masses of the fields $\phi_a$, $a = 1, \cdots, N-1$, and $\chi = \phi_N - v$, are given by

$$m_{\phi_a}^2 = 0, \qquad m_\chi^2 = -2\mu^2. \tag{6.94}$$

By defining

$$m^2 = -2\mu^2, \tag{6.95}$$

we can write the potential as a function of the new fields

$$V = \frac{m^4}{16\lambda} + \frac{1}{2}m^2\chi^2 + \sqrt{\frac{m^2\lambda}{2}}\chi\left(\sum_{a=1}^{N-1}\phi_a^2 + \chi^2\right) + \frac{\lambda}{4}\left(\sum_{a=1}^{N-1}\phi_a^2 + \chi^2\right)^2.$$
(6.96)

In this form the original symmetry $O(N)$ is broken. However a residual symmetry $O(N-1)$ is left. In fact, $V$ depends only on the combination $\sum_{a=1}^{N-1}\phi_a^2$, and this is invariant under rotations around the axis we have chosen as representative for the fundamental state, $(0,\cdots,v)$. However, it must be stressed that this is not the most general potential invariant under $O(N-1)$. In fact the most general potential (up to the fourth order in the fields) describing $N$ scalar fields with a symmetry $O(N-1)$ is a polynomial in $\sum_{a=1}^{N-1}\phi_a^2$ and $\chi$ depending on 7 coupling constants, whereas the one we have obtained depends only on the two original parameters $m$ and $\lambda$. Therefore spontaneous symmetry breaking puts heavy constraints on the dynamics of the system. We have also seen that there are $N-1$ massless scalars. Clearly the rotations along the first $N-1$ directions leave the potential invariant, whereas the $N-1$ rotations on the planes $(a,N), a = 1,\cdots,N-1$ move away from the surface of the minima. This can also be seen in terms of generators. Since the field we have chosen as representative of the ground state is $\phi_i|_{\min} = v\delta_{iN}$, we have, using eq. (6.14),

$$T_{ij}^{ab}\phi_j|_{\min} = i(\delta_i^a\delta_j^b - \delta_i^b\delta_j^a)v\delta_{jN} = 0,\qquad(6.97)$$

since $a,b = 1,\cdots,N-1$, and

$$T_{ij}^{aN}\phi_j|_{\min} = i(\delta_i^a\delta_j^N - \delta_i^N\delta_j^a)v\delta_{jN} = iv\delta_i^a \neq 0.\qquad(6.98)$$

Therefore we have $N-1$ broken symmetries and $N-1$ massless scalars. The generators of $O(N)$ divide up naturally in the generators of the vacuum symmetry (here $O(N-1)$), and in the so-called broken generators, each of them corresponding to a massless Goldstone boson. In general, if the original symmetry group $G$ of the theory is spontaneously broken down to a subgroup $H$ (which is the symmetry of the vacuum), the Goldstone bosons correspond to the generators of $G$ which are left after subtracting the generators of $H$. Intuitively one can understand the origin of the massless particles noticing that the broken generators allow transitions from a possible vacuum to another. Since these states are degenerate the operation does not cost any energy. From the relativistic dispersion relation this implies that we must have massless particles. One can say that Goldstone bosons correspond to flat directions in the potential.

## 6.4 QED as a gauge theory

Many field theories possess global symmetries. These are transformations leaving invariant the action of the system and are characterized by a certain number of parameters which are independent on the space-time point. As a prototype we can consider the free Dirac Lagrangian density

$$\mathcal{L}_0 = \bar{\psi}(x)[i\not{\partial} - m]\psi(x), \tag{6.99}$$

which is invariant under the global phase transformation

$$\psi(x) \to \psi'(x) = e^{-iQ\alpha}\psi(x), \tag{6.100}$$

where $Q$ is the electric charge of the field in units of $e$ which, only for this Chapter, will be assumed as the charge of the proton. If there are many charged fields, $Q$ is a diagonal matrix with eigenvalues equal to the charges of the different fields (always measured in units of $e$). For instance, a term as $\bar{\psi}_2\psi_1\phi$, with $\phi$ a scalar field, is invariant by choosing $Q(\psi_1) = Q(\phi) = 1$, and $Q(\psi_2) = 2$. This is said to be an **abelian symmetry** because different transformations commute among themselves

$$e^{-i\alpha Q}e^{-i\beta Q} = e^{-i(\alpha+\beta)Q} = e^{-i\beta Q}e^{-i\alpha Q}. \tag{6.101}$$

It is also referred to as a $U(1)$ symmetry. The physical meaning of this invariance lies in the possibility of assigning the phases of fields in arbitrary way, without changing the observable quantities. This way of thinking is in some sort of contradiction with causality, since it requires to assign the phase of the fields simultaneously in all space-time points. It looks more physical to be able to assign the phase arbitrarily in each space-time point. If the theory does not depend by this arbitrary choice of the phases we say that it is **gauge invariant**[3]. The free Lagrangian (6.99) cannot be invariant under the transformation (6.100), with a parameter $\alpha$ depending on the space-time point, due to the derivative present in the kinetic term. Then, the idea is simply to generalize the derivative $\partial_\mu$ to a so-called covariant derivative $D_\mu$ having the property that $D_\mu\psi$ transforms as $\psi$, that is

$$D_\mu\psi(x) \to [D_\mu\psi(x)]' = e^{-iQ\alpha(x)}D_\mu\psi(x). \tag{6.102}$$

In this case the term

$$\bar{\psi}D_\mu\psi \tag{6.103}$$

will be invariant as the mass term under the local phase transformation. To construct the covariant derivative, we need to enlarge the field content

---

[3]This was the main idea underlying the work of [Yang and Mills (1954)] that lead to the formulation of non-abelian gauge symmetries, see later.

of the theory, by introducing a vector field, the **gauge field** $A_\mu$, in the following way

$$D_\mu = \partial_\mu + ieQA_\mu. \tag{6.104}$$

The transformation law of $A_\mu$ is obtained from eq. (6.102)

$$[(\partial_\mu + ieQA_\mu)\psi]' = (\partial_\mu + ieQA'_\mu)\psi'(x)$$

$$= (\partial_\mu + ieQA'_\mu)e^{-iQ\alpha(x)}\psi$$

$$= e^{-iQ\alpha(x)}\left[\partial_\mu + ieQ(A'_\mu - \frac{1}{e}\partial_\mu\alpha)\right]\psi, \tag{6.105}$$

from which

$$A'_\mu = A_\mu + \frac{1}{e}\partial_\mu\alpha. \tag{6.106}$$

The Lagrangian density

$$\mathcal{L}_\psi = \bar\psi[i\slashed{D} - m]\psi = \bar\psi[i\gamma^\mu(\partial_\mu + ieQA_\mu) - m]\psi = \mathcal{L}_0 - e\bar\psi Q\gamma^\mu\psi A_\mu \tag{6.107}$$

is then invariant under gauge transformations or under the local group $U(1)$. We see that the requirement of local invariance gives rise to the electromagnetic interaction as obtained through the minimal substitution we discussed before.

In order to determine the kinetic term for the vector field $A_\mu$ we notice that eq. (6.102) implies that under a gauge transformation, the covariant derivative transforms according to the unitary transformation

$$D_\mu \to D_\mu' = e^{-iQ\alpha(x)}D_\mu e^{iQ\alpha(x)}. \tag{6.108}$$

Then, also the commutator of two covariant derivatives

$$[D_\mu, D_\nu] = [\partial_\mu + ieQA_\mu, \partial_\nu + ieQA_\nu] = ieQF_{\mu\nu}, \tag{6.109}$$

with

$$F_{\mu\nu} = \partial_\mu A_\nu - \partial_\nu A_\mu, \tag{6.110}$$

transforms exactly in the same way

$$F_{\mu\nu} \to e^{-iQ\alpha(x)}F_{\mu\nu}e^{iQ\alpha(x)} = F_{\mu\nu}. \tag{6.111}$$

The last equality follows from the commutativity of $F_{\mu\nu}$ with the phase factor. The complete invariant Lagrangian density can be written as

$$\mathcal{L} = \mathcal{L}_\psi + \mathcal{L}_A = \bar\psi[i\gamma^\mu(\partial_\mu + ieQA_\mu) - m]\psi - \frac{1}{4}F_{\mu\nu}F^{\mu\nu}. \tag{6.112}$$

The gauge principle has automatically generated an interaction between the gauge field and the charged field. We notice also that gauge invariance prevents any mass term for the vector field $A_\mu$. As a consequence the photon field is necessarily massless. Also, since the local invariance implies the global one, by using the Noether theorem we find the conserved current

$$j_\mu = \frac{\partial\mathcal{L}}{\partial\psi_{,\mu}}\delta\psi = \bar\psi\gamma_\mu(Q\alpha)\psi, \tag{6.113}$$

from which, eliminating the infinitesimal parameter $\alpha$,

$$J_\mu = \bar\psi\gamma_\mu Q\psi. \tag{6.114}$$

## 6.5  Non-abelian gauge theories

The approach of the previous Section can be easily extended to non-abelian local symmetries. We will consider the case of $N$ Dirac fields. The free Lagrangian density

$$\mathcal{L}_0 = \sum_{a=1}^{N} \bar{\psi}_a(i\partial\!\!\!/ - m)\psi_a \tag{6.115}$$

is invariant under the global transformation

$$\Psi(x) \to \Psi'(x) = A\Psi(x), \tag{6.116}$$

where $A$ is a unitary $N \times N$ matrix, and we have denoted by $\Psi$ the column vector with components $\psi_a$. Notice that $U(N)$ is the maximal symmetry group of eq. (6.115). More generally the actual symmetry could be a subgroup of $U(N)$. For instance, when the masses are not all equal. So we will consider here the gauging of a subgroup $G$ of $U(N)$. The fields $\psi_a(x)$ will belong, in general, to some reducible representation of $G$. Denoting by $U$ the generic element of $G$, we will write the corresponding matrix $U_{ab}$ acting upon the fields $\psi_a$ as

$$U = e^{-i\alpha_A T^A}, \qquad U \in G, \tag{6.117}$$

where $T^A$ denote the generators of the Lie algebra associated to $G$, $\mathrm{Lie}(G)$, (that is the vector space spanned by the infinitesimal generators of the group) in the fermion representation. The generators $T^A$ satisfy the algebra

$$[T^A, T^B] = if_C^{AB} T^C, \tag{6.118}$$

where $f_C^{AB}$ are the structure constants of $\mathrm{Lie}(G)$. For instance, if $G = SU(2)$, and we take the fermions in the fundamental representation,

$$\Psi = \begin{pmatrix} \psi_1 \\ \psi_2 \end{pmatrix}, \tag{6.119}$$

we have

$$T^A = \frac{\sigma^A}{2}, \qquad A = 1, 2, 3, \tag{6.120}$$

where $\sigma^A$ are the Pauli matrices. In the general case the $T^A$'s are $N \times N$ hermitian matrices that we will choose normalized in such a way that

$$Tr(T^A T^B) = \frac{1}{2}\delta^{AB}. \tag{6.121}$$

To make local the transformation (6.117), means to promote the parameters $\alpha_A$ to space-time functions

$$\alpha_A \to \alpha_A(x). \tag{6.122}$$

Notice, that the group does not need to be abelian, and therefore, in general

$$e^{-i\alpha_A T^A} e^{-i\beta_A T^A} \neq e^{-i\beta_A T^A} e^{-i\alpha_A T^A}. \tag{6.123}$$

In the case of a local symmetry we will define a covariant derivative in strict analogy to what we have done in the abelian case. That is we will require the following transformation properties

$$D_\mu \Psi(x) \to [D_\mu \Psi(x)]' = U(x)[D_\mu \psi(x)]. \tag{6.124}$$

We will put again

$$D_\mu = \partial_\mu + igB_\mu, \tag{6.125}$$

where $B_\mu$ is an $N \times N$ matrix acting upon $\Psi(x)$. In components

$$D_{ab}^\mu = \delta_{ab}\partial^\mu + ig(B^\mu)_{ab}. \tag{6.126}$$

Then eq. (6.124) implies

$$\begin{aligned}
D_\mu \Psi &\to (\partial_\mu + igB'_\mu)U(x)\Psi \\
&= U(x)\partial_\mu\Psi + U(x)[U^{-1}(x)igB'_\mu U(x)]\Psi + (\partial_\mu U(x))\Psi \\
&= U(x)[\partial_\mu + U^{-1}(x)igB'_\mu U(x) + U^{-1}(x)\partial_\mu U(x)]\Psi, \quad (6.127)
\end{aligned}$$

from which

$$U^{-1}(x)igB'_\mu U(x) + U^{-1}(x)\partial_\mu U(x) = igB_\mu \tag{6.128}$$

and

$$B'_\mu(x) = U(x)B_\mu(x)U^{-1}(x) + \frac{i}{g}(\partial_\mu U(x))U^{-1}(x). \tag{6.129}$$

For an infinitesimal transformation

$$U(x) \approx 1 - i\alpha_A(x)T^A, \tag{6.130}$$

we get

$$\delta B_\mu(x) = -i\alpha_A(x)[T^A, B_\mu(x)] + \frac{1}{g}(\partial_\mu\alpha_A(x))T^A. \tag{6.131}$$

Since $B_\mu(x)$ acquires a term proportional to $T^A$, the transformation law is consistent with $B_\mu$ being linear in the generators of the Lie algebra, that is

$$(B^\mu)_{ab} \equiv A_A^\mu(T^A)_{ab}. \tag{6.132}$$

The fields $A_A^\mu$ are called non-abelian gauge fields and their transformation law is

$$\delta A_C^\mu = f_C^{AB} \alpha_A A_B^\mu + \frac{1}{g} \partial^\mu \alpha_C. \tag{6.133}$$

The difference with respect to the abelian case is that the transformation acts on the fields not only with the inhomogeneous piece, but also with a homogeneous one depending on the non-trivial commutation relations among the generators of the Lie algebra of the non-abelian group. In fact, for an abelian group the structure constants are zero and we reproduce the abelian transformation.

The kinetic term for the gauge fields is constructed as in the abelian case. In fact the quantity

$$[D_\mu, D_\nu] \tag{6.134}$$

transforms as

$$[D_\mu, D_\nu] \to U(x)[D_\mu, D_\nu]U^{-1}(x), \tag{6.135}$$

by virtue of eq. (6.124), that implies

$$D_\mu \to U(x)D_\mu U^{-1}(x). \tag{6.136}$$

Therefore, defining as in the abelian case

$$[D_\mu, D_\nu] \equiv ig F_{\mu\nu}, \tag{6.137}$$

we see that

$$F'_{\mu\nu} = U(x)F_{\mu\nu}U^{-1}(x). \tag{6.138}$$

This time the tensor $F_{\mu\nu}$ is not invariant but transforms homogeneously, since it does not commute with the gauge transformation as in the abelian case.

To construct invariant quantities from objects transforming homogeneously, it is enough to take the trace. The invariant kinetic term is then

$$\mathcal{L}_A = -\frac{1}{2} Tr[F_{\mu\nu} F^{\mu\nu}]. \tag{6.139}$$

Let us now evaluate $F_{\mu\nu}$

$$\begin{aligned} ig F_{\mu\nu} = [D_\mu, D_\nu] &= [\partial_\mu + ig B_\mu, \partial_\nu + ig B_\nu] \\ &= ig(\partial_\mu B_\nu - \partial_\nu B_\mu) - g^2[B_\mu, B_\nu], \end{aligned} \tag{6.140}$$

or

$$F_{\mu\nu} = (\partial_\mu B_\nu - \partial_\nu B_\mu) + ig[B_\mu, B_\nu]. \tag{6.141}$$

In components

$$F^{\mu\nu} = F_C^{\mu\nu} T^C, \tag{6.142}$$

with

$$F_C^{\mu\nu} = \partial^\mu A_C^\nu - \partial^\nu A_C^\mu - g f_C^{AB} A_A^\mu A_B^\nu. \tag{6.143}$$

The main feature of the non-abelian gauge theories is the bilinear term in the previous expression. Such a term originates from the fact that $G$ is not abelian, corresponding to $f_C^{AB} \neq 0$. The kinetic term for the gauge fields, expressed in components, is given by

$$\mathcal{L}_A = -\frac{1}{4} \sum_A F_{\mu\nu A} F_A^{\mu\nu}. \tag{6.144}$$

Whereas in the abelian case $\mathcal{L}_A$ is a free Lagrangian (it contains only quadratic terms), in the case of non-abelian symmetries it contains interaction terms both cubic and quartic in the fields. The physical motivation lies in the fact that the gauge fields couple to all the fields transforming in a non-trivial way under the gauge group. Therefore they couple also to themselves (remember the homogeneous piece of transformation).

To derive the equations of motion for the gauge fields, let us consider the total action

$$\int_V d^4x \left[ \bar{\Psi}(i\slashed{\partial} - m)\Psi - g\bar{\Psi}\gamma_\mu B^\mu \Psi \right] + S_A, \tag{6.145}$$

where

$$S_A = -\frac{1}{2} \int_V d^4x \, Tr(F_{\mu\nu} F^{\mu\nu}). \tag{6.146}$$

The variation of $S_A$ is

$$\delta S_A = -\int_V d^4x \, Tr(F_{\mu\nu} \delta F^{\mu\nu}). \tag{6.147}$$

Using the definition (6.141) for the field strength we get

$$\delta F_{\mu\nu} = \partial_\mu \delta B_\nu + ig(\delta B_\mu)B_\nu + igB_\mu\delta(B_\nu) - (\mu \leftrightarrow \nu), \tag{6.148}$$

from which

$$\delta S_A = -2 \int_V d^4x \, Tr[F^{\mu\nu}(\partial_\mu \delta B_\nu + ig(\delta B_\mu)B_\nu + igB_\mu(\delta B_\nu))], \tag{6.149}$$

where we have taken into account the antisymmetry properties of $F_{\mu\nu}$. Integrating by parts we obtain

$$\delta S_A = -2 \int_V d^4x Tr[-(\partial_\mu F^{\mu\nu})\delta B_\nu - igB_\mu F^{\mu\nu}\delta B_\nu + igF^{\mu\nu}B_\mu\delta B_\nu]$$

$$= 2 \int_V d^4x Tr\left[ (\partial_\mu F^{\mu\nu} + ig[B_\mu, F^{\mu\nu}]) \delta B_\nu \right]$$

$$= \int_V d^4x \, (\partial_\mu F^{\mu\nu} + ig[B_\mu, F^{\mu\nu}])_A \, \delta A_{\nu A}. \tag{6.150}$$

Here we have made use of the cyclic property of the trace. By taking into account also the free term for the Dirac fields and the interaction we find the equations of motion

$$\partial_\mu F^{\mu\nu A} + ig[B_\mu, F^{\mu\nu}]^A = g\bar{\Psi}\gamma^\nu T^A\Psi,$$
$$(i\slashed{\partial} - m)\Psi = g\gamma_\mu B^\mu\Psi. \tag{6.151}$$

From the first equation we see that the currents $\bar{\Psi}\gamma_\mu T^A\Psi$ are not conserved. In fact, the conserved currents turn out to be

$$J_\nu^A = \bar{\Psi}\gamma_\nu T^A\Psi - i[B^\mu, F_{\mu\nu}]^A, \tag{6.152}$$

as it follows from the first equation in (6.151). The reason is that under a global transformation of the symmetry group, the gauge fields are not invariant, in other words they are <u>charged</u> fields with respect to the gauge fields. In fact, we can verify immediately that the previous currents are precisely the Noether currents. Under a global variation we have

$$\delta A_C^\mu = f_C^{AB}\alpha_A A_B^\mu, \qquad \delta\Psi = -i\alpha_A T^A\Psi \tag{6.153}$$

and we get

$$j^\mu = \frac{\partial\mathcal{L}}{\partial\Psi_{,\mu}}\delta\Psi + \frac{\partial\mathcal{L}}{\partial A_{\nu,\mu C}}\delta A_{\nu C}, \tag{6.154}$$

from which

$$j^\mu = \bar{\Psi}\gamma^\mu\alpha_A T^A\Psi - F_C^{\mu\nu}f_C^{AB}\alpha_A A_{\nu B}. \tag{6.155}$$

In the case of simple compact Lie groups one can define $f^{ABC} = f_C^{AB}$ with the property $f^{ABC} = f^{BCA}$. It follows

$$F_C^{\mu\nu}f^{ABC}A_{\nu B}T^A = i[B^\nu, F_{\nu\mu}] \equiv i[B^\nu, F_{\nu\mu}]^A T^A. \tag{6.156}$$

Therefore

$$j^\mu = \bar{\Psi}\gamma^\mu\alpha_A T^A\Psi - i[B_\nu, F^{\nu\mu}]^A\alpha_A. \tag{6.157}$$

After division by $\alpha_A$ we get the Noether currents (6.152). The contribution of the gauge fields to the currents is also crucial for their conservation. In fact, the divergence of the fermionic contribution is given by

$$\partial^\mu(\bar{\Psi}\gamma_\mu T^A\Psi) = -ig\bar{\Psi}\gamma_\mu T^A B^\mu\Psi + ig\bar{\Psi}\gamma_\mu B^\mu T^A\Psi = -ig\bar{\Psi}\gamma_\mu[T^A, B^\mu]\Psi, \tag{6.158}$$

which vanishes for abelian gauge fields, whereas it is compensated by the gauge fields contribution in the non-abelian case.

## 6.6   The Higgs mechanism

We have seen that in the case of continuous symmetries, the spontaneous symmetry breaking mechanism leads to massless scalar particles, the Goldstone bosons. Also gauge theories lead to massless vector bosons. In fact, as in the electromagnetic case, gauge invariance forbids the presence in the Lagrangian of terms quadratic in the gauge fields. Unfortunately in nature the only massless particles we know are the photons[4]. But once spontaneous symmetry breaking and gauge symmetry are present at the same time, things change and one gets completely different results. In fact, if we look back at the hypotheses underlying a gauge theory, it turns out that the Goldstone theorem does not hold in this context. The reason is that it is impossible to quantize a gauge theory in a way which is at the same time manifestly covariant and with a Hilbert space with positive definite metric. In fact, remember that for the electromagnetic field one has to choose the gauge before quantization. What happens is that, if one chooses a physical gauge, as the Coulomb gauge, in order to have a Hilbert space spanned by only the physical states, then the theory loses the manifest covariance. If one goes to a covariant gauge, as the Lorenz one, the theory is covariant but one has to work with a bigger Hilbert space with non-definite positive metric, and where the physical states are obtained through a supplementary condition. This situation is common to all the gauge theories.

The way in which the Goldstone theorem is evaded is that the Goldstone bosons disappear and, at the same time, the gauge bosons corresponding to the broken symmetries acquire mass. This is the famous **Higgs mechanism**.

Let us start with a scalar theory invariant under $O(2)$

$$\mathcal{L} = \frac{1}{2}\partial_\mu \vec{\phi} \cdot \partial^\mu \vec{\phi} - \frac{\mu^2}{2}\vec{\phi} \cdot \vec{\phi} - \frac{\lambda}{4}(\vec{\phi} \cdot \vec{\phi})^2 \qquad (6.159)$$

where $\vec{\phi} = (\phi_1, \phi_2)$. Let us now analyze the spontaneous symmetry breaking mechanism. In the case $\mu^2 < 0$ the symmetry is broken and we can choose the vacuum as the state

$$\vec{\phi}_0 = (v, 0), \qquad v = \sqrt{\frac{-\mu^2}{\lambda}}. \qquad (6.160)$$

After the translation $\phi_1 = \chi + v$, with $\langle 0|\chi|0\rangle = 0$, we get the potential

---

[4]In 1998 the SuperKamiokande collaboration has observed neutrino oscillations, proving that also the neutrinos, that for a very long time were supposed to be massless, have indeed a mass.

$(m^2 = -2\mu^2)$

$$V = \frac{m^4}{16\lambda} + \frac{1}{2}m^2\chi^2 + \sqrt{\frac{m^2\lambda}{2}}\chi(\phi_2^2 + \chi^2) + \frac{\lambda}{4}(\phi_2^2 + \chi^2)^2. \qquad (6.161)$$

Correspondingly the group $O(2)$ is completely broken (except for the discrete symmetry $\phi_2 \to -\phi_2$). The Goldstone particle is described by $\phi_2$. This field has a peculiar way of transforming under $O(2)$. In fact, the original fields transform as

$$\delta\phi_1 = -\alpha\phi_2, \qquad \delta\phi_2 = \alpha\phi_1, \qquad (6.162)$$

from which

$$\delta\chi = -\alpha\phi_2, \qquad \delta\phi_2 = \alpha\chi + \alpha v. \qquad (6.163)$$

We see that the Goldstone field, $\phi_2$, under the transformation, performs both a rotation and a translation, $\alpha v$. This is the main reason why the Goldstone particle is massless. In fact, one can have invariance under translations of the field, only if the potential is flat in the corresponding direction. This is what happens when one moves in a way which is tangent to the surface of the degenerate vacua (in this case a circle). How do things change if our theory is gauge invariant? In that case the translation becomes an arbitrary function of the space-time point and it should be possible to eliminate the Goldstone field from the theory. This is better seen by using polar coordinates for the fields, that is

$$\rho = \sqrt{\phi_1^2 + \phi_2^2}, \qquad \sin\theta = \frac{\phi_2}{\sqrt{\phi_1^2 + \phi_2^2}}. \qquad (6.164)$$

Under a finite rotation, the new fields transform as

$$\rho \to \rho, \qquad \theta \to \theta + \alpha. \qquad (6.165)$$

Notice that the two coordinate systems coincide when we are close to the vacuum. In fact, in that case we can perform the following expansion

$$\rho = \sqrt{\phi_2^2 + \chi^2 + 2\chi v + v^2} \approx v + \chi, \qquad \theta \approx \frac{\phi_2}{v + \chi} \approx \frac{\phi_2}{v}. \qquad (6.166)$$

Therefore, if we make the theory invariant under a local transformation, we will have invariance under

$$\theta(x) \to \theta(x) + \alpha(x). \qquad (6.167)$$

By choosing $\alpha(x) = -\theta(x)$ we can eliminate the $\theta$ field from the theory. The only remaining degree of freedom in the scalar sector is $\rho(x)$.

Let us study the gauging of this model. It is convenient to introduce complex variables

$$\phi = \frac{1}{\sqrt{2}}(\phi_1 + i\phi_2), \qquad \phi^\dagger = \frac{1}{\sqrt{2}}(\phi_1 - i\phi_2). \qquad (6.168)$$

The $O(2)$ transformation becomes a phase transformation on $\phi$

$$\phi \to e^{i\alpha}\phi \qquad (6.169)$$

and the Lagrangian density (6.159) can be written as

$$\mathcal{L} = \partial_\mu \phi^\dagger \partial^\mu \phi - \mu^2 \phi^\dagger \phi - \lambda(\phi^\dagger \phi)^2. \qquad (6.170)$$

We know that it is possible to promote a global symmetry to a local one by introducing the covariant derivative

$$\partial_\mu \phi \to (\partial_\mu + igA_\mu)\phi, \qquad (6.171)$$

from which

$$\mathcal{L} = (\partial_\mu - igA_\mu)\phi^\dagger(\partial^\mu + igA^\mu)\phi - \mu^2 \phi^\dagger \phi - \lambda(\phi^\dagger \phi)^2 - \frac{1}{4}F_{\mu\nu}F^{\mu\nu}. \quad (6.172)$$

In terms of the polar coordinates $(\rho, \theta)$ we have

$$\phi = \frac{1}{\sqrt{2}}\rho e^{i\theta}, \qquad \phi^\dagger = \frac{1}{\sqrt{2}}\rho e^{-i\theta}. \qquad (6.173)$$

By performing the following gauge transformation on the scalars

$$\phi \to \phi' = \phi e^{-i\theta} \qquad (6.174)$$

and the corresponding transformation on the gauge fields

$$A_\mu \to A'_\mu = A_\mu + \frac{1}{g}\partial_\mu \theta, \qquad (6.175)$$

the Lagrangian will depend only on the fields $\rho$ and $A'_\mu$ (we will put $A'_\mu = A_\mu$ in the following)

$$\mathcal{L} = \frac{1}{2}(\partial_\mu - igA_\mu)\rho(\partial^\mu + igA_\mu)\rho - \frac{\mu^2}{2}\rho^2 - \frac{\lambda}{4}\rho^4 - \frac{1}{4}F_{\mu\nu}F^{\mu\nu}. \quad (6.176)$$

In this way the Goldstone boson disappears. We have now to translate the field $\rho$

$$\rho = \chi + v, \qquad \langle 0|\chi|0\rangle = 0 \qquad (6.177)$$

and we see that this generates a bilinear term in $A_\mu$, coming from the covariant derivative, given by

$$\frac{1}{2}g^2v^2 A_\mu A^\mu. \qquad (6.178)$$

Therefore the gauge field acquires a mass

$$m_A^2 = g^2 v^2. \tag{6.179}$$

It is instructive to count the degrees of freedom in the symmetric and in the broken case. In the symmetric case we had 4 degrees of freedom, two from the scalar fields and two from the gauge field. In the broken case, after the gauge transformation, we have only one degree of freedom from the scalar sector and three degrees of freedom from the vector field which is now massive. This is as it should be, since the number of degrees of freedom cannot change simply changing the phase. Notice that the reason why we may read clearly the number of degrees of freedom only after the gauge transformation is that before performing the transformation the Lagrangian contains a mixing term

$$A_\mu \partial^\mu \theta, \tag{6.180}$$

between the Goldstone field and the gauge vector which forbids to read directly the mass of the states from the Lagrangian. The previous gauge transformation realizes the purpose of making that term vanish and, at the same time, makes diagonal the mass matrix of the fields $(\rho, A_\mu)$. The gauge in which such a thing happens is called the **unitary gauge**.

We will consider now the example of a symmetry $O(N)$. The Lagrangian density, invariant under local transformations, is

$$\mathcal{L} = \frac{1}{2}(D_\mu)_{ij}\phi_j (D^\mu)_{ik}\phi_k - \frac{\mu^2}{2}\phi_i\phi_i - \frac{\lambda}{4}(\phi_i\phi_i)^2, \tag{6.181}$$

where

$$(D_\mu)_{ij} = \delta_{ij}\partial_\mu + i\frac{g}{2}(T^{AB})_{ij}W_\mu^{AB}, \tag{6.182}$$

where $(T^{AB})_{lm} = i(\delta_l^A\delta_m^B - \delta_m^A\delta_l^B)$. In the instance of broken symmetry $(\mu^2 < 0)$, we choose again the vacuum along the direction $N$, with $v$ defined as in (6.160)

$$\phi_i|_{\min} = v\delta_{iN}. \tag{6.183}$$

Recalling that

$$T_{ij}^{ab}\phi_j\Big|_{\min} = 0, \quad T_{ij}^{aN}\phi_j\Big|_{\min} = iv\delta_i^a, \quad a,b = 1,\cdots,N-1, \tag{6.184}$$

the mass term for the gauge field is given by

$$-\frac{1}{8}g^2 T_{ij}^{AB}\phi_j\Big|_{\min} (T^{CD})_{ik}\phi_k\Big|_{\min} W_\mu^{AB}W^{\mu CD}$$

$$= -\frac{1}{2}g^2 T_{ij}^{aN}\phi_j\Big|_{\min} (T^{bN})_{ik}\phi_k\Big|_{\min} W_\mu^{aN}W^{\mu bN}$$

$$= \frac{1}{2}g^2 v^2 \delta_i^a \delta_i^b W_\mu^{aN}W^{\mu bN} = \frac{1}{2}g^2 v^2 W_\mu^{aN}W^{\mu aN}. \tag{6.185}$$

Therefore, the fields $W_\mu^{aN}$ associated to the broken directions $T^{aN}$ acquire a mass $g^2v^2$, whereas the fields $W_\mu^{ab}$, associated to the unbroken symmetry $O(N-1)$, remain massless.

In general, if $G$ is the global symmetry group of the Lagrangian, $H$ the subgroup of $G$ leaving invariant the vacuum, and $G_W$ the group of local (gauge) symmetries, $G_W \in G$, one can divide up the broken generators in two categories. In the first category we put the broken generators lying in $G_W$; they have associated massive vector bosons. In the second category we put the other broken generators; they have associated massless Goldstone bosons. Finally the gauge fields associated to generators of $G_W$ lying in $H$ remain massless. From the previous derivation this follows noticing that the generators of $H$ annihilate the minimum of the fields, leaving the corresponding gauge bosons massless, whereas the non-zero action of the broken generators generate a mass term for the corresponding gauge fields. The situation is represented in Fig. 6.1. Now let us show how to eliminate

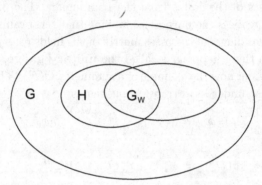

Fig. 6.1   This figure shows the various groups, $G$, the global symmetry of the Lagrangian, $H \in G$, the symmetry of the vacuum, and $G_W$, the group of local symmetries. The broken generators in $G_W$ correspond to massive vector bosons. The broken generators not belonging to $G_W$ correspond to massless Goldstone bosons. The unbroken generators in $G_W$ correspond to massless vector bosons.

the Goldstone bosons corresponding to the broken generators in the gauge group. In fact, we can define new fields $\xi_a$ and $\chi$ as

$$\phi_i = \left(e^{-iT^{aN}\xi_a}\right)_{iN}(\chi+v), \qquad (6.186)$$

where $a = 1, \cdots, N-1$, that is with the sum restricted to the broken directions. The other degree of freedom is in the second factor. The correspondence among the fields $\vec{\phi}$ and $(\xi_a, \chi)$ can be seen easily by expanding around the vacuum

$$\left(e^{-iT^{aN}\xi_a}\right)_{iN} \approx \delta_{iN} - i(T^{aN})_{iN}\xi_a = \delta_{iN} + \delta_i^a \xi_a, \qquad (6.187)$$

from which

$$\phi_i \approx (v\xi_a, \chi + v) \qquad (6.188)$$

showing that the $\xi_a$'s are really the Goldstone fields. The unitary gauge is defined through the transformation

$$\phi_i \to \left(e^{iT^{aN}\xi_a}\right)_{ij} \phi_j = \delta_{iN}(\chi + v), \qquad (6.189)$$

$$W_\mu \to e^{iT^{aN}\xi_a} W_\mu e^{-iT^{aN}\xi_a} - \frac{i}{g}\left(\partial_\mu e^{iT^{aN}\xi_a}\right) e^{-iT^{aN}\xi_a}. \qquad (6.190)$$

This transformation eliminates the Goldstone degrees of freedom and the resulting Lagrangian depends on the field $\chi$, on the massive vector fields $W_\mu^{aN}$ and on the massless fields $W_\mu^{ab}$. Notice again the counting of the degrees of freedom: $N$ scalar fields $+ 2N(N-1)/2$ massless vector fields $= N^2$ fields in the symmetric case, and 1 massive scalar field $+ 3(N-1)$ massive vector fields $+ 2(N-1)(N-2)/2$ massless vector fields $= N^2$ fields after the breaking of the symmetry.

## 6.7 Exercises

(1) Consider the Lagrangian density

$$\partial_\mu \phi^\dagger \partial^\mu \phi - m^2 \phi^\dagger \phi \qquad (6.191)$$

with

$$\begin{pmatrix} \phi_1 \\ \phi_2 \end{pmatrix} \qquad (6.192)$$

a doublet of $SU(2)$. Show the invariance under $SU(2)$ and evaluate the conserved currents.

(2) Verify the commutation relations given in eq. (6.15).

(3) Derive the commutation relations among the non-abelian charges of eq. (6.22), using the definition (6.21) and the canonical commutation relations for the fields.

(4) Derive the potential given in eq. (6.96).

(5) Using the equations of motion for the fields verify explicitly that the currents in eq. (6.152) have zero divergence.

(6) Consider the electromagnetic tensor $F_{\mu\nu}$. Show that the gauge invariant quantity

$$A = \epsilon_{\mu\nu\rho\lambda} F^{\mu\nu} F^{\rho\lambda} \qquad (6.193)$$

is the divergence of a four-vector $K_\mu$. Find the expression of $K_\mu$.

(7) Show that the Dirac equation is invariant under the chiral transformation

$$\psi \to e^{i\theta\gamma_5}\psi, \qquad (6.194)$$

if and only if the mass of the Dirac field vanishes. For $m \neq 0$ evaluate the Noether current and its divergence.

(8) Consider the $O(4)$ $\sigma$-model interacting with a doublet of Dirac fields $\Psi$:

$$\mathcal{L} = \frac{1}{2}\left[\partial_\mu\sigma\partial^\mu\sigma + \partial_\mu\vec{\pi}\partial^\mu\vec{\pi}\right] + i\bar{\Psi}\hat{\partial}\Psi + \bar{\Psi}(\sigma + i\vec{\tau}\cdot\vec{\pi})\Psi$$
$$- \frac{m^2}{2}(\sigma^2 + \vec{\pi}^2) + \frac{\lambda}{4}(\sigma^2 + \vec{\pi}^2)^2. \qquad (6.195)$$

Check the invariance of $\mathcal{L}$ under the following $SU(2)$ transformations (see (6.41))

$$\delta\sigma = 0, \qquad \delta\vec{\pi} = \vec{\theta}\wedge\vec{\pi}, \qquad (6.196)$$

$$\delta\Psi = -i\frac{\vec{\theta}\cdot\vec{\tau}}{2}\Psi. \qquad (6.197)$$

Then, evaluate the Noether currents.

(9) Consider the following dilation transformation

$$x_\mu \to x'_\mu = e^\lambda x^\mu, \qquad \phi(x) \to \phi'(x') = e^\lambda\phi(x). \qquad (6.198)$$

Show that the classical Lagrangian for a massless scalar field is invariant under this transformation. Evaluate the infinitesimal transformation and derive the Noether current.

# Chapter 7

# Time ordered products

## 7.1 Time ordered products and propagators

In the next Section we will discuss perturbation theory in the context of quantum fields. We will see that one of the most relevant quantities will be the propagator, that is the vacuum expectation value of a time ordered product of two fields. To introduce this quantity from a physical point of view we will begin considering a charged Klein-Gordon field. As we know from Section 3.6, the field $\phi$ destroys a particle of charge $+1$ and creates a particle of charge $-1$. In any case the net variation of the charge is $-1$. In analogous way the field $\phi^\dagger$ gives rise to a net variation of the charge equal to $+1$. Let us now construct a state with charge $+1$ applying $\phi^\dagger$ to the vacuum

$$|\psi(\vec{y}, t)\rangle = \phi^\dagger(y)|0\rangle = \int \frac{d^3k}{\sqrt{2\omega_k (2\pi)^3}} e^{iky} |\vec{k}, m, 1\rangle, \qquad (7.1)$$

where $|\vec{k}, m, 1\rangle$ is the single particle state with charge $+1$, momentum $\vec{k}$ and mass $m$. We want to evaluate the probability amplitude for the state, $|\psi(\vec{y}, t)\rangle$, to propagate to the same state, $|\psi(\vec{x}, t')\rangle$, at a later time $t' > t$. This is given by the matrix element

$$\theta(t' - t)\langle\psi(\vec{x}, t')|\psi(\vec{y}, t)\rangle = \theta(t' - t)\langle 0|\phi(\vec{x}, t')\phi^\dagger(\vec{y}, t)|0\rangle. \qquad (7.2)$$

It is convenient to think of this matrix element as the one corresponding to the creation of a charge $+1$ at the point $\vec{y}$ and time $t$, and to its annihilation at the point $\vec{x}$ and time $t'$. This interpretation is correct since the state $|\psi(\vec{y}, t)\rangle$ is an eigenstate of the normal ordered charge density operator, $j_0(\vec{x}, t)$, given in eq. (3.174)

$$\rho(\vec{x}, t)|\psi(\vec{y}, t)\rangle = [\rho(\vec{x}, t), \phi^\dagger(\vec{y}, t)]|0\rangle$$
$$= +\delta^3(\vec{x} - \vec{y})\phi^\dagger(\vec{y}, t)|0\rangle = \delta^3(\vec{x} - \vec{y})|\psi(\vec{y}, t)\rangle, \qquad (7.3)$$

159

where we have used eq. (3.168) and the fact that the vacuum has zero charge.

However, for $t' < t$ we could reach the same result by creating a particle of charge $-1$ at $(\vec{x}, t')$, and annihilating it at $(\vec{y}, t)$. The corresponding amplitude is

$$\theta(t - t')\langle 0|\phi^\dagger(\vec{y}, t)\phi(\vec{x}, t')|0\rangle. \tag{7.4}$$

The situation is represented in Fig. 7.1, where we have considered the case of a charged particle exchanged between a proton (charge $+1$) and a neutron (charge $0$). The total amplitude is obtained by adding the two contributions. We define the vacuum expectation value of the time ordered product ($T$ product) of two fields as

$$
\begin{aligned}
-i\Delta_F(x - y) &\equiv \langle 0|T(\phi(x)\phi^\dagger(y)|0\rangle = \langle 0|T(\phi^\dagger(y)\phi(x))|0\rangle \\
&= \theta(x_0 - y_0)\langle 0|\phi(x)\phi^\dagger(y)|0\rangle \\
&\quad + \theta(y_0 - x_0)\langle 0|\phi(y)^\dagger\phi(x)|0\rangle.
\end{aligned}
\tag{7.5}
$$

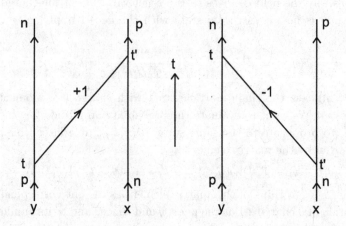

Fig. 7.1   The two probability amplitudes contributing to the process $np \to np$.

The function $\Delta_F(x - y)$ is called the Feynman propagator, and we will show immediately that it depends indeed on the difference of the two coordinates $x$ and $y$. Using the expressions (3.184) for the fields, we find

$$-i\Delta_F(x-y) = \int d^3k\, d^3k' \left[\theta(x_0-y_0)f_{\vec{k}}^*(y)f_{\vec{k}'}(x)\langle 0|a(\vec{k}')a^\dagger(\vec{k})|0\rangle \right.$$
$$\left. +\theta(y_0-x_0)f_{\vec{k}}(y)f_{\vec{k}'}^*(x)\langle 0|b(\vec{k}')b^\dagger(\vec{k})|0\rangle\right]$$
$$= \int \frac{d^3k}{(2\pi)^3}\frac{1}{2\omega_k}\left[\theta(x_0-y_0)e^{-ik(x-y)}\right.$$
$$\left. +\theta(y_0-x_0)e^{ik(x-y)}\right], \tag{7.6}$$

where, we recall that $k_0 = \omega_k$. This expression can be written in a more convenient way by using the following integral representation of the step function

$$\theta(t) = \lim_{\eta\to 0^+} \frac{i}{2\pi}\int_{-\infty}^{+\infty} d\omega\, \frac{e^{-i\omega t}}{\omega + i\eta}. \tag{7.7}$$

This representation can be verified immediately by noticing that for $t < 0$ the integral is convergent in the upper complex half-plane of $\omega$. Since there are no singularities in this region (the integral has a pole at $\omega = -i\eta$), we see that the integral vanishes. In the case $t > 0$ the integral is convergent in the lower half-plane. Then we pick up the contribution of the pole (in a clockwise direction) and we find 1 as a result. By substituting this representation in eq. (7.6) we get (omitting the limit that, however, should be taken at the end of the integration)

$$-i\Delta_F(x-y) = i\int \frac{d^3k}{(2\pi)^4}\int d\omega \frac{1}{2\omega_k}\left[\frac{e^{-i\omega(x_0-y_0)}}{\omega+i\eta}e^{-ik(x-y)}\right.$$
$$\left. +\frac{e^{i\omega(x_0-y_0)}}{\omega+i\eta}e^{ik(x-y)}\right]. \tag{7.8}$$

By performing the change of variable $k_0 = \omega + \omega_k$, we find

$$-i\Delta_F(x-y) = i\int \frac{d^4k}{(2\pi)^4}\frac{1}{2\omega_k}\left[\frac{e^{-ik(x-y)}}{k_0-\omega_k+i\eta}+\frac{e^{ik(x-y)}}{k_0-\omega_k+i\eta}\right]$$
$$= i\int \frac{d^4k}{(2\pi)^4}\frac{1}{2\omega_k}e^{-ik(x-y)}\left[\frac{1}{k_0-\omega_k+i\eta}-\frac{1}{k_0+\omega_k-i\eta}\right]$$
$$= i\int \frac{d^4k}{(2\pi)^4}\frac{e^{-ik(x-y)}}{k^2-m^2+i\epsilon}, \tag{7.9}$$

where we have defined $\epsilon = 2\eta\omega_k$. Notice that $\epsilon$ is a positive quantity and, again, we omit to write explicitly the limit. Then

$$\Delta_F(x-y) = -\int \frac{d^4k}{(2\pi)^4}\frac{e^{-ik(x-y)}}{k^2-m^2+i\epsilon}. \tag{7.10}$$

From this representation it follows that $\Delta_F(x)$ is a Green function for the Klein-Gordon operator

$$(\Box + m^2)\Delta_F(x) = \delta^4(x). \tag{7.11}$$

This property of the $T$-product is a simple consequence of its very definition. In fact, by using the canonical commutators

$$\begin{aligned}
(\Box + m^2)_x \langle 0|T(\phi(x)\phi^\dagger(y))|0\rangle &= \partial_0^2 \langle 0|T(\phi(x)\phi^\dagger(y))|0\rangle \\
&\quad + \langle 0|T((-\vec{\nabla}_x^2 + m^2)\phi(x)\phi^\dagger(y))|0\rangle \\
&= \partial_0 \langle 0|\delta(x_0 - y_0)[\phi(x), \phi^\dagger(y)]|0\rangle + \partial_0 \langle 0|T(\dot\phi(x)\phi^\dagger(y))|0\rangle \\
&\quad + \langle 0|T((-\vec{\nabla}_x^2 + m^2)\phi(x)\phi^\dagger(y))|0\rangle \\
&= \langle 0|\delta(x_0 - y_0)[\dot\phi(x), \phi^\dagger(y)]|0\rangle + \langle 0|T((\Box_x + m^2)\phi(x)\phi^\dagger(y))|0\rangle \\
&= -i\delta^4(x - y). \tag{7.12}
\end{aligned}$$

It is easily seen that an analogous result holds for the hermitian Klein-Gordon field, $\phi^\dagger = \phi$, that is

$$\Delta_F(x - y) = i\langle 0|T(\phi(x)\phi(y))|0\rangle. \tag{7.13}$$

It is possible to specify different Green's functions by choosing in a convenient way the integration path of the following integral

$$\Delta_C(x) = \int_C \frac{d^4k}{(2\pi)^4} \frac{e^{-ikx}}{k^2 - m^2}. \tag{7.14}$$

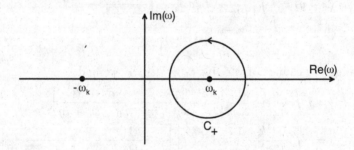

Fig. 7.2   The integration path $C_+$.

In fact, the integrand has two poles in $k_0$, which are located at $k_0 = \pm\omega_k$, with $\omega_k = (|\vec{k}|^2 + m^2)^{1/2}$. To define the integral we need to specify how the integration path goes around the poles. The result is different by taking paths closed around the poles, rather than paths extending from $-\infty$ to

$+\infty$. In fact, one gets solutions for the homogeneous Klein-Gordon equation in the first case and Green's functions in the second one. Let us consider the first case and let us define the two integration paths $C_\pm$ given in Figs. 7.2 and 7.3. Correspondingly we define the integrals

$$\Delta^{(\pm)}(x) = -i \int_{C_\pm} \frac{d^4k}{(2\pi)^4} \frac{e^{-ikx}}{k^2 - m^2}. \tag{7.15}$$

We find

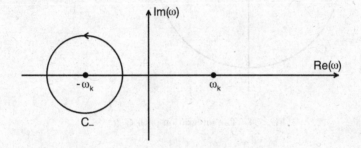

Fig. 7.3  The integration path $C_-$.

$$\Delta^{(+)}(x) = -i \int_{C_+} \frac{d^4k}{(2\pi)^4} \frac{e^{-ikx}}{(k_0 - \omega_k)(k_0 + \omega_k)} = \int \frac{d^3k}{(2\pi)^3} \frac{e^{-ikx}}{2\omega_k}$$
$$= [\phi^{(+)}(x), \phi^{(-)}(y)], \tag{7.16}$$

where we have used eq. (3.140). Also

$$\Delta^{(-)}(x) = -\int \frac{d^3k}{(2\pi)^3} \frac{e^{ikx}}{2\omega_k} = -\Delta^{(+)\dagger}(x) = [\phi^{(-)}(x), \phi^{(+)}(y)]. \tag{7.17}$$

It follows

$$\Delta^{(+)}(x) + \Delta^{(-)}(x) = i\Delta(x) \tag{7.18}$$

and using eq. (3.138), $i\Delta(x - y) = [\phi(x), \phi(y)]$. Therefore the commutator can be represented as

$$\Delta(x) = -\int_C \frac{d^4k}{(2\pi)^4} \frac{e^{-ikx}}{k^2 - m^2}, \tag{7.19}$$

with $C$ given in Fig. 7.4. Quite clearly $\Delta(x)$ satisfies the homogeneous Klein-Gordon equation.

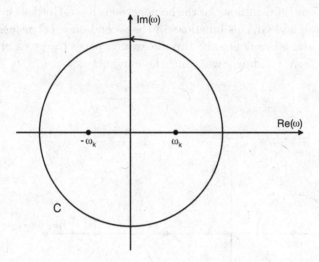

Fig. 7.4   The integration path $C$.

In the case of $\Delta_F$, the position of the poles is the one in Fig. 7.5. Then, the integral can be defined by shifting the poles to the real axis and choosing an integration path $C_F$, as specified in Fig. 7.6. That is

$$\Delta_F(x) = -\int_{C_F} \frac{d^4k}{(2\pi)^4} \frac{e^{-ikx}}{k^2 - m^2}. \tag{7.20}$$

Using eqs. (7.6), (7.16) and (7.17), we see that

$$\Delta_F(x) = i\theta(x_0)\Delta^{(+)}(x) - i\theta(-x_0)\Delta^{(-)}(x). \tag{7.21}$$

The reason why the $\Delta_C$ invariant functions, defined on a closed path $C$, satisfy the homogeneous Klein-Gordon equation, is a simple consequence of the action of the Klein-Gordon operator. This action removes the singularities from the integrand, leaving the integral of an analytic function on a closed path, which vanishes due to the Cauchy theorem.

For a free Dirac field, the Feynman propagator has a similar definition

$$S_F(x - y)_{\alpha\beta} = -i\langle 0|T(\psi_\alpha(x)\bar{\psi}_\beta(y))|0\rangle, \tag{7.22}$$

but with the $T$-product defined as follows

$$T(\psi_\alpha(x)\bar{\psi}_\beta(y)) = \theta(x_0 - y_0)\psi_\alpha(x)\bar{\psi}_\beta(y) - \theta(y_0 - x_0)\bar{\psi}_\beta(y)\psi_\alpha(x). \tag{7.23}$$

Notice that

$$T(\psi_\alpha(x)\bar{\psi}_\beta(y)) = -T(\bar{\psi}_\beta(y)\psi_\alpha(x)), \tag{7.24}$$

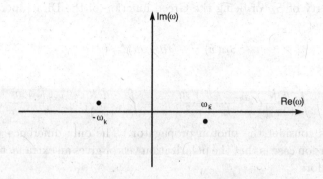

Fig. 7.5  The position of the poles in the definition of $\Delta_F(x)$.

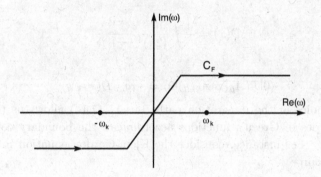

Fig. 7.6  The integration path $C_F$.

(the analogous property holds for the Klein-Gordon case, but with a plus sign, see eq. (7.5)). The minus sign, introduced in the definition of the $T$-product for the Dirac field, is needed because only in this way it may represent the Green function for the Dirac operator. In fact,

$$(i\hat{\partial}_x - m)_{\alpha\beta} T(\psi_\beta(x)\bar{\psi}_\gamma(y)) = i(\gamma_0)_{\alpha\beta}\delta(x_0 - y_0)\left[\psi_\beta(x), \bar{\psi}_\gamma(y)\right]_+$$
$$= i\delta(x_0 - y_0)(\gamma_0)_{\alpha\beta}(\gamma_0)_{\beta\gamma}\delta^3(\vec{x} - \vec{y})$$
$$= i\delta^4(x - y)\delta_{\alpha\gamma}, \qquad (7.25)$$

that is

$$(i\hat{\partial} - m)S_F(x) = \delta^4(x). \qquad (7.26)$$

It should be clear that the choice of sign in the definition of the $T$-product is related to the way in which we perform the canonical quantization. From the property of $S_F$ of being the Green function of the Dirac operator, we can see that

$$S_F(x) = -(i\hat{\partial} + m)\Delta_F(x) \tag{7.27}$$

and

$$S_F(x) = \int \frac{d^4k}{(2\pi)^4} e^{-ikx} \frac{\hat{k} + m}{k^2 - m^2 + i\epsilon} = \int_{C_F} \frac{d^4k}{(2\pi)^4} e^{-ikx} \frac{\hat{k} + m}{k^2 - m^2}. \tag{7.28}$$

Let us consider the photon propagator. The only difference with the Klein-Gordon case is that the polarization vector gives an extra factor $-g_{\mu\nu}$, and therefore

$$\langle 0|T(A_\mu(x)A_\nu(y)|0\rangle = -ig_{\mu\nu} \int \frac{d^4k}{(2\pi)^4} \frac{e^{-ikx}}{k^2 + i\epsilon}. \tag{7.29}$$

Defining

$$D(x) = -\int \frac{d^4k}{(2\pi)^4} \frac{e^{-ikx}}{k^2 + i\epsilon}, \tag{7.30}$$

we get

$$\langle 0|T(A_\mu(x)A_\nu(y)|0\rangle = +ig_{\mu\nu}D(x - y). \tag{7.31}$$

The choice of the integration paths in eq. (7.14) allows us to define various types of Green's functions according to the boundary conditions we require. For instance, consider the Klein-Gordon equation in a given external source

$$(\Box + m^2)\phi(x) = j(x). \tag{7.32}$$

The solution can be given in terms of the Green's function defined by

$$(\Box + m^2)G(x) = \delta^4(x). \tag{7.33}$$

In fact, the solution can be written as

$$\phi(x) = \phi^{(0)}(x) + \int d^4y \, G(x - y)j(y), \tag{7.34}$$

where $\phi^{(0)}(x)$ satisfies the homogeneous Klein-Gordon equation and it is chosen in such a way that $\phi(x)$ satisfies the boundary conditions of the problem. For instance, if we give the function $\phi(x)$ at $t = -\infty$ we require

$$\lim_{t \to -\infty} \phi(x) = \lim_{t \to -\infty} \phi^{(0)}(x). \tag{7.35}$$

Then, in order to satisfy the boundary conditions it is enough to choose for $G(x)$ the retarded solution defined by

$$G_{\text{ret}}(\vec{x}, x_0 < 0) = 0. \tag{7.36}$$

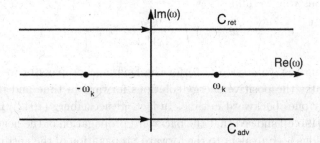

Fig. 7.7   The integration paths for $G_\text{ret}$ and $G_\text{adv}$.

Such a solution can be found easily by applying the method we have illustrated in the previous Section. By choosing the integration path as in Fig. 7.7, that is, leaving both poles below the path we get

$$G_\text{ret}(x) = -\int_{C_\text{ret}} \frac{d^4k}{(2\pi)^4} \frac{e^{-ikx}}{k^2 - m^2}. \tag{7.37}$$

Clearly $G_\text{ret}(x)$ vanishes for $x_0 < 0$. In fact, in this case, we can close the path on the half-plane, Im $\omega > 0$, without surrounding any singularity. Therefore the function vanishes. In analogous way we can define a function $G_\text{adv}(x)$ vanishing for $x_0 > 0$, by choosing a path below the poles (see Fig. 7.7). By integrating explicitly over $\omega$ one sees easily that the retarded solution propagates forward in time both the positive and negative energy solutions, whereas the advanced one propagates both solutions backward in time. Both $G_\text{ret}(x)$ and $G_\text{adv}(x)$ can be connected to the invariant $\Delta$ functions defined previously. In particular, using Fig. 7.6, we can establish the following relations between paths in the frequency complex plane

$$C_\text{ret} = C_F - C_-. \tag{7.38}$$

Then, from eqs. (7.37) and (7.20)

$$G_\text{ret}(x) = \Delta_F(x) + \int_{C_-} \frac{d^4}{(2\pi)^4} \frac{e^{-ikx}}{k^2 - m^2} = \Delta_F(x) + i\Delta^{(-)}(x), \tag{7.39}$$

where we have used eq. (7.15). Then, from (7.21)

$$
\begin{aligned}
G_\text{ret}(x) &= i\theta(x_0)\Delta^{(+)}(x) - i\theta(-x_0)\Delta^{(-)}(x) + i\Delta^{(-)}(x) \\
&= i\theta(x_0)\left(\Delta^{(+)}(x) + \Delta^{(-)}(x)\right) \\
&= i\theta(x_0)\int \frac{d^3k}{(2\pi)^3} \frac{1}{2\omega_k}\left(e^{-ikx} - e^{ikx}\right).
\end{aligned}
\tag{7.40}
$$

In analogous way we find

$$G_{\mathrm{adv}} = -i\theta(-x_0) \int \frac{d^3k}{(2\pi)^3} \frac{1}{2\omega_k} \left(e^{-ikx} - e^{ikx}\right). \qquad (7.41)$$

By using the expression (7.6) for the Feynman propagator we see that it propagates the positive energy solutions forward in time and the negative energy ones backward in time. In fact, [Stueckelberg (1942); Feynman (1948c, 1949b,c)] showed that the backward propagation of the negative energy solutions is equivalent to the forward propagation of the anti-particles.

## 7.2   A physical application of the propagators

The Feynman propagator acquires its full meaning only in quantum theory where, as we shall prove, it represents the central element of the perturbation theory.

In order to stress the relevance of the Feynman propagator we will consider now a simple application of what discussed so far. Let us consider two static point-like electric charges placed at $\vec{x}_1$ and $\vec{x}_2$. They can be described by the charge density

$$j_0(\vec{x}) = \sum_{m=1,2} e_m \delta^3(\vec{x} - \vec{x}_m), \qquad (7.42)$$

where the $e_m$'s are the values of the electric charges. The current density vanishes because we have supposed the charges to be static. Therefore the interaction Hamiltonian is (see eq. (4.171))

$$H_{\mathrm{int}} = -L_{int} = \int d^3x \, j_\mu(x) A^\mu(x) = \sum_{m=1,2} e_m A_0(\vec{x}_m, 0). \qquad (7.43)$$

In this equation we have taken the electromagnetic field operator $A_0$ at the time $t = 0$, since we will make use of the Schrödinger representation. So far, we have worked in the Heisenberg representation, since it is more convenient from the point of view of the relativistic covariance of the formalism, but the problem we are interested here is the evaluation of the interaction energy between two static electric charges. We recall that the relation between an operator, $A_H(t)$ in the Heisenberg representation and the same operator in the Schrödinger representation, $A_S$, is

$$A_H(t) = e^{iHt} A_S e^{-iHt}, \qquad (7.44)$$

Therefore the two operators are the same at $t = 0$. To evaluate the interaction energy we can do a perturbative calculation by determining the energy shift induced by the interaction Hamiltonian. We will think to the static electric charges as classical objects. Therefore we will quantize only the photon field $A_0(\vec{x}, 0)$. Furthermore we will consider the effect of the perturbation on the vacuum state (the state without photons). Since

$$\langle 0|H_{\text{int}}|0\rangle = 0, \qquad (7.45)$$

we must evaluate the energy shift at the second order in perturbation theory. We have

$$\Delta E = \sum_n \frac{\langle 0|H_{\text{int}}|n\rangle \langle n|H_{\text{int}}|0\rangle}{E_0 - E_n} = \int \frac{d^3k}{-\omega_k} \langle 0|H_{\text{int}}|\vec{k}\rangle \langle \vec{k}|H_{\text{int}}|0\rangle. \qquad (7.46)$$

The only state which contributes in the sum is the state with a single photon of energy $\omega_k$. By using the expression for $H_{\text{int}}$ and the following representation for $1/\omega_k$

$$\frac{1}{\omega_k} = \lim_{\epsilon \to 0^+} i \int_0^\infty dt \, e^{-i(\omega_k - i\epsilon)t}, \qquad (7.47)$$

we can write $\Delta E$ in the form (notice that we have suppressed the limit because the resulting expression is regular for $\epsilon \to 0^+$):

$$\begin{aligned}
\Delta E &= -i \sum_{m,n} e_m e_n \int d^3k \int_0^\infty dt \\
&\quad \times e^{-i\omega_k t} \langle 0|A_0(\vec{x}_n, 0)|\vec{k}\rangle \langle \vec{k}|A_0(\vec{x}_m, 0)|0\rangle \\
&= -i \sum_{m,n} e_m e_n \int d^3k \int_{-\infty}^{+\infty} dt \, \theta(t) \\
&\quad \times \langle 0|e^{iH_0 t} A_0(\vec{x}_n, 0) e^{-iH_0 t}|\vec{k}\rangle \langle \vec{k}|A_0(\vec{x}_m, 0)|0\rangle \\
&= -i \sum_{m,n} e_m e_n \int_{-\infty}^{+\infty} dt \, \theta(t) \langle 0|A_0(\vec{x}_n, t) A_0(\vec{x}_m, 0)|0\rangle, \qquad (7.48)
\end{aligned}$$

where $H_0$ is the free Hamiltonian for the electromagnetic field. To get this expression we have used the completeness (recalling again that $A_0$ couples only states that differ by a photon). Then, notice that the operators appearing in the last line can be interpreted as operators in the Heisenberg representation. By writing explicitly the various terms in the sum we get

$$\begin{aligned}
\Delta E = -i \int_{-\infty}^{+\infty} dt\theta(t) \Big[ & e_1^2 \langle 0|A_0(\vec{x}_1, t) A_0(\vec{x}_1, 0)|0\rangle \\
&+ e_1 e_2 \langle 0|A_0(\vec{x}_1, t) A_0(\vec{x}_2, 0)|0\rangle \\
&+ e_1 e_2 \langle 0|A_0(\vec{x}_2, t) A_0(\vec{x}_1, 0)|0\rangle \\
&+ e_2^2 \langle 0|A_0(\vec{x}_2, t) A_0(\vec{x}_2, 0)|0\rangle \Big].
\end{aligned} \qquad (7.49)$$

The intermediate states description of these four contributions is given in Fig. 7.8. Notice that the first and the fourth diagrams describe a correction to the intrinsic properties of the charges, and since we are interested in the evaluation of the interaction energy we can omit these terms from our calculation. Then, the energy interaction $\Delta E_{12}$ is given by

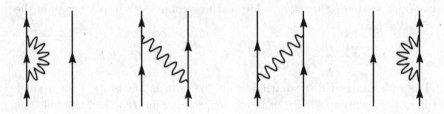

Fig. 7.8   The graphical description of eq. (7.49).

$$\Delta E_{12} = -ie_1 e_2 \int_{-\infty}^{+\infty} dt \; \theta(t) \Big[ \langle 0|A_0(\vec{x}_1, t) A_0(\vec{x}_2, 0)|0 \rangle$$

$$+ \langle 0|A_0(\vec{x}_2, t) A_0(\vec{x}_1, 0)|0 \rangle \Big]. \tag{7.50}$$

Sending $t \to -t$ in the second term, we get

$$\Delta E_{12} = -ie_1 e_2 \int_{-\infty}^{+\infty} dt \Big[ \theta(t) \langle 0|A_0(\vec{x}_1, t) A_0(\vec{x}_2, 0)|0 \rangle$$

$$+ \theta(-t) \langle 0|A_0(\vec{x}_2, -t) A_0(\vec{x}_1, 0)|0 \rangle \Big] \tag{7.51}$$

and using $H_0|0\rangle = 0$,

$$\langle 0|A_0(\vec{x}_2, -t) A_0(\vec{x}_1, 0)|0 \rangle = \langle 0|e^{-iH_0 t} A_0(\vec{x}_2, 0) e^{iH_0 t} A_0(\vec{x}_1, 0) e^{-iH_0 t}|0 \rangle$$

$$= \langle 0|A_0(\vec{x}_2, 0) A_0(\vec{x}_1, t)|0 \rangle. \tag{7.52}$$

Therefore our final result is

$$\Delta E_{12} = -ie_1 e_2 \int_{-\infty}^{+\infty} dt \; \langle 0|T(A_0(\vec{x}_1, t) A_0(\vec{x}_2, 0))|0 \rangle. \tag{7.53}$$

We see that the energy interaction is expressed in terms of the Feynman propagator. Recalling eqs. (7.30) and (7.31) we get (here we put $x_1^0 = t$ and $x_2^0 = 0$)

$$\Delta E_{12} = -e_1 e_2 \int_{-\infty}^{+\infty} dt \int \frac{d^4 k}{(2\pi)^4} \frac{e^{-ik(x_1 - x_2)}}{k^2 + i\epsilon}$$

$$= e_1 e_2 \int \frac{d^3 k}{(2\pi)^3} \frac{e^{i\vec{k}(\vec{x}_1 - \vec{x}_2)}}{\vec{k}^2} = \frac{e_1 e_2}{4\pi} \frac{1}{|\vec{x}_1 - \vec{x}_2|}. \tag{7.54}$$

For the final equality, see exercise (4) in this Chapter. This result tells us that the interaction energy between two classical charges evaluated at the lowest order in quantum theory coincides with the Coulomb energy. Therefore we can expect quantum corrections from higher orders. We shall see in the following that, in fact, this is the case.

## 7.3 Exercises

(1) Consider a Klein-Gordon field interacting with a classical source $j(x)$ described by the following Lagrangian density

$$\mathcal{L} = \frac{1}{2}\partial_\mu \phi(x)\partial^\mu \phi(x) - \frac{1}{2}m^2\phi^2(x) + j(x)\phi(x). \qquad (7.55)$$

a) Find the solutions to the equations of motion in the limit $x_0 \to \infty$ and show that in this limit the solution can be written in the form

$$\phi(x) = \int \frac{d^3k}{\sqrt{(2\pi)^3}} \frac{1}{\sqrt{2\omega_k}} \left[ \left( a(k) + \frac{i}{\sqrt{2\omega_k}}\tilde{j}(k) \right) e^{-ikx} \right.$$
$$\left. + \left( a^\dagger(k) - \frac{i}{\sqrt{2\omega_k}}\tilde{j}(-k) \right) e^{ikx} \right], \qquad (7.56)$$

where

$$\tilde{j}(\pm k) = \int \frac{d^4y}{\sqrt{(2\pi)^3}} e^{\mp ikx} j(x), \quad k^2 = m^2. \qquad (7.57)$$

(Hint: Write $\phi(x) = \phi^{(0)} + \int G_{\text{ret}}(x-y)j(y)d^4y$ with $\phi^{(0)}$ a free field.)
b) Evaluate the v.e.v.'s of the Hamiltonian and of the number of particles operator.

(2) Prove that the Feynman propagator of a massless scalar field is given by

$$\Delta_F(x) = \frac{1}{4\pi^2}\delta(x^2) - \frac{i}{4\pi^2}\mathcal{P}\frac{1}{x^2}, \qquad (7.58)$$

where $\mathcal{P}$ denotes the principal part.

(3) Evaluate the Feynman propagator for a massive vector field and show that

$$\langle 0|T(V_\mu(x)V_\nu(y))|0\rangle = -i \int \frac{d^4k}{(2\pi)^4} e^{-ik(x-y)} \left( \frac{g_{\mu\nu} - k_\mu k_\nu/m^2}{k^2 - m^2 + i\epsilon} \right). \qquad (7.59)$$

(4) Prove eq. (7.54). In order to perform the integration over $\vec{k}$ the following integral is useful

$$\int_0^\infty \frac{\sin y}{y}dy = \frac{\pi}{2}. \qquad (7.60)$$

(5) Repeat the calculation made in Section 7.2 in the case of a fixed scalar charge interacting with a massive scalar field, $\phi$ according to the Hamiltonian

$$H_{\text{int}} = \int d^3\vec{x}j(x)\phi(x) = \sum_{m=1,2} g\phi(\vec{x}_m, 0). \qquad (7.61)$$

Show that the interaction energy between the two sources is given by the Yukawa potential

$$\Delta E_{12} = \frac{g^2}{4\pi} \frac{1}{|\vec{x}_1 - \vec{x}_2|} e^{-m|\vec{x}|}, \qquad (7.62)$$

with $m$ the mass of the field.

(6) Consider the Lagrangian density for the electromagnetic field with the gauge fixing of eq. (5.45):

$$\mathcal{L} = -\frac{1}{4}F_{\mu\nu}F^{\mu\nu} - \frac{\lambda}{2}(\partial_\mu A^\mu)^2. \qquad (7.63)$$

Evaluate the Feynman propagator.

# Chapter 8

# Perturbation theory

Relativistic quantum field theory does not offer simple calculable examples and therefore one has to develop perturbative methods in order to discuss interacting fields. In the following we will be interested in describing scattering processes, since they are the typical processes appearing in the experiments in particle physics. The methods we will use for introducing the perturbation theory are quite general but, for the sake of simplicity and for the particular physical interest, we will discuss in detail only the case of the electromagnetic interactions of charged spin 1/2 particles described by a Dirac field. The results we will obtain can be easily generalized to other interacting theories.

## 8.1 The electromagnetic interaction

As we have already discussed, the electromagnetic interaction of an arbitrary charged particle is obtained through the minimal substitution or, equivalently, by invoking the gauge principle

$$\partial_\mu \to \partial_\mu + ieA_\mu. \tag{8.1}$$

For a charged Klein-Gordon particle, following this prescription, we get the following Lagrangian density

$$\mathcal{L}_{\text{free}} = \partial_\mu \phi^\dagger \partial^\mu \phi - m^2 \phi^\dagger \phi - \frac{1}{4} F_{\mu\nu} F^{\mu\nu}$$

$$\to [(\partial_\mu + ieA_\mu)\phi)]^\dagger [(\partial^\mu + ieA^\mu)\phi] - m^2 \phi^\dagger \phi - \frac{1}{4} F_{\mu\nu} F^{\mu\nu}. \tag{8.2}$$

The interacting part is given by the following two terms

$$\mathcal{L}_{\text{int.}} = -ie \left[ \phi^\dagger \partial_\mu \phi - (\partial_\mu \phi^\dagger)\phi \right] A^\mu + e^2 A^2 \phi^\dagger \phi. \tag{8.3}$$

In the first term the gauge field is coupled to the current

$$j_\mu = ie\left[\phi^\dagger \partial_\mu \phi - (\partial_\mu \phi^\dagger)\phi\right], \tag{8.4}$$

but another interacting term appears. This term is a straight consequence of the gauge invariance. In fact, the current $j_\mu$, which is conserved in the absence of the interaction, is neither conserved, nor gauge invariant, when the electromagnetic field is turned on. In fact, consider the infinitesimal gauge transformation

$$\delta\phi(x) = -ie\Lambda(x)\phi(x), \qquad \delta A_\mu(x) = \partial_\mu \Lambda(x), \tag{8.5}$$

then

$$\delta\mathcal{L}_{\text{free}} = ie\Lambda_{,\mu}\phi^\dagger \partial^\mu \phi - ie\partial_\mu \phi^\dagger \Lambda^{,\mu}\phi = j^\mu \partial_\mu \Lambda \tag{8.6}$$

and writing $\mathcal{L}_{\text{int}}$ in the form

$$\mathcal{L}_{\text{int}} = -j_\mu A^\mu + e^2 A^2 \phi^\dagger \phi, \tag{8.7}$$

we find

$$\delta\mathcal{L}_{\text{int}} = -j_\mu \partial^\mu \Lambda - (\delta j_\mu)A^\mu + 2e^2 \Lambda_{,\mu} A^\mu \phi^\dagger \phi. \tag{8.8}$$

The first term cancels with the variation of $\mathcal{L}_{\text{free}}$, whereas the other two terms cancel among themselves. In fact,

$$\delta j_\mu = ie\left[\phi^\dagger(-ie)\Lambda_{,\mu}\phi - ie\Lambda_{,\mu}\phi^\dagger\phi\right] = 2e^2 \Lambda_{,\mu}\phi^\dagger\phi. \tag{8.9}$$

This shows that the $A^2$ term is necessary to compensate the non-invariance of the current under gauge transformations. In fact, the conserved and gauge invariant current can be obtained through the Noether theorem

$$J_\mu = ie\left[\phi^\dagger(\partial_\mu + ieA_\mu)\phi - (\partial_\mu - ieA_\mu)\phi^\dagger)\phi\right] = j_\mu - 2e^2 A_\mu \phi^\dagger \phi. \tag{8.10}$$

The situation is much simpler in the case of the Dirac equation where

$$\mathcal{L}_{\text{free}} = \bar{\psi}(i\hat{\partial} - m)\psi - \frac{1}{4}F_{\mu\nu}F^{\mu\nu} \to \bar{\psi}(i\hat{\partial} - e\hat{A} - m)\psi - \frac{1}{4}F_{\mu\nu}F^{\mu\nu}, \tag{8.11}$$

giving the interaction term

$$\mathcal{L}_{\text{int}} = -e\bar{\psi}\gamma_\mu \psi A^\mu. \tag{8.12}$$

Here the gauge field is coupled to a conserved and gauge invariant current. As a consequence $-j_\mu A^\mu$ is the only interaction term. The gauge variations of the free and of the interacting Lagrangian densities are respectively

$$\delta\mathcal{L}_{\text{free}} = e\bar{\psi}\gamma_\mu \psi \Lambda^{,\mu} = j_\mu \Lambda^{,\mu} \tag{8.13}$$

and

$$\delta\mathcal{L}_{\text{int}} = -j_\mu \delta A^\mu = -j_\mu \Lambda^{,\mu}. \tag{8.14}$$

The canonical quantization for an interacting system follows the same procedure as in the non-interacting case. We require canonical commutation and/or anticommutation relations at equal times for the various fields. For different fields we require equal time vanishing commutation (anticommutation) relations for spin integer (half-integer) fields, whereas we require zero commutation relations among fields of integer spin and fields of half-integer spin. Usually the canonical commutation relations among the fields are not changed by the interactions with respect to the free case. However, this is not the case if the interaction term involves derivatives of the fields. This follows from the definition of the canonical momentum densities

$$\Pi^i = \frac{\partial \mathcal{L}}{\partial \dot\phi_i} = \frac{\partial \mathcal{L}_{\text{free}}}{\partial \dot\phi_i} + \frac{\partial \mathcal{L}_{\text{int}}}{\partial \dot\phi_i}. \tag{8.15}$$

For instance, for the charged scalar field we get

$$\Pi \equiv \Pi_\phi = \frac{\partial \mathcal{L}}{\partial \dot\phi} = \dot\phi^\dagger - ie\phi^\dagger A^0, \quad \Pi^\dagger = \Pi_{\phi^\dagger} = \dot\phi + ie\phi A^0. \tag{8.16}$$

Since the canonical momenta contain the time component of the gauge field, one can verify that the canonical commutators among the fields and their derivatives are changed by the interaction. Also the propagators are modified. However, we will not insist on this point, because in practice it has no consequences on the perturbation theory (see later). When derivative interactions are not present, the canonical momentum densities coincide with the free ones, and we get

$$\mathcal{H} = \Pi\dot\phi - \mathcal{L} = \Pi\dot\phi - \mathcal{L}_{\text{free}} - \mathcal{L}_{\text{int}} = \mathcal{H}_{\text{free}} - \mathcal{L}_{\text{int}} \tag{8.17}$$

and therefore

$$\mathcal{H}_{\text{int}} = -\mathcal{L}_{\text{int}}. \tag{8.18}$$

This is what happens for the electromagnetic interaction between a Dirac field and the electromagnetic one. The corresponding theory is called QED (Quantum Electro Dynamics). We recall also that in general the Hamiltonian and the electromagnetic current are normal ordered in such a way that the vacuum is an eigenstate of the energy and of the charge operators with vanishing eigenvalues. Therefore the interaction term is written as

$$\mathcal{L}_{\text{int}} = -e : \bar\psi\gamma_\mu\psi : A^\mu. \tag{8.19}$$

We can verify that this is equivalent to write

$$\mathcal{L}_{\text{int}} = -\frac{e}{2}[\bar{\psi}, \gamma_\mu \psi]A^\mu. \tag{8.20}$$

For instance, if we consider the electric charge as defined indirectly in eq. (8.20), we get

$$Q = \frac{e}{2} \int d^3x (\psi^\dagger \psi - \gamma_0 \psi \psi^\dagger \gamma_0)$$

$$= \frac{e}{2} \sum_{\pm n} \int d^3p \Big[ b^\dagger(p,n)b(p,n) + d(p,n)d^\dagger(p,n)$$

$$- b(p,n)b^\dagger(p,n) - d^\dagger(p,n)d(p,n) \Big]$$

$$= \frac{e}{2} \sum_{\pm n} \int d^3p \Big[ 2b^\dagger(p,n)b(p,n) - 2d^\dagger(p,n)d(p,n)$$

$$- \big[b^\dagger(p,n),b(p,n)\big]_+ + \big[d(p,n),d^\dagger(p,n)\big]_+ \Big] =: Q :, \tag{8.21}$$

since the two anticommutators cancel among themselves.

## 8.2 The scattering matrix

The scattering processes are a central element in the study of the elementary particles. In the typical scattering process the incoming particles are prepared in a state of definite momentum. These particles are sent either on a fixed target, or into another beam moving in the opposite direction. After the collision has taken place, one looks at the final states produced by the scattering process. In ordinary quantum mechanics this situation is well described by using free wave functions for the initial and final states. This description is certainly correct if one has to do with short-range potentials. However, in field theory, this representation is not really correct since, also in the absence of reciprocal interactions, the particles have self interactions as we have already observed in Section 7.2. For instance, a real electron can be thought of as being surrounded by a cloud of photons which can be emitted and reabsorbed. The point is that, in general, the number of particles is not a conserved quantity. Particles can be created and annihilated and this can happen also outside the strict region of interaction. A rigorous treatment of these problems is non-trivial and it goes beyond the scope of this course. Therefore we will confine ourselves to a rather intuitive treatment of the problem. On the other hand the limitations of the method will be rather obvious so it may well give the basis for a more refined treatment. To simplify the matter we will make use of the **adiabatic hypothesis**. This consists in looking at a scattering process in the following way. At time $t = -\infty$ we will suppose that our system can be described in terms of free particles, that is with the interaction turned off. Between $t = -\infty$ and a time $t = -T$, much before the scattering process takes place, we let the coupling describing the interaction to grow from zero to its actual value. In the interval $-T < t < +T$, the coupling stays constant, and then from $t = +T$ and $t = +\infty$ the coupling goes again to zero. In practice, one defines the interacting part of the Hamiltonian as

$$H_{\text{int}}(t, \epsilon) = e^{-\epsilon|t|} H_{\text{int}}, \qquad (8.22)$$

performing all the calculations and taking the limit $\epsilon \to 0^+$ at the end. The consistency of this procedure has been shown by various authors and a detailed discussion can be found, for instance, in the book by [Jauch and Rohrlich (1980)].

By using the adiabatic hypothesis we can now discuss the perturbative calculation of the scattering amplitudes. The perturbative expansion will be possible only if the interaction term is small. For instance in QED the perturbative expansion gives rise to a series of powers in the fine structure constant $e^2/4\pi \approx 1/137$. Therefore, if the coefficients of the expansion do not grow too much, the expansion is justified. Let us start with the equation of motion for the states in the Schrödinger representation

$$i\frac{\partial |\Phi_S(t)\rangle}{\partial t} = H_S |\Phi_S(t)\rangle. \qquad (8.23)$$

Suppose also that we have two interacting fields $A$ and $B$. Then we can write

$$H_S = H_S^0 + H_S^I, \qquad (8.24)$$

with

$$H_S^0 = H_S^0(A) + H_S^0(B) \qquad (8.25)$$

and

$$H_S^I \equiv H_S^I(A, B), \qquad (8.26)$$

where $H_S^0(A)$ and $H_S^0(B)$ are the free Hamiltonians for the fields $A$ and $B$, and $H_S^I$ is the interaction Hamiltonian. It turns out convenient to introduce a new representation for the vectors of state, the <u>interaction representation</u>. This is defined by the following unitary transformation upon the states and the operators in the Schrödinger representation

$$|\Phi(t)\rangle = e^{iH_S^0 t} |\Phi_S(t)\rangle, \qquad O(t) = e^{iH_S^0 t} O_S e^{-iH_S^0 t}. \qquad (8.27)$$

Of course the matrix elements of any operator in the interaction representation are the same as in the Schrödinger representation

$$\langle \Phi'(t)|O(t)|\Phi(t)\rangle = \langle \Phi'_S(t)|O_S|\Phi_S(t)\rangle. \qquad (8.28)$$

We also have $H_S^0 = H^0$, where $H^0$ is the free Hamiltonian in the interaction representation. Notice also that the interaction representation coincides with the Heisenberg representation when we switch off the interaction. In the interaction representation the time evolution of the states is dictated by the interaction Hamiltonian

$$i\frac{\partial|\Phi(t)\rangle}{\partial t} = -H_S^0 e^{iH_S^0 t}|\Phi_S(t)\rangle + e^{iH_S^0 t}(H_S^0 + H_S^I)|\Phi_S(t)\rangle$$

$$= e^{iH_S^0 t}H_S^I e^{-iH_S^0 t}|\Phi(t)\rangle, \qquad (8.29)$$

from which

$$i\frac{\partial|\Phi(t)\rangle}{\partial t} = H^I|\Phi(t)\rangle, \qquad (8.30)$$

where $H^I$ is the interaction Hamiltonian in the interaction representation. On the other hand the operators evolve with the free Hamiltonian. Therefore, in the interaction representation they coincide with the Heisenberg operators of the non-interacting case.

In order to describe a scattering process we will assign to the vector of state a condition at $t = -\infty$

$$|\Phi(-\infty)\rangle \equiv |\Phi_i\rangle, \qquad (8.31)$$

where the state $\Phi_i$ will be specified by assigning the set of incoming free particles in terms of eigenstates of momentum, spin and other possible quantum numbers. For instance, in QED we will have to specify how many electrons, positrons and photons are in the initial state and we will have to specify their momenta, the spin projection of the fermions and the polarization of the photons. The equations of motion will tell us how this state evolves in time and it will be possible to evaluate the state at $t = +\infty$, where, ideally, we will detect the final states. In practice the preparation and the detection processes are made at some finite times. It follows that our ideal description will be correct only if these times are much bigger than the typical interaction time of the scattering process. Once we know $|\Phi(+\infty)\rangle$, we are interested to evaluate the probability amplitude of detecting at $t = +\infty$ a given set of free particles (see the adiabatic hypothesis) specified by a vector of state $|\Phi_f\rangle$. Therefore the probability amplitude (or scattering amplitude) we are interested in is

$$S_{fi} = \langle \Phi_f|\Phi(+\infty)\rangle. \qquad (8.32)$$

We will define the $S$ matrix as the operator giving $|\Phi(+\infty)\rangle$ when applied to $|\Phi(-\infty)\rangle$

$$|\Phi(+\infty)\rangle = S|\Phi(-\infty)\rangle. \tag{8.33}$$

The amplitude $S_{fi}$ is then

$$S_{fi} = \langle \Phi_f|S|\Phi_i\rangle. \tag{8.34}$$

We see that $S_{fi}$ is the $S$ matrix element between free states. To evaluate the $S$ matrix we first transform the Schrödinger equation in the interaction representation into an integral equation

$$|\Phi(t)\rangle = |\Phi(-\infty)\rangle - i \int_{-\infty}^{t} dt_1\, H^I(t_1)|\Phi(t_1)\rangle. \tag{8.35}$$

The right-hand side of this equation satisfies both the Schrödinger equation and the boundary condition at $t = -\infty$. The perturbative expansion consists in evaluating $|\Phi(t)\rangle$ by iterating this integral equation

$$|\Phi(t)\rangle = |\Phi(-\infty)\rangle$$
$$-i \int_{-\infty}^{t} dt_1\, H^I(t_1)\Big[|\Phi(-\infty)\rangle$$
$$-i \int_{-\infty}^{t_1} dt_2\, H^I(t_2)|\Phi(t_2)\rangle\Big]. \tag{8.36}$$

Continuing the iteration we get

$$|\Phi(t)\rangle = \Big[1 - i \int_{-\infty}^{t} dt_1\, H^I(t_1) + (-i)^2 \int_{-\infty}^{t} dt_1 \int_{-\infty}^{t_1} dt_2\, H^I(t_1)H^I(t_2)$$
$$+ \cdots \Big]|\Phi(-\infty)\rangle. \tag{8.37}$$

Of course this is meaningful only if the expansion is convergent. By taking the limit for $t = +\infty$ we get the perturbative expansion of the $S$ matrix

$$S = 1 + \sum_{n=1}^{\infty} (-i)^n \int_{-\infty}^{+\infty} dt_1 \int_{-\infty}^{t_1} dt_2 \cdots \int_{-\infty}^{t_{n-1}} dt_n$$
$$\times \big[ H^I(t_1)H^I(t_2) \cdots H^I(t_n) \big]. \tag{8.38}$$

We can rewrite this expression in terms of $T$-products

$$S = 1 + \sum_{n=1}^{\infty} \frac{(-i)^n}{n!} \int_{-\infty}^{+\infty} dt_1 \cdots \int_{-\infty}^{+\infty} dt_n\, T\big(H^I(t_1) \cdots H^I(t_n)\big). \tag{8.39}$$

The $T$-product of $n$ terms means that the factors have to be written from the left to the right with decreasing times. For instance, if $t_1 \geq t_2 \geq \cdots \geq t_n$, then

$$T\big(H^I(t_1) \cdots H^I(t_n)\big) = H^I(t_1) \cdots H^I(t_n). \tag{8.40}$$

The equality of the two expressions (8.38) and (8.39) holds term by term. As an example, consider $n = 2$ in eq. (8.39) for finite times $t_1$ and $t_2$. This term can be written as

$$A = \int_{t_1}^{t_2} \int_{t_1}^{t_2} dt\, ds\, T\left(H^I(t)H^I(s)\right)$$

$$= \int_{t_1}^{t_2} dt\, H^I(t) \left( \int_{t_1}^{t} ds\, H^I(s) \right)$$

$$+ \int_{t_1}^{t_2} dt \left( \int_{t}^{t_2} ds\, H^I(s) \right) H^I(t). \tag{8.41}$$

By looking at Figs. 8.1 and 8.2 one sees easily that exchanging the integrations on $s$ and $t$ one gets

$$\int_{t_1}^{t_2} dt \int_{t}^{t_2} ds = \int_{t_1}^{t_2} ds \int_{t_1}^{s} dt. \tag{8.42}$$

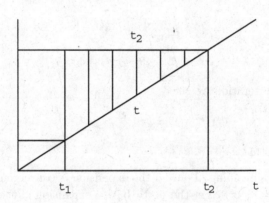

Fig. 8.1   The figure represents schematically the integral $\int_{t_1}^{t_2} dt \int_{t}^{t_2} ds$.

Therefore

$$A = \int_{t_1}^{t_2} dt \int_{t_1}^{t} ds\, H^I(t)H^I(s) + \int_{t_1}^{t_2} ds \int_{t_1}^{s} dt\, H^I(s)H^I(t) \tag{8.43}$$

and exchanging $s \leftrightarrow t$ in the second integral,

$$A = 2 \int_{t_1}^{t_2} dt \int_{t_1}^{t} ds\, H^I(t)H^I(s). \tag{8.44}$$

By letting $t_2 \to +\infty$ and $t_1 \to -\infty$, we prove the equality of the $n = 2$ terms in (8.38) and (8.39).

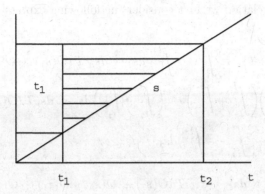

Fig. 8.2 The figure represents schematically the integral $\int_{t_1}^{t_2} ds \int_{t_1}^{s} dt$.

The result for the $n^{th}$ term in the series can be obtained in a completely analogous way.

Since the $S$ matrix connects the set of free states at $t = -\infty$ with a set of free states at $t = +\infty$, it should represent simply a change of basis and as such it should be unitary. From this point of view the unitarity property of the $S$ matrix is very important because it has to do with the fundamental properties of quantum mechanics. So we have to check that at least formally[1] the expression (8.39) represents a unitary operator. In order to prove this point we rewrite $S$ in the form

$$S = T\left(e^{-i\int_{-\infty}^{+\infty} dt\ H^I(t)}\right). \tag{8.45}$$

This expression is a symbolic one and it is really defined by its series expansion

$$S = \sum_{n=0}^{\infty} \frac{(-i)^n}{n!} T\left(\int_{-\infty}^{+\infty} dt\ H^I(t)\right)^n$$

$$= \sum_{n=0}^{\infty} \frac{(-i)^n}{n!} \int_{-\infty}^{+\infty} dt_1 \cdots dt_n T\left(H^I(t_1)\cdots H^I(t_n)\right). \tag{8.46}$$

The motivation for introducing the $T$-ordered exponential is that it satisfies the following factorization property

$$T\left(e^{\int_{t_1}^{t_3} O(t)dt}\right) = T\left(e^{\int_{t_2}^{t_3} O(t)dt}\right) T\left(e^{\int_{t_1}^{t_2} O(t)dt}\right). \tag{8.47}$$

---

[1]This observation refers to the fact that we don't really know if the series which we have found for the $S$ matrix is convergent or not.

To prove this relation we first consider the following expression ($t_1 \leq t_2 \leq t_3$)

$$
T\left(\int_{t_1}^{t_3} O(t)dt\right)^n \equiv \int_{t_1}^{t_3}\cdots\int_{t_1}^{t_3} ds_1\cdots ds_n\, T\left(O(s_1)\cdots O(s_n)\right)
$$

$$
= \left(\int_{t_2}^{t_3}+\int_{t_1}^{t_2}\right)\left(\int_{t_2}^{t_3}+\int_{t_1}^{t_2}\right)\cdots\left(\int_{t_2}^{t_3}+\int_{t_1}^{t_2}\right)ds_1\cdots ds_n T\left(O(s_1)\cdots O(s_n)\right)
$$

$$
= \sum_{k=0}^{n}\frac{n!}{(n-k)!k!}\int_{t_2}^{t_3}\cdots\int_{t_2}^{t_3} ds_1\cdots ds_{n-k}
$$

$$
\times \int_{t_1}^{t_2}\cdots\int_{t_1}^{t_2} dz_1\cdots dz_k\, T\left(O(s_1)\cdots O(s_{n-k})O(z_1)\cdots O(z_k)\right)
$$

$$
= \sum_{k=0}^{n}\frac{n!}{(n-k)!k!}\int_{t_2}^{t_3}\cdots\int_{t_2}^{t_3} ds_1\cdots ds_{n-k} T\left(O(s_1)\cdots O(s_{n-k})\right)
$$

$$
\times \int_{t_1}^{t_2}\cdots\int_{t_1}^{t_2} dz_1\cdots dz_k\, T\left(O(z_1)\cdots O(z_k)\right). \tag{8.48}
$$

In the last term we have used the fact that all the times $z_i$ are smaller than the times $s_i$. Therefore we have shown the relation

$$
T\left(\int_{t_1}^{t_3} dt\, O(t)\right)^n = \sum_{k=0}^{n}\frac{n!}{(n-k)!k!}
$$

$$
\times T\left(\int_{t_2}^{t_3} dt\, O(t)\right)^{n-k} T\left(\int_{t_1}^{t_2} dt\, O(t)\right)^k. \tag{8.49}
$$

The factorization property (8.47) follows immediately if we remember that the analogous property for the ordinary exponential,

$$
e^{a+b} = e^a e^b, \tag{8.50}
$$

follows from the binomial expansion

$$
e^{a+b} = \sum_{n=0}^{\infty}\frac{1}{n!}(a+b)^n = \sum_{n=0}^{\infty}\frac{1}{n!}\sum_{k=0}^{n}\frac{n!}{(n-k)!k!}a^{n-k}b^k, \tag{8.51}
$$

using

$$
\sum_{n=0}^{\infty}\sum_{k=0}^{n} = \sum_{k=0}^{\infty}\sum_{n=k}^{\infty} \tag{8.52}
$$

and replacing $k$ with $h = n - k$ in eq. (8.51). Since eq. (8.49) generalizes the binomial formula to $T$-products of powers of time integrals of operators, by the same token we get the formula (8.47). With this property we can now prove the unitarity of any operator of the form

$$U = T\left(e^{-i\int_{t_i}^{t_f} dt\, O(t)}\right), \tag{8.53}$$

with $O(t)$ a hermitian operator. To this end let us divide the time interval $(t_i, t_f)$ in $N$ infinitesimal intervals $\Delta t$ with

$$t_i \equiv t_1 \leq t_2 \leq \cdots \leq t_N = t_f, \tag{8.54}$$

then we can write

$$U = \lim_{N \to \infty} e^{-i\Delta t O(t_N)} e^{-i\Delta t O(t_{N-1})} \cdots e^{-i\Delta t O(t_1)}, \tag{8.55}$$

from which

$$U^\dagger = \lim_{N \to \infty} e^{+i\Delta t O(t_1)} e^{+i\Delta t O(t_2)} \cdots e^{+i\Delta t O(t_N)} \tag{8.56}$$

and the unitarity follows immediately.

Let us notice that if there are no derivative interactions we have

$$S = T\left(e^{-i\int_{-\infty}^{+\infty} dt\, H^I(t)}\right) = T\left(e^{+i\int d^4x\, \mathcal{L}_{\text{int}}}\right). \tag{8.57}$$

From this expression we can prove that the Lorentz invariance of the theory implies the invariance of the $S$ matrix. Since in a relativistic theory the Lagrangian is Lorentz invariant, the previous statement would be trivial but for the presence of the $T$-product in the expression of the $S$ matrix. For simplicity, let us consider the second order contribution (but the argument is valid at any order)

$$\int d^4x_1\, d^4x_2 \left[\theta(x_1^0 - x_2^0)\mathcal{L}_{\text{int}}(x_1)\mathcal{L}_{\text{int}}(x_2) + \theta(x_2^0 - x_1^0)\mathcal{L}_{\text{int}}(x_2)\mathcal{L}_{\text{int}}(x_1)\right]. \tag{8.58}$$

The $T$-product is Lorentz invariant for time-like separations of the two operators (remember that a proper Lorentz transformation cannot change the sign of the time component of a four-vector). On the other hand we integrate over $x_1$ and $x_2$ and the separation between the two points can be either space-like or time-like. In the case of space-like separations, and assuming that the Lagrangian density is a local function of the fields, it follows that

$$[\mathcal{L}_{\text{int}}(x_1), \mathcal{L}_{\text{int}}(x_2)] = 0 \quad \text{for} \quad (x_1 - x_2)^2 < 0 \tag{8.59}$$

and therefore for $(x_1 - x_2)^2 < 0$ we get

$$T(\mathcal{L}_{\text{int}}(x_1)\mathcal{L}_{\text{int}}(x_2)) = \mathcal{L}_{\text{int}}(x_1)\mathcal{L}_{\text{int}}(x_2). \tag{8.60}$$

This shows that the $T$-product of local invariant Lorentz operators is Lorentz invariant. We saw an example of this property when we have evaluated the propagator for a scalar field.

One could think that for theories with derivative interactions the Lorentz invariance is lost. However, it is possible to show that also in these theories the $S$ matrix is given by the expression on the right-hand side of eq. (8.57) (see, for instance, [Itzykson and Zuber (1980)]), that is it depends on $\mathcal{L}_{\text{int}}$.

## 8.3  Wick's theorem

In the previous Section we have shown that the $S$ matrix can be evaluated in terms of matrix elements of $\mathcal{T}$-products. As we shall see in the applications, the matrix elements of the $S$ matrix between free particle states can in turn be expressed as vacuum expectation values (VEV's) of $T$-products. These VEV's satisfy an important theorem due to Wick that states that the VEV of a $T$-product of an arbitrary number of free fields (the ones that appear in the interaction representation) can be expressed as combinations of VEV's of $T$-products among two fields, that is in terms of Feynman propagators. In order to prove the theorem we will use the technique of generating functionals. That is we will start by proving the following identity

$$T\left( \exp(-i \int d^4x \, j(x)\phi(x)) \right) =: \exp\left(-i \int d^4x \, j(x)\phi(x)\right):$$

$$\times \exp\left(-\frac{1}{2} \int d^4x \, d^4y \, j(x)j(y)\langle 0|T(\phi(x)\phi(y))|0\rangle\right), \qquad (8.61)$$

where $\phi(x)$ is a free real scalar field and $j(x)$ an ordinary real function. This formula can be easily extended to charged scalar, fermion and photon fields. The Wick theorem is then obtained by expanding both sides of this equation in powers of $j(x)$ and taking the VEV of both sides. Let us start by expanding the left-hand side of (8.61), by using the factorization property in eq. (8.47). Let us also define

$$O(t) = \int d^3x \, j(x)\phi(x) \qquad (8.62)$$

and notice that for free fields $[O(t), O(t')]$ is just an ordinary number (called a $c$-number, to be contrasted with operators which are called $q$-numbers). Dividing again the interval $(t_i, t_f)$ in $N$ pieces of amplitude $\Delta t$ as in the previous Section, we get

$$T\Big( \exp\big(-i \int_{t_i}^{t_f} dt\, O(t)\big) = \lim_{N\to\infty} \exp\left(-i\Delta t O(t_N)\right) \exp\left(-i\Delta t O(t_{N-1})\right)$$
$$\cdots \times \exp\left(-i\Delta t O(t_1)\right) \qquad (8.63)$$

and using

$$\exp\left(A\right)\exp\left(B\right) = \exp\left(A+B\right)\exp\left(+\frac{1}{2}[A,B]\right), \qquad (8.64)$$

valid if $[A, B]$ commutes with $A$ and $B$, we get

$$T\Big( \exp\big(-i \int_{t_i}^{t_f} dt\, O(t)\big)\Big) = \lim_{N\to\infty} \prod_{i=3}^{N} \exp\left(-i\Delta t O(t_i)\right)$$
$$\times \exp\left(-i\Delta t(O(t_2)+O(t_1))\right) \exp\left(-\frac{1}{2}\Delta t^2[O(t_2),O(t_1)]\right)$$
$$= \lim_{N\to\infty} \prod_{i=4}^{N} \exp\left(-i\Delta t O(t_i)\right)\exp\left(-i\Delta t(O(t_3)+O(t_2)+O(t_1))\right)$$
$$\times \exp\left(-\frac{1}{2}\Delta t^2[O(t_3),O(t_2)+O(t_1)] - \frac{1}{2}\Delta t^2[O(t_2),O(t_1)]\right)$$
$$= \lim_{N\to\infty} \exp\left(-i\Delta t \sum_{i=1}^{N} O(t_i)\right)\exp\left(-\frac{1}{2}\Delta t^2 \sum_{1\le i\le j\le N} [O(t_j),O(t_i)]\right)$$
$$= \exp\left(-i \int_{t_i}^{t_f} dt\, O(t)\right)$$
$$\times \exp\left(-\frac{1}{2}\int_{t_i}^{t_f} dt_1\, dt_2\, \theta(t_1-t_2)[O(t_1),O(t_2)]\right), \qquad (8.65)$$

from which

$$T\Big( \exp\big(-i \int d^4x\, j(x)\phi(x)\big)\Big) = \exp\left(-i\int d^4x\, j(x)\phi(x)\right)$$
$$\times \exp\left(-\frac{1}{2}\int d^4x\, d^4y\, j(x)j(y)\theta(x_0-y_0)[\phi(x),\phi(y)]\right). \qquad (8.66)$$

The next step is to expand the first exponential on the right-hand side of this equation in normal products. We have

$$\exp\left(-i\int d^4x\, j(x)\phi(x)\right) = \exp\left(-i\int d^4x\, j(x)(\phi^{(+)}(x)+\phi^{(-)}(x))\right)$$
$$= \exp\left(-i\int d^4x\, j(x)\phi^{(-)}(x)\right)\exp\left(-i\int d^4x\, j(x)\phi^{(+)}(x)\right)$$
$$\times \exp\left(+\frac{1}{2}\int d^4x\, d^4y\, j(x)j(y)[\phi^{(-)}(x),\phi^{(+)}(y)]\right), \qquad (8.67)$$

where we have used again eq. (8.64). Therefore

$$\exp\left(-i\int d^4x\, j(x)\phi(x)\right) =:\exp\left(-i\int d^4x\, j(x)\phi(x)\right): $$
$$\times \exp\left(+\frac{1}{2}\int d^4x\, d^4y\, j(x)j(y)[(\phi^{(-)}(x),\phi^{(+)}(y)]\right). \qquad (8.68)$$

Substituting in eq. (8.66)

$$T\left(\exp\left(-i\int d^4x\, j(x)\phi(x)\right)\right) =:\exp\left(-i\int d^4x\, j(x)\phi(x)\right): $$
$$\times \exp\left(+\frac{1}{2}\int d^4x\, d^4y\, j(x)j(y)A(x,y)\right), \qquad (8.69)$$

where

$$A(x,y) = [\phi^{(-)}(x),\phi^{(+)}(y)] - \theta(x_0 - y_0)[\phi(x),\phi(y)]. \qquad (8.70)$$

This is a *c*-number because it contains commutators of free fields, and therefore it can be evaluated by taking its VEV

$$A(x,y) = \langle 0|[\phi^{(-)}(x),\phi^{(+)}(y)] - \theta(x_0 - y_0)[\phi(x),\phi(y)]|0\rangle. \qquad (8.71)$$

Since $\phi^{(+)}|0\rangle = 0$, we can write

$$A(x,y) = -\langle 0|\left[\phi(y)\phi(x) + \theta(x_0 - y_0)\phi(x)\phi(y) - \theta(x_0 - y_0)\phi(y)\phi(x)\right]|0\rangle$$
$$= -\langle 0|\left[\theta(y_0 - x_0)\phi(y)\phi(x) + \theta(x_0 - y_0)\phi(x)\phi(y)|0\rangle\right.$$
$$= -\langle 0|T(\phi(x)\phi(y))|0\rangle. \qquad (8.72)$$

This proves our identity (8.61). Let us now expand both sides of this identity in a series of $j(x)$ and compare term by term. We will use the simplified notation $\phi_i \equiv \phi(x_i)$. We get

$$T(\phi) = \;\; :\phi:, \qquad (8.73)$$

$$T(\phi_1\phi_2) = \;\; :\phi_1\phi_2: + \langle 0|T(\phi_1\phi_2)|0\rangle, \qquad (8.74)$$

$$T(\phi_1\phi_2\phi_3) = \;\; :\phi_1\phi_2\phi_3: + \sum_{i\neq j\neq k=1}^{3} :\phi_i: \langle 0|T(\phi_j\phi_k)|0\rangle, \qquad (8.75)$$

$$T(\phi_1\phi_2\phi_3\phi_4) = \;\; :\phi_1\phi_2\phi_3\phi_4: + \sum_{i\neq j\neq k\neq l=1}^{4}\Big[ :\phi_i\phi_j: \langle 0|T(\phi_k\phi_l)|0\rangle$$
$$+ \langle 0|T(\phi_i\phi_j)|0\rangle\langle 0|T(\phi_k\phi_l)|0\rangle\Big] \qquad (8.76)$$

and analogous expressions for higher order monomials in the scalar field. By taking the VEV of these expressions, and recalling that the VEV of a normal product is zero, we get the Wick theorem. The $T$-product of two field operators is sometimes called the **contraction** of the two operators. Therefore to evaluate the VEV of a $T$-product of an arbitrary number of free fields, it is sufficient to consider all the possible contractions of the fields appearing in the $T$-product. For instance, from the last of the previous relations we get

$$\langle 0|T(\phi_1\phi_2\phi_3\phi_4)|0\rangle = \sum_{i\neq j\neq k\neq l=1}^{4} \langle 0|T(\phi_i\phi_j)|0\rangle\langle 0|T(\phi_k\phi_l)|0\rangle. \qquad (8.77)$$

An analogous theorem holds for the photon field. For the fermions one has to remember that the $T$-product is defined in a slightly different way. This gives a minus sign any time we have an odd permutation of the fermionic fields with respect to the original ordering. As an illustration the previous formula, in the case of fermions, becomes

$$\langle 0|T(\psi_1\psi_2\psi_3\psi_4)|0\rangle = \sum_{i\neq j\neq k\neq l=1}^{4} \sigma_P\langle 0|T(\psi_i\psi_j)|0\rangle\langle 0|T(\psi_k\psi_l)|0\rangle, \qquad (8.78)$$

where $\sigma_P = \pm 1$ is the sign of the permutation $(i, j, k, l)$ with respect to the fundamental one $(1, 2, 3, 4)$ appearing on the left-hand side. More explicitly

$$\begin{aligned}\langle 0|T(\psi_1\psi_2\psi_3\psi_4)|0\rangle = &+\langle 0|T(\psi_1\psi_2)|0\rangle\langle 0|T(\psi_3\psi_4)|0\rangle \\ &-\langle 0|T(\psi_1\psi_3)|0\rangle\langle 0|T(\psi_2\psi_4)|0\rangle \\ &+\langle 0|T(\psi_1\psi_4)|0\rangle\langle 0|T(\psi_2\psi_3)|0\rangle. \qquad (8.79)\end{aligned}$$

## 8.4 Evaluation of the $S$ matrix at second order in QED

In the case of QED the $S$ matrix is given by

$$S = 1 + \sum_{n=1}^{\infty} \frac{(+i)^n}{n!} \int \cdots \int d^4x_1 \cdots d^4x_n T\left(\mathcal{L}_I(x_1)\cdots\mathcal{L}_I(x_n)\right), \qquad (8.80)$$

with (see eq. (8.19))

$$\mathcal{L}_I = -e : \bar{\psi}\hat{A}\psi : . \qquad (8.81)$$

We have now to understand how to use the Wick theorem in the actual situation with normal ordered operators inside the $T$-product. Consider, for simplicity, two scalar fields. From eq. (8.74) we get (at equal times the $T$-product and the usual product coincides)

$$\phi_a(x)\phi_b(x) =: \phi_a(x)\phi_b(x): +\langle 0|T(\phi_a(x)\phi_b(x))|0\rangle, \qquad (8.82)$$

from which

$$: \phi_a(x)\phi_b(x) := \phi_a(x)\phi_b(x) - \langle 0|T(\phi_a(x)\phi_b(x))|0\rangle. \qquad (8.83)$$

Therefore

$$T(: \phi_a(x)\phi_b(x): \phi_1(x_1)\cdots\phi_n(x_n)) = T(\phi_a(x)\phi_b(x)\phi_1(x_1)\cdots\phi_n(x_n))$$
$$-\langle 0|T(\phi_a(x)\phi_b(x))|0\rangle T(\phi_1(x_1)\cdots\phi_n(x_n)). \qquad (8.84)$$

Since the second term subtracts the contraction between the two operators taken at the same point, we can generalize the Wick expansion by saying that, when normal products are contained inside a $T$-product, the Wick expansion applies with the further rule that the contractions of operators at the same point, inside the normal product, vanish. With this convention we can write

$$T(: A(x_1)B(x_1)\cdots: \cdots: A(x_n)B(x_n)\cdots:)$$
$$= T(A(x_1)B(x_1)\cdots A(x_n)B(x_n)\cdots). \qquad (8.85)$$

In the case of QED one can be convinced more easily by recalling that (see eqs. (8.19) and (8.20))

$$: \bar{\psi}\gamma_\mu\psi := \frac{1}{2}[\bar{\psi}, \gamma_\mu\psi] \qquad (8.86)$$

and noticing that, inside a $T$-product, the fields can be freely commuted except for taking into account their statistics. Therefore

$$T(: \bar{\psi}\gamma_\mu\psi :) = \frac{1}{2}T([\bar{\psi}, \gamma_\mu\psi]) = T(\bar{\psi}\gamma_\mu\psi). \qquad (8.87)$$

For the following analysis it is useful to remember how the various field operators act on the kets

$$
\begin{array}{llllll}
\psi^+ & \text{annihilates} & e^-, & \psi^- & \text{creates} & e^+, \\
\bar{\psi}^+ & \text{annihilates} & e^+, & \bar{\psi}^- & \text{creates} & e^-, \\
A^+ & \text{annihilates} & \gamma, & A^- & \text{creates} & \gamma.
\end{array} \qquad (8.88)
$$

Decomposing $\mathcal{L}_I$ in positive and negative frequency components[2]

$$\mathcal{L}_I = -e : (\bar{\psi}^+ + \bar{\psi}^-)(A_\mu^+ + A_\mu^-)\gamma^\mu(\psi^+ + \psi^-) :, \qquad (8.89)$$

---

[2]Remember that $e$ is the electric charge of the particle with its own sign. Therefore, for the electron, $e < 0$.

we get 8 terms with non-vanishing matrix elements. For instance,

$$: \bar{\psi}_\alpha^+ \hat{A}_{\alpha\beta}^- \psi_\beta^- := -\psi_\beta^- \hat{A}_{\alpha\beta}^- \bar{\psi}_\alpha^+ \tag{8.90}$$

has the following non-vanishing matrix element

$$\langle e^+ \gamma | \psi^- \hat{A}^- \bar{\psi}^+ | e^+ \rangle. \tag{8.91}$$

This process corresponds to a positron emitting a photon. This and the other seven processes, described by the $S$ matrix at first order in perturbation theory

$$S^{(1)} = -ie \int d^4x \; : \bar{\psi}(x)\hat{A}(x)\psi(x) : \tag{8.92}$$

are represented by the diagrams of Fig. 8.3.

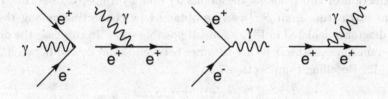

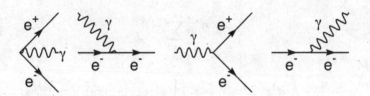

Fig. 8.3 Diagrams for the processes described by the $S$ matrix at first order.

However none of these contributions corresponds to a physically possible process since the four-momentum is not conserved. We will show later that the conservation of the four-momentum is a consequence of the theory. For the moment we will assume it and we will show that for real particles (that is for particles on mass shell, $p^2 = m^2$) these processes cannot happen. For instance, consider

$$e^-(p) \to e^-(p') + \gamma(k), \tag{8.93}$$

If the four-momentum is conserved

$$p' = p - k, \tag{8.94}$$

from which

$$m^2 = m^2 - 2p \cdot k, \tag{8.95}$$

where we have used $k^2 = 0$ for the photon. In the rest frame of the initial electron we get $mk_0 = 0$. Therefore the process is possible only for a photon with vanishing four-momentum. Let us now consider the 2$^{nd}$ order contribution

$$S^{(2)} = \frac{(-ie)^2}{2!} \int d^4x_1 \, d^4x_2 \, T\left(\bar{\psi}(x_1)\hat{A}(x_1)\psi(x_1)\bar{\psi}(x_2)\hat{A}(x_2)\psi(x_2)\right). \tag{8.96}$$

We can expand $S^{(2)}$ using Wick's theorem, and classifying the various contributions according to the number of contractions. If we associate to a contraction of two fields at the points $x_1$ and $x_2$ a line, we see that the terms originating from $S^{(2)}$ can be obtained by connecting among them the diagrams depicted in Fig. 8.3 in all possible ways. In this case the only non-vanishing contractions are the ones between $\psi$ and $\bar{\psi}$, and between $A_\mu$ and $A_\nu$. Recalling from Section 7.1

$$\langle 0|T(\psi_\alpha(x)\bar{\psi}_\beta(y))|0\rangle = iS_F(x-y)_{\alpha\beta}$$
$$\langle 0|T(A_\mu(x)A_\nu(y))|0\rangle = ig_{\mu\nu}D(x-y), \tag{8.97}$$

we get

$$S^{(2)} = \sum_{i=1}^{6} S_i^{(2)}, \tag{8.98}$$

where

$$S_1^{(2)} = \frac{(-ie)^2}{2!} \int d^4x_1 \, d^4x_2 \; : \bar{\psi}(x_1)\hat{A}(x_1)\psi(x_1)\bar{\psi}(x_2)\hat{A}(x_2)\psi(x_2) :, \tag{8.99}$$

$$S_2^{(2)} = \frac{(-ie)^2}{2!} \int d^4x_1 \, d^4x_2 \; : \bar{\psi}(x_1)\hat{A}(x_1)iS_F(x_1-x_2)\hat{A}(x_2)\psi(x_2) :$$
$$+ \frac{(-ie)^2}{2!} \int d^4x_1 \, d^4x_2 \; : \bar{\psi}(x_2)\hat{A}(x_2)iS_F(x_2-x_1)\hat{A}(x_1)\psi(x_1) :$$
$$= (-ie)^2 \int d^4x_1 \, d^4x_2$$
$$\times \; : \bar{\psi}(x_1)\hat{A}(x_1)iS_F(x_1-x_2)\hat{A}(x_2)\psi(x_2) :, \tag{8.100}$$

$$S_3^{(2)} = \frac{(-ie)^2}{2!} \int d^4x_1 \, d^4x_2 \; : \bar{\psi}(x_1)\gamma_\mu\psi(x_1)ig^{\mu\nu}D(x_1-x_2)\bar{\psi}(x_2)\gamma_\nu\psi(x_2) :, \tag{8.101}$$

$$S_4^{(2)} = \frac{(-ie)^2}{2!} \int d^4x_1 \, d^4x_2$$
$$\times : \bar{\psi}(x_1)\gamma_\mu iS_F(x_1 - x_2)ig^{\mu\nu}D(x_1 - x_2)\gamma_\nu\psi(x_2) :$$
$$+\frac{(-ie)^2}{2!}\int d^4x_1 \, d^4x_2 : \bar{\psi}(x_2)\gamma_\mu iS_F(x_2 - x_1)ig^{\mu\nu}D(x_2 - x_1)\gamma_\nu\psi(x_1):$$
$$= (-ie)^2 \int d^4x_1 \, d^4x_2$$
$$\times : \bar{\psi}(x_1)\gamma_\mu iS_F(x_1 - x_2)ig^{\mu\nu}D(x_1 - x_2)\gamma_\nu\psi(x_2) :, \qquad (8.102)$$

$$S_5^{(2)} = \frac{(-ie)^2}{2!} \int d^4x_1 \, d^4x_2(-1)$$
$$\times : Tr[iS_F(x_1 - x_2)\hat{A}(x_2)iS_F(x_2 - x_1)\hat{A}(x_1)] :, \qquad (8.103)$$

$$S_6^{(2)} = \frac{(-ie)^2}{2!} \int d^4x_1 \, d^4x_2(-1)$$
$$\times Tr[iS_F(x_1 - x_2)\gamma_\mu iS_F(x_2 - x_1)\gamma_\nu ig^{\mu\nu}D(x_2 - x_1)]. \, (8.104)$$

The term $S_1^{(2)}$ is nothing but the product of two processes of type $S^{(1)}$ and it does not give rise to real processes. The term $S_2^{(2)}$ is obtained by contracting two fermionic fields, meaning that we are connecting a fermion line with two of the vertices of Fig. 8.3. The possible external particles are two $\gamma$, two $e^-$, two $e^+$, or a pair $e^+e^-$. Selecting the external states we can get different physical processes. One of these processes is the Compton scattering $\gamma + e^- \to \gamma + e^-$. In this case we must select in $S_2^{(2)}$, $\psi^+(x_2)$ to destroy the initial electron and $\bar{\psi}^-(x_1)$ to create the final electron. As the photons are concerned, since $A_\mu$ is a real field, we can destroy the initial photon both in $x_2$ and $x_1$ and create the final photon in the other point. Therefore we get two contributions

$$S_2^{(2)}(\gamma e^- \to \gamma e^-) = S_a + S_b, \qquad (8.105)$$

with

$$S_a = (-ie)^2 \int d^4x_1 \, d^4x_2 \, \bar{\psi}^-(x_1)\gamma^\mu iS_F(x_1 - x_2)\gamma^\nu A_\mu^-(x_1)A_\nu^+(x_2)\psi^+(x_2) \qquad (8.106)$$

and

$$S_b = (-ie)^2 \int d^4x_1 \, d^4x_2 \, \bar{\psi}^-(x_1)\gamma^\mu iS_F(x_1 - x_2)\gamma^\nu A_\nu^-(x_2)A_\mu^+(x_1)\psi^+(x_2). \qquad (8.107)$$

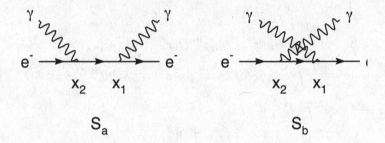

Fig. 8.4  Diagrams for the Compton scattering.

The corresponding diagrams are given in Fig. 8.4.

The terms corresponding to the Compton scattering for a positron are obtained from the previous ones by substituting $\psi^+$ (annihilates an electron) with $\psi^-$ (creates a positron) and $\bar{\psi}^-$ (creates an electron) with $\bar{\psi}^+$ (annihilates a positron). The other two processes coming from $S_2^{(2)}$ are $2\gamma \to e^+ e^-$ (pair creation) and $e^+ e^- \to 2\gamma$ (pair annihilation). The $S$ matrix element for the pair creation is given by

$$S_2^{(2)}(2\gamma \to e^+ e^-) = (-ie)^2 \int d^4 x_1 \, d^4 x_2$$

$$\times \bar{\psi}^-(x_1)\gamma^\mu i S_F(x_1 - x_2)\gamma^\nu A_\mu^+(x_1) A_\nu^+(x_2)\psi^-(x_2). \qquad (8.108)$$

Notice that in evaluating

$$A_\mu^+(x_1) A_\nu^+(x_2)|\gamma(k_1)\gamma(k_2)\rangle, \qquad (8.109)$$

we get two contributions since one of the fields $A_\mu^+$ can annihilate any one of the two external photons. The diagrams for these two contributions are given in Fig. 8.5.

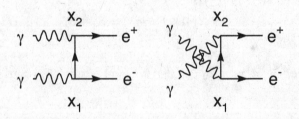

Fig. 8.5  Diagrams for the pair creation.

We have analogous contributions and diagrams for the pair annihilation process.

The next processes we consider are the ones generated by $S_3^{(2)}$ in which we have contracted the photon fields. They are: electron scattering $e^-e^- \to e^-e^-$, positron scattering $e^+e^+ \to e^+e^+$ and electron-positron scattering $e^+e^- \to e^+e^-$. For the electron scattering we have

$$S_3^{(2)}(2e^- \to 2e^-) = \frac{(-ie)^2}{2!} \int d^4x_1 \, d^4x_2$$
$$\times \, : \bar{\psi}^-(x_1)\gamma_\mu\psi^+(x_1)ig^{\mu\nu}D(x_1 - x_2)\bar{\psi}^-(x_2)\gamma_\nu\psi^+(x_2) : . \quad (8.110)$$

The term

$$\psi^+(x_1)\psi^+(x_2)|e^-(p_1)e^-(p_2)\rangle \qquad (8.111)$$

gives rise to two contributions, and other two come from the final state. The corresponding diagrams are given in Fig. 8.6.

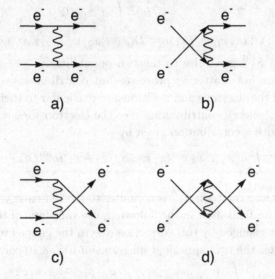

Fig. 8.6 Diagrams for electron scattering.

The terms a) and d) differ only for the exchange $x_1 \leftrightarrow x_2$ and therefore they are equal after having exchanged the integration variables. The same is true for the terms b) and c). In this way we get a factor 2 which cancels the 2! in the denominator. This is the same kind of cancellation we have seen for the term $S_2^{(2)}$. At the order $n$ one has $n!$ equivalent diagrams which cancel the factor $n!$ coming from the expansion of the $S$ matrix. This means that it is enough to draw all the inequivalent diagrams. For the electron scattering we have two such diagrams differing for a minus sign due to the exchange of the fermion lines. This happens because field theory takes automatically into account the statistics of the particles.

For the process $e^+e^- \to e^+e^-$ we get the diagrams of Fig. 8.7. The inequivalent diagrams are those of Fig. 8.8 and correspond to the following contributions

$$S_3^{(2)}(e^+e^- \to e^+e^-) = S_a(e^+e^- \to e^+e^-) + S_b(e^+e^- \to e^+e^-), \quad (8.112)$$

with

$$S_a(e^+e^- \to e^+e^-) = (-ie)^2 \int d^4x_1 \, d^4x_2$$
$$\times : \bar{\psi}^-(x_1)\gamma_\mu\psi^+(x_1)ig^{\mu\nu}D(x_1-x_2)\bar{\psi}^+(x_2)\gamma_\nu\psi^-(x_2) : \quad (8.113)$$

and

$$S_b(e^+e^- \to e^+e^-) = (-ie)^2 \int d^4x_1 \, d^4x_2$$
$$\times : \bar{\psi}^-(x_1)\gamma_\mu\psi^-(x_1)ig^{\mu\nu}D(x_1-x_2)\bar{\psi}^+(x_2)\gamma_\nu\psi^+(x_2) : . \quad (8.114)$$

The term $S_4^{(2)}$ gives rise to the two possibilities $e^- \to e^-$ and $e^+ \to e^+$. These are not scattering processes but contributions to the intrinsic properties of the electrons and positrons, in particular to their mass. These are called self-energy contributions. For the electron we have the diagram of Fig. 8.9 with a contribution given by

$$S_4^{(2)} = (-ie)^2 \int d^4x_1 \, d^4x_2 \, \bar{\psi}^-(x_1)\gamma_\mu iS_F(x_1-x_2)ig^{\mu\nu}D(x_1-x_2)\gamma_\nu\psi^+(x_2).$$
$$(8.115)$$

In analogous way the term $S_2^{(5)}$ contributes to the self-energy of the photon. However, as we shall show in the following, the vanishing of the mass of the photon is not changed by this correction due to the gauge invariance of the theory. We get the two equivalent diagrams of Fig. 8.10 corresponding to

$$S_5^{(2)} = (-ie)^2 \int d^4x_1 \, d^4x_2(-1)Tr[iS_F(x_1-x_2)\gamma^\mu iS_F(x_2-x_1)\gamma^\nu]$$
$$\times A_\mu^-(x_1)A_\nu^+(x_2). \quad (8.116)$$

The minus sign arises because in a fermionic loop we have to exchange two fermion fields inside the $T$-product. The last term is the one where all the fields are contracted. There are no external particles and the corresponding diagram of Fig. 8.11 is called a vacuum diagram. We can ignore it because it is possible to show that it contributes to a phase factor for the vacuum state.

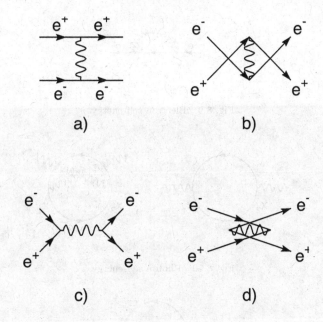

Fig. 8.7   Diagrams for the process $e^+e^- \to e^+e^-$.

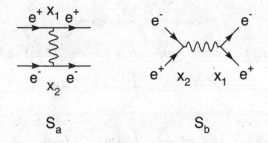

Fig. 8.8   Inequivalent diagrams for the process $e^+e^- \to e^+e^-$.

## 8.5   Feynman diagrams in momentum space

As we have already discussed, in a typical experiment in particle physics one or two beams of particles with definite momenta are prepared. Furthermore, the momenta of the final states are measured. For this reason it is convenient to work in momentum space. Let us recall the expressions for

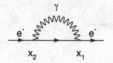

Fig. 8.9   Electron self-energy.

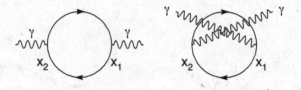

Fig. 8.10   Photon self-energy.

Fig. 8.11   Vacuum diagram.

the fermion and the photon propagators (see eqs. (7.28) and (7.30))

$$S_F(x) = \frac{1}{(2\pi)^4} \int d^4p \, e^{-ipx} S_F(p) \tag{8.117}$$

and

$$D(x) = \frac{1}{(2\pi)^4} \int d^4p \, e^{-ipx} D(p), \tag{8.118}$$

where

$$S_F(p) = \frac{1}{\hat{p} - m + i\epsilon} \tag{8.119}$$

and

$$D(p) = -\frac{1}{p^2 + i\epsilon}. \tag{8.120}$$

From the expansion of the fields in terms of creation and annihilation operators, one can evaluate the action of the positive frequency part of $\psi$, $\bar{\psi}$

and $A_\mu$ on the one particle states. We get

$$\psi^+(x)|e^-(p,r)\rangle = \sum_{\pm s} \int d^3k \sqrt{\frac{m}{(2\pi)^3 E_k}} e^{-ikx} u(k,s) b(k,s) b^\dagger(p,r)|0\rangle,$$

$$(8.121)$$

where the indices $r,s$ stands for the polarization of the fermions. Then, using

$$\left[b(k,s), b^\dagger(p,r)\right]_+ = \delta_{rs}\delta^3(\vec{p} - \vec{k}),$$ $$(8.122)$$

it follows

$$\psi^+(x)|e^-(p,r)\rangle = \sqrt{\frac{m}{(2\pi)^3 E_k}} u(p,r) e^{-ipx}|0\rangle.$$ $$(8.123)$$

For later applications it is more convenient to use the normalization in a box than the continuous one. This amounts to the substitution

$$\frac{1}{\sqrt{(2\pi)^3}} \to \frac{1}{\sqrt{V}}$$ $$(8.124)$$

and to the use of discrete momenta $\vec{p} = (2\pi/L)^3 \vec{n}$. We get

$$\psi^+(x)|e^-(p,r)\rangle = \sqrt{\frac{m}{V E_p}} u(p,r) e^{-ipx}|0\rangle,$$ $$(8.125)$$

$$\bar\psi^+(x)|e^+(p,r)\rangle = \sqrt{\frac{m}{V E_p}} \bar{v}(p,r) e^{-ipx}|0\rangle$$ $$(8.126)$$

and for the photon

$$A_\mu^+(x)|\gamma(k,\lambda)\rangle = \sqrt{\frac{1}{2V E_k}} \epsilon_\mu^{(\lambda)}(k) e^{-ikx}|0\rangle.$$ $$(8.127)$$

By conjugating these expressions we obtain the action of the negative frequency operators on the bra vectors.

As an example let us consider a process associated to $S^{(1)}$

$$|i\rangle = |e^-(p)\rangle \to |f\rangle = |e^-(p'), \gamma(k)\rangle.$$ $$(8.128)$$

From (8.92) we get[3]

$$\langle f|S^{(1)}|i\rangle = -ie \int d^4x \,\langle e^-(p'), \gamma(k)|\bar\psi^-(x) A_\mu^-(x) \gamma^\mu \psi^+(x)|e^-(p)\rangle$$

$$= -ie \int d^4x \left(\sqrt{\frac{m}{V E_{p'}}} \bar{u}(p') e^{ip'x}\right)$$

$$\times \left(\sqrt{\frac{1}{2V E_k}} \hat\epsilon^*(k) e^{ikx}\right) \left(\sqrt{\frac{m}{V E_p}} u(p) e^{-ipx}\right)$$

$$= \frac{-iem}{\sqrt{V^3 E_p E_{p'} 2E_k}} (2\pi)^4 \delta^4(p' + k - p) \bar{u}(p')\hat\epsilon^*(k) u(p). \quad (8.129)$$

---

[3]Notice that the polarization vectors of the photons can be real, as for linear polarization, or complex, as for circular polarization.

This expression can be written as

$$\langle f|S^{(1)}|i\rangle = (2\pi)^4\delta^4(p' + k - p)\sqrt{\frac{m}{VE_p}}\sqrt{\frac{m}{VE_{p'}}}\sqrt{\frac{1}{2VE_k}}\mathcal{M}, \qquad (8.130)$$

where the quantity

$$\mathcal{M} = -ie\bar{u}(p')\hat{\epsilon}^*(k)u(p) \qquad (8.131)$$

is called the Feynman, or the invariant amplitude for the process. Notice that $\mathcal{M}$ is a Lorentz invariant quantity. The term $(2\pi)^4\delta^4(p' + k - p)$ gives the conservation of the four-momentum in the process, whereas the other factors are associated to the various external particles (incoming and outgoing). As we said previously this process is not physically possible since it does not respect the four-momentum conservation.

This structure is quite general and for any process a delta-function expressing the four-momentum conservation will appear. Furthermore there will be factors as $\sqrt{m/VE_p}$ for each external fermion and $\sqrt{1/2VE_k}$ for each boson. Now we have to investigate the rules for evaluating the invariant amplitude $\mathcal{M}$. To this end let us start considering the Compton scattering

$$|i\rangle = |e^-(p), \gamma(k)\rangle \rightarrow |f\rangle = |e^-(p'), \gamma(k')\rangle. \qquad (8.132)$$

The $S$ matrix element for the Compton scattering is given in eqs. (8.105), (8.106) and (8.107)

$$S_2^{(2)}(\gamma e^- \rightarrow \gamma e^-) = S_a + S_b \qquad (8.133)$$

and

$$\langle f|S_a|i\rangle = (-ie)^2 \int d^4x_1\, d^4x_2 \sqrt{\frac{m}{VE_p}}\sqrt{\frac{m}{VE_{p'}}}\sqrt{\frac{1}{2VE_k}}\sqrt{\frac{1}{2VE_{k'}}}$$

$$\times \bar{u}(p')e^{ip'x_1}\hat{\epsilon}^*(k')e^{ik'x_1}\frac{i}{(2\pi)^4}\int d^4q e^{-iq(x_1-x_2)}S_F(q)$$

$$\times \hat{\epsilon}(k)e^{-ikx_2}u(p)e^{-ipx_2}$$

$$= (-ie)^2 \int d^4x_1 d^4x_2 \int \frac{d^4q}{(2\pi)^4}\sqrt{\frac{m}{VE_p}}\sqrt{\frac{m}{VE_{p'}}}\sqrt{\frac{1}{2VE_k}}\sqrt{\frac{1}{2VE_{k'}}}$$

$$\times e^{i(p'+k'-q)x_1}e^{-i(p+k-q)x_2}\epsilon_\mu^*(k')\epsilon_\nu(k)$$

$$\times \bar{u}(p')\gamma^\mu\frac{i}{\hat{q} - m + i\epsilon}\gamma^\nu u(p)$$

$$= (2\pi)^4\delta^4(p' + k' - p - k)\sqrt{\frac{m}{VE_p}}\sqrt{\frac{m}{VE_{p'}}}\sqrt{\frac{1}{2VE_k}}\sqrt{\frac{1}{2VE_{k'}}}$$

$$\times \epsilon_\mu^*(k')\epsilon_\nu(k)\bar{u}(p')(-ie\gamma^\mu)\frac{i}{\hat{p} + \hat{k} - m + i\epsilon}(-ie\gamma^\nu)u(p). \qquad (8.134)$$

Also in this case we can write

$$\langle f|S_a|i\rangle = (2\pi)^4\delta^4(p'+k'-p-k)\sqrt{\frac{m}{VE_p}}\sqrt{\frac{m}{VE_{p'}}}\sqrt{\frac{1}{2VE_k}}\sqrt{\frac{1}{2VE_{k'}}}\mathcal{M}_a,$$

(8.135)

with

$$\mathcal{M}_a = \epsilon_\mu^*(k')\epsilon_\nu(k)\bar{u}(p')(-ie\gamma^\mu)\frac{i}{\hat{q}-m+i\epsilon}(-ie\gamma^\nu)u(p), \quad q = p+k.$$

(8.136)

We will associate to this expression the diagram in Fig. 8.12.

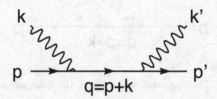

Fig. 8.12 Contribution to the Compton scattering.

The four-momentum $q$ is determined by the conservation of the four-momentum at the vertices: $q = p + k = p' + k'$. However, notice that in general $q^2 = m^2 + 2p \cdot k \neq m^2$, that is the exchanged particle (described by the propagator) is not a real particle but a **virtual** one. Looking at the previous expression one understands immediately how the various pieces are connected to the graphical elements in the diagram. In fact, we have the following rules for any given diagram (Feynman diagram):

• write the amplitude starting from the outgoing particles,
• for each vertex there is a factor $-ie\gamma_\mu$,
• for each internal fermion line there is a factor $iS_F(p)$ (fermion propagator),
• for each incoming and/or outgoing fermion line there is a factor $u(p)$ and/or $\bar{u}(p)$,
• for each incoming and/or outgoing photon line there is a factor $\epsilon_\mu^\lambda$ (or its complex conjugate for outgoing photons if we consider complex polarization as the circular one).

Notice also that the spinorial factors start from the final states and end up with the initial ones. The further contribution to the Compton scattering (given by $S_b$) corresponds to the diagram in Fig. 8.13 and, using the previous rules, we get

$$\mathcal{M}_b = \bar{u}(p')(-ie\gamma_\mu)\frac{i}{\hat{q} - m + i\epsilon}(-ie\gamma_\nu)u(p)\epsilon^\mu(k)\epsilon^\nu(k'), \quad q = p - k'.$$
$$(8.137)$$

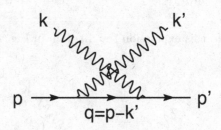

Fig. 8.13   The crossed contribution to the Compton scattering.

If we write down the Compton amplitude for the positrons we see that we must associate $\bar{v}(p)$, $v(p)$ to the initial and final states respectively. This is seen from eq. (8.126) which shows that the annihilation operator for the positrons is associated to $\bar{v}(p)$. For this reason, when drawing the diagrams in momentum space it is often convenient to invert the direction of the positron lines, in such a way to follow the rule of associating unbarred spinors to incoming fermion lines and barred spinors to outgoing lines (see Fig. 8.14).

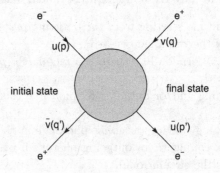

Fig. 8.14   The spinor conventions for antifermions.

In this case one has to be careful with the direction of momenta which are flowing in a direction opposite to the arrow in the case of antiparticles. As for the internal lines, there is no distinction between particles and antiparticles. Therefore, in general, one draws the arrows in a way consistent with the flow of the momenta. Consider now $e^- e^-$ scattering. The diagrams for this process are in Fig. 8.15, and the matrix element is given by

$$\langle f|S_3^{(2)}|i\rangle = (2\pi)^4 \delta^4(p_1' + p_2' - p_1 - p_2) \prod_{i=1}^{4} \sqrt{\frac{m}{VE_i}} \mathcal{M}, \qquad (8.138)$$

where $\mathcal{M} = \mathcal{M}_a + \mathcal{M}_b$, with

$$M_a = \bar{u}(p_1')(-ie\gamma_\mu)u(p_1)ig^{\mu\nu}D(p_2 - p_2')\bar{u}(p_2')(-ie\gamma_\nu)u(p_2) \qquad (8.139)$$

and

$$M_b = -\bar{u}(p_2')(-ie\gamma_\mu)u(p_1)ig^{\mu\nu}D(p_2 - p_1')\bar{u}(p_1')(-ie\gamma_\nu)u(p_2). \qquad (8.140)$$

The relative minus sign comes from the exchange of the two electrons in the final state, that is from the Fermi statistics. From this example we

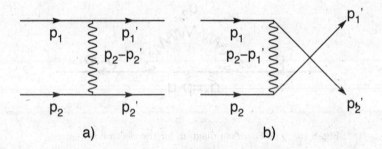

Fig. 8.15 The Feynman diagrams for the scattering $e^- e^- \to e^- e^-$.

get the further rule for the Feynman diagrams

- - for each internal photon line there is a factor $ig^{\mu\nu}D(p)$ (propagator).

As a last example let us consider the electron self-energy

$$|i\rangle = |e^-(p)\rangle \to |f\rangle = |e^-(p')\rangle, \qquad (8.141)$$

the $S$ matrix element is

$$\langle f|S_4^{(2)}|i\rangle = -e^2 \int d^4x_1\, d^4x_2\, \sqrt{\frac{m}{VE_{p'}}}\sqrt{\frac{m}{VE_p}}\bar{u}(p')e^{ip'x_1}]$$

$$\times\gamma_\mu \int \frac{d^4q_1}{(2\pi)^4}e^{-iq_1(x_1-x_2)}iS_F(q_1)$$

$$\times\gamma_\nu \int \frac{d^4q_2}{(2\pi)^4}e^{-iq_2(x_1-x_2)}iD(q_2)u(p)e^{-ipx_2}$$

$$= \sqrt{\frac{m}{VE_{p'}}}\sqrt{\frac{m}{VE_p}}\int \frac{d^4q_1}{(2\pi)^4}\frac{d^4q_2}{(2\pi)^4}$$

$$\times(2\pi)^4\delta^4(p'-q_1-q_2)(2\pi)^4\delta^4(p-q_1-q_2)$$

$$\times\bar{u}(p')(-ie\gamma_\mu)iS_F(q_1)(-ie\gamma_\nu)u(p)ig^{\mu\nu}D(q_2)$$

$$= (2\pi)^4\delta^4(p'-p)\sqrt{\frac{m}{VE_{p'}}}\sqrt{\frac{m}{VE_p}}\mathcal{M}, \qquad (8.142)$$

where in the last term we have integrated over $q_1$. $\mathcal{M}$ is given by

$$\mathcal{M} = \int \frac{d^4q_2}{(2\pi)^4}\bar{u}(p')(-ie\gamma_\mu)iS_F(p-q_2)(-ie\gamma_\nu)u(p)ig^{\mu\nu}D(q_2), \quad (8.143)$$

corresponding to the diagram in Fig. 8.16.

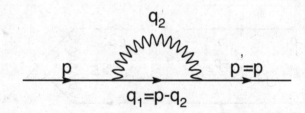

Fig. 8.16   The Feynman diagram for the electron self-energy.

We see that the rule of conservation of the four-momentum is always valid and also, that we have to integrate with measure $d^4q/(2\pi)^4$ over all the momenta which are not determined by the conservation law. For the general case it is more convenient to formulate this rule in the following way:

- for each vertex there is an explicit factor $(2\pi)^4\delta^4(\sum_{\text{ent}} p_{\text{ent}})$, where $p_{in}$ are the momenta entering the vertex,
- integrate all the internal momenta, $p_{\text{int}}$, with a measure $d^4p_{\text{int}}/(2\pi)^4$.

In this way the factor $(2\pi)^4 \delta^4(\sum p_{\text{ext}})$ is automatically produced. We can verify the previous rule in the self-energy case

$$\int \frac{d^4 q_1}{(2\pi)^4} \int \frac{d^4 q_2}{(2\pi)^4} (2\pi)^4 \delta^4(q_1 + q_2 - p)(2\pi)^4 \delta^4(p' - q_1 - q_2)$$

$$= (2\pi)^4 \delta^4(p' - p) \int \frac{d^4 q_2}{(2\pi)^4}. \tag{8.144}$$

As a last rule we recall

- a factor $(-1)$ for each fermionic loop.

## 8.6  Exercises

(1) Consider a scalar field of mass $M$, coupled to the Dirac field with mass $m$, with an interaction given by

$$\mathcal{L}_I = g \bar{\psi} \psi \phi. \tag{8.145}$$

Discuss under which conditions the $S$-matrix, at first order in perturbation theory, is non-vanishing.

(2) Draw the Feynman diagrams and evaluate the corresponding Feynman amplitudes for the Bhabha scattering $(e^+ e^- \to e^+ e^-)$ at the 2$^{\text{nd}}$ order in perturbation theory.

(3) Repeat the previous exercise for the scattering Møller $(e^- e^- \to e^- e^-)$, the pair creation $\gamma\gamma \to e^+ e^-$ and for the pair annihilation $e^+ e^- \to \gamma\gamma$.

(4) Write the amplitude $\mathcal{M}$ for the Compton scattering at the second order in the electric charge and show that it is invariant under the following substitution on any of the photon polarization vectors

$$\epsilon_\mu(k) + \lambda k_\mu, \tag{8.146}$$

for any value of $\lambda$. What is the relation with gauge invariance?

(5) The result of the previous exercise holds for any amplitude involving external photons. That is substituting a polarization vector with the corresponding photon four-momentum the result vanishes. Using this result and eq. (5.60) for the sum over the physical photon polarizations, show the following relation (assuming real polarization vectors)

$$\sum_{\lambda=1,2} \epsilon_\mu^{(\lambda)} \mathcal{M}^\mu (\epsilon_\nu^{(\lambda)} \mathcal{M}^\nu)^* = -\mathcal{M}^\mu \mathcal{M}^{\nu*} g_{\mu\nu} \tag{8.147}$$

where $\mathcal{M}^\mu$ is a generic amplitude involving at least one external photon. This result is very useful when evaluating the probability for emission or absorption of unpolarized photons where one has to sum or average over the photon polarizations. Of course, this result can be used for performing the polarization sum over any number of external photons.

# Chapter 9

# Applications

## 9.1  The cross-section

Let us consider a scattering process with a set of initial particles with four-momenta $p_i = (E_i, \vec{p}_i)$ which collide and produce a set of final particles with four-momenta $p_f = (E_f, \vec{p}_f)$. From the rules of the previous Chapter we know that each external photon line contributes with a factor $(1/2VE)^{1/2}$, whereas each external fermionic line contributes with $(m/VE)^{1/2}$. Furthermore, the conservation of the total four-momentum gives a term $(2\pi)^4 \delta^4(\sum_i p_i - \sum_f p_f)$. If we separate in the $S$ matrix the term $\delta_{fi}$ corresponding to no scattering events we can write

$$S_{fi} = \delta_{fi} + (2\pi)^4 \delta^4(\sum_i p_i - \sum_f p_f)$$

$$\times \prod_{\text{ferm}} \left(\frac{m}{VE}\right)^{1/2} \prod_{\text{bos}} \left(\frac{1}{2VE}\right)^{1/2} \mathcal{M}, \tag{9.1}$$

where $\mathcal{M}$ is the Feynman amplitude which can be evaluated by drawing the corresponding Feynman diagrams and using the rules of the previous Chapter.

Let us consider the typical case of a two-particle collision giving rise to $N$ particles in the final state. Since we are interested in a situation with the final state different from the initial one, the probability for the transition will be the modulus square of the second term in eq. (9.1). In doing this operation we encounter the square of the Dirac delta which is not a definite quantity. However, we should recall that we are quantizing the theory in a box, and we are considering the system for a finite, although large, time interval that we parametrize as $(-T/2, T/2)$. Therefore we are not dealing with the delta function but rather with $(P_{i(f)} = \sum_{i(f)} p_{i(f)})$

$$(2\pi)^4 \delta^4 (P_f - P_i) \to \int_V d^3x \int_{-T/2}^{T/2} dt\, e^{i(P_f - P_i)x}. \tag{9.2}$$

Consider one of the factors appearing in this equation, for instance the time integral. By performing the integration we get $(\Delta E = E_f - E_i)$

$$\int_{-T/2}^{T/2} dt\, e^{i\Delta Et} = \frac{2\sin(T\Delta E/2)}{\Delta E}. \tag{9.3}$$

And evaluating the modulus square

$$|(2\pi)\delta(E_f - E_i)|^2 \to \frac{4\sin^2(T\Delta E/2)}{(\Delta E)^2}. \tag{9.4}$$

On the right-hand side we have a function of $\Delta E$, whose integral holds $2\pi T$, and has a peak at $\Delta E = 0$. Therefore in the $T \to \infty$ limit we have a delta-convergent sequence

$$\lim_{T\to\infty} \frac{4\sin^2(T\Delta E/2)}{(\Delta E)^2} = 2\pi T \delta(E_f - E_i). \tag{9.5}$$

By doing the same operations also for the space integrals we get

$$\left|(2\pi)^4 \delta^4 (P_f - P_i)\right|^2 \to (2\pi)^4 L^3 T \delta^4(P_f - P_i), \tag{9.6}$$

where $L$ is the side of the volume $V = L^3$. Therefore, the probability per unit time of the transition is

$$w = V(2\pi)^4 \delta^4(P_f - P_i) \prod_i \frac{1}{2VE_i} \prod_f \frac{1}{2VE_f} \prod_{\text{ferm}} (2m)|\mathcal{M}|^2. \tag{9.7}$$

Here, for reasons of convenience, we have written

$$\frac{m}{VE} = (2m)\frac{1}{2VE}. \tag{9.8}$$

$w$ gives the probability per unit time of a transition toward a state with well-defined quantum numbers, but we are rather interested in final states having momenta between $\vec{p}_f$ and $\vec{p}_f + d\vec{p}_f$. Since in the volume $V$ the momentum is given by $\vec{p} = 2\pi\vec{n}/L$, the number of final states is

$$\left(\frac{L}{2\pi}\right)^3 d^3p. \tag{9.9}$$

The cross-section is defined as the probability per unit time divided by the flux of the incoming particles (number of particles per unit surface and unit of time), and has the dimensions of a squared length

$$[\text{cross-section}] = [\text{Probability per unit time/Flux}]$$
$$= [t^{-1}/(\ell^{-2}t^{-1})] = [\ell^2]. \tag{9.10}$$

The flux is given by the relative velocity of the two beams, or the velocity of the beam in the instance of scattering over a fixed target, times the density of particles, $\rho$:

$$v_{\text{rel}}\rho = v_{\text{rel}}/V, \tag{9.11}$$

since in our normalization we have one particle in the volume $V$, and $v_{\text{rel}}$ is the relative velocity of the incoming particles. For bosons this follows from the normalization condition (3.37). For fermions recall that, in the box normalization, the wave function is

$$\psi(x) = \sqrt{\frac{m}{EV}}u(p)e^{-ipx}, \tag{9.12}$$

from which we get that the total number of particles in the volume $V$ is one:

$$\int_V d^3x|\psi(x)|^2 = \int_V d^3x \frac{m}{VE}u^\dagger(p)u(p) = 1. \tag{9.13}$$

The cross-section for getting the final states with momenta between $\vec{p}_f$ and $\vec{p}_f + d\vec{p}_f$ is given

$$d\sigma = w\frac{V}{v_{\text{rel}}}dN_F = w\frac{V}{v_{\text{rel}}}\prod_f \frac{Vd^3p_f}{(2\pi)^3}. \tag{9.14}$$

We obtain

$$d\sigma = \frac{V}{v_{\text{rel}}}\left(\prod_f \frac{Vd^3p_f}{(2\pi)^3}\right)V(2\pi)^4\delta^4(P_f - P_i)\frac{1}{4V^2E_1E_2}$$

$$\times \prod_f \frac{1}{2VE_f}\prod_{\text{ferm}}(2m)|\mathcal{M}|^2$$

$$= (2\pi)^4\delta^4(P_f - P_i)\frac{1}{4E_1E_2v_{\text{rel}}}\prod_f \frac{d^3p_f}{(2\pi)^32E_f}\prod_{\text{ferm}}(2m)|\mathcal{M}|^2. \tag{9.15}$$

Notice that the dependence on the quantization volume $V$ disappears, as it should be, in the final equation. Furthermore, the total cross-section, which is obtained by integrating the previous expression over all the final momenta, is Lorentz invariant. In fact, both the factors $\mathcal{M}$ and $d^3p/2E$ are invariant. Furthermore

$$\vec{v}_{\text{rel}} = \vec{v}_1 - \vec{v}_2 = \frac{\vec{p}_1}{E_1} - \frac{\vec{p}_2}{E_2}, \tag{9.16}$$

but in the frame where the particle 2 is at rest (laboratory frame) we have $p_2 = (m_2, \vec{0})$ and $\vec{v}_{\text{rel}} = \vec{v}_1$, from which

$$E_1E_2|\vec{v}_{\text{rel}}| = E_1m_2\frac{|\vec{p}_1|}{E_1} = m_2|\vec{p}_1| = m_2\sqrt{E_1^2 - m_1^2}$$

$$= \sqrt{m_2^2E_1^2 - m_1^2m_2^2} = \sqrt{(p_1 \cdot p_2)^2 - m_1^2m_2^2}. \tag{9.17}$$

We see that also this factor is Lorentz invariant.

## 9.2   The scattering $e^+e^- \to \mu^+\mu^-$

In order to exemplify the previous techniques we will study the process $e^+e^- \to \mu^+\mu^-$. The Lagrangian density describing the interaction is[1]

$$\mathcal{L}_I = -e \left[ \bar{\psi}_e \gamma^\lambda \psi_e + \bar{\psi}_\mu \gamma^\lambda \psi_\mu \right] A_\lambda, \qquad (9.18)$$

where $\psi_e$ and $\psi_\mu$, are the electron and the muon fields[2]. Figure 9.1 describes the Feynman diagram for this process at the second order (in this diagram the arrows are oriented according to the direction of the momenta). Notice that in contrast to the process $e^+e^- \to e^+e^-$ the diagram in the crossed channel, Fig. 9.2, is now missing.

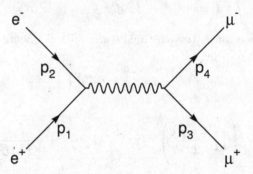

Fig. 9.1   The Feynman diagram for the scattering $e^+e^- \to \mu^+\mu^-$.

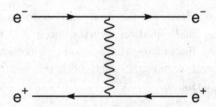

Fig. 9.2   The crossed diagram for the scattering $e^+e^- \to e^+e^-$.

---

[1] Here we will consider only the electromagnetic interaction disregarding the contribution due to the exchange of the $Z$ vector boson.

[2] The muon is a particle with the same spin and electric charge of the electron, but its mass is 105.66 MeV, about 200 times the mass of the electron.

The Feynman amplitude is

$$\mathcal{M} = \bar{u}(p_4, r_4)(-ie\gamma_\mu)v(p_3, r_3)\frac{-ig^{\mu\nu}}{(p_1 + p_2)^2}\bar{v}(p_1, r_1)(-ie\gamma_\nu)u(p_2, r_2)$$

$$= ie^2\bar{u}(p_4, r_4)\gamma_\mu v(p_3, r_3)\frac{1}{(p_1 + p_2)^2}\bar{v}(p_1, r_1)\gamma^\mu u(p_2, r_2), \qquad (9.19)$$

where we have introduced the polarization of the fermions, $r_i$ (the direction of the spin in the rest frame). Often one is interested in unpolarized cross-sections. In that case one has to sum the cross-section over the final polarizations and to average over the initial ones. In other words, we are assuming that both the initial and the final states are in a statistical mixture. Correspondingly we need to evaluate the following quantity

$$X = \frac{1}{4}\sum_{r_i}|\mathcal{M}|^2. \qquad (9.20)$$

Using

$$\gamma_0\gamma_\mu^\dagger\gamma_0 = \gamma_\mu, \qquad (9.21)$$

we can write

$$\mathcal{M}^\star = -ie^2\bar{v}(p_3, r_3)\gamma_\mu u(p_4, r_4)\frac{1}{(p_1 + p_2)^2}\bar{u}(p_2, r_2)\gamma^\mu v(p_1, r_1). \qquad (9.22)$$

The amplitude $\mathcal{M}$ can be expressed in the following form

$$\mathcal{M} = \frac{ie^2}{(p_1 + p_2)^2}A_\mu^{\text{muons}}(r_3, r_4)A_{\text{electrons}}^\mu(r_1, r_2), \qquad (9.23)$$

with

$$A_\mu^{\text{muons}} = \bar{u}(p_4, r_4)\gamma_\mu v(p_3, r_3),$$
$$A_\mu^{\text{electrons}} = \bar{v}(p_1, r_1)\gamma_\mu u(p_2, r_2). \qquad (9.24)$$

Therefore

$$X = \frac{1}{4}\frac{e^4}{(p_1 + p_2)^4}\sum_{r_1, r_2}\left(A_\mu^{\text{electrons}}A_\nu^{\star\ \text{electrons}}\right)\sum_{r_3, r_4}\left(A_{\text{muons}}^\mu A_{\text{muons}}^{\star\ \nu}\right). \qquad (9.25)$$

Defining the quantity

$$A_{\mu\nu}^{\text{electrons}} = \sum_{r_1, r_2}\left(A_\mu^{\text{electrons}}A_\nu^{\star\ \text{electrons}}\right)$$

$$= \sum_{r_1, r_2}\bar{v}(p_1, r_1)\gamma_\mu u(p_2, r_2)\bar{u}(p_2, r_2)\gamma_\nu v(p_1, r_1) \qquad (9.26)$$

and using eqs. (4.279) for the positive and negative energy projectors, we obtain

$$A_{\mu\nu}^{\text{electrons}} = Tr'\left[\frac{\hat{p}_1 - m_e}{2m_e}\gamma_\mu\frac{\hat{p}_2 + m_e}{2m_e}\gamma_\nu\right] \tag{9.27}$$

and the analogous quantity for the muons

$$A_{\mu\nu}^{\text{muons}} = Tr\left[\frac{\hat{p}_4 + m_\mu}{2m_\mu}\gamma_\mu\frac{\hat{p}_3 - m_\mu}{2m_\mu}\gamma_\nu\right]. \tag{9.28}$$

To evaluate the trace of Dirac matrices we will make use of their algebraic properties. Let us start by showing that the trace of an odd number of gamma matrices is zero. In fact for odd $n$

$$Tr\,[\hat{a}_1\cdots\hat{a}_n] = Tr\,[\hat{a}_1\cdots\hat{a}_n\gamma_5\gamma_5] = Tr\,[\gamma_5\hat{a}_1\cdots\hat{a}_n\gamma_5]$$
$$= (-1)^n Tr\,[\hat{a}_1\cdots\hat{a}_n]\,, \tag{9.29}$$

where we have used the cyclic property of the trace and the anticommutativity of $\gamma_5$ and $\gamma_\mu$. Obviously we have

$$Tr[1] = 4. \tag{9.30}$$

Furthermore

$$Tr[\hat{a}\hat{b}] = \frac{1}{2}Tr[\hat{a}\hat{b} + \hat{b}\hat{a}] = \frac{1}{2}a_\mu b_\nu Tr([\gamma^\mu,\gamma^\nu]_+) = 4a_\mu b_\nu g^{\mu\nu} = 4a\cdot b. \tag{9.31}$$

Then, using $\hat{a}\hat{b} = -\hat{b}\hat{a} + 2a\cdot b$ we can evaluate

$$Tr[\hat{a}_1\hat{a}_2\hat{a}_3\hat{a}_4] = Tr[(-\hat{a}_2\hat{a}_1 + 2a_1\cdot a_2)\hat{a}_3\hat{a}_4]$$
$$= -Tr[\hat{a}_2(-\hat{a}_3\hat{a}_1 + 2a_1\cdot a_3)\hat{a}_4] + 8(a_1\cdot a_2)(a_3\cdot a_4)$$
$$= Tr[\hat{a}_2\hat{a}_3(-\hat{a}_4\hat{a}_1 + 2a_1\cdot a_4)] - 8(a_1\cdot a_3)(a_2\cdot a_4) + 8(a_1\cdot a_2)(a_3\cdot a_4)$$
$$= -Tr[\hat{a}_2\hat{a}_3\hat{a}_4\hat{a}_1] + 8(a_1\cdot a_4)(a_2\cdot a_3)$$
$$- 8(a_1\cdot a_3)(a_2\cdot a_4) + 8(a_1\cdot a_2)(a_3\cdot a_4), \tag{9.32}$$

that is

$$Tr[\hat{a}_1\hat{a}_2\hat{a}_3\hat{a}_4] = 4[(a_1\cdot a_2)(a_3\cdot a_4) - (a_1\cdot a_3)(a_2\cdot a_4) + (a_1\cdot a_4)(a_2\cdot a_3)]. \tag{9.33}$$

This relation can be easily extended, by induction, to the trace of $2n$ $\gamma$-matrices. Other useful properties are

$$\gamma_\mu\gamma^\mu = 4, \tag{9.34}$$

$$\gamma_\mu\hat{a}\gamma^\mu = (-\hat{a}\gamma_\mu + 2a_\mu)\gamma^\mu = -2\hat{a} \tag{9.35}$$

and

$$\gamma_\mu\hat{a}\hat{b}\gamma^\mu = 4a\cdot b. \tag{9.36}$$

Let us go back to our process. Evaluating the trace we get

$$A^{\mu\nu}_{\text{electrons}} = \frac{1}{4m_e^2} Tr[(\hat{p}_1 \gamma^\mu \hat{p}_2 \gamma^\nu) - m_e^2 \gamma^\mu \gamma^\nu]$$

$$= \frac{1}{m_e^2} [p_1^\mu p_2^\nu + p_1^\nu p_2^\mu - g^{\mu\nu}(p_1 \cdot p_2 + m_e^2)]. \tag{9.37}$$

Analogously we get

$$A^{\mu\nu}_{\text{muons}} = \frac{1}{m_\mu^2} [p_3^\mu p_4^\nu + p_3^\nu p_4^\mu - g^{\mu\nu}(p_3 \cdot p_4 + m_\mu^2)]. \tag{9.38}$$

Substituting into $X$ we find

$$X = \frac{e^4}{2m_e^2 m_\mu^2 (p_1 + p_2)^4} [(p_1 \cdot p_3)(p_2 \cdot p_4) + (p_1 \cdot p_4)(p_2 \cdot p_3)$$

$$+ m_\mu^2 (p_1 \cdot p_2) + m_e^2 (p_3 \cdot p_4) + 2m_e^2 m_\mu^2]. \tag{9.39}$$

This process is studied in machines, called colliders, where two beams, one of electrons and the other of positrons with the same energy are made to collide, and one looks for a final pair $\mu^+ - \mu^-$. Therefore it is convenient to use the frame of the center of mass for the pair $e^+ - e^-$. We will choose the momentum variables as in Fig. 9.3 with

$$p_1 = (E, \vec{p}), \quad p_2 = (E, -\vec{p}), \quad p_3 = (E, \vec{p}\,'), \quad p_4 = (E, -\vec{p}\,'). \tag{9.40}$$

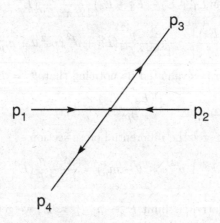

Fig. 9.3 The kinematic for the scattering $e^+ e^- \to \mu^+ \mu^-$.

In this frame the scalar products can be expressed as

$$p_1 \cdot p_3 = p_2 \cdot p_4 = E^2 - pp' \cos\theta, \quad p_1 \cdot p_4 = p_2 \cdot p_3 = E^2 + pp' \cos\theta, \tag{9.41}$$

$$p_1 \cdot p_2 = E^2 + p^2, \qquad p_3 \cdot p_4 = E^2 + p'^2, \qquad (p_1 + p_2)^2 = 4E^2, \qquad (9.42)$$

where $p = |\vec{p}|$ and $p' = |\vec{p}'|$. In order for the process to be kinematically possible we must have $E > m_\mu \approx 200 \, m_e$. We can then neglect $m_e$ with respect to $E$ and $m_\mu$, obtaining

$$X = \frac{e^4}{2m_e^2 m_\mu^2} \frac{1}{16E^4} \left[ (E^2 - pp' \cos\theta)^2 + (E^2 + pp' \cos\theta)^2 + m_\mu^2 (E^2 + p^2) \right]$$

$$= \frac{e^4}{16m_e^2 m_\mu^2} \frac{1}{E^2} [E^2 + p'^2 \cos^2\theta + m_\mu^2], \qquad (9.43)$$

from which

$$d\sigma = \frac{e^4}{16m_e^2 m_\mu^2} \frac{1}{E^2} [E^2 + p'^2 \cos^2\theta + m_\mu^2] (2\pi)^4 \delta^4(p_1 + p_2 - p_3 - p_4)$$

$$\times \frac{1}{4(E^2 + p^2)} (2m_e)^2 (2m_\mu)^2 \frac{d^3p_3 d^3p_4}{(2\pi)^6 4E^2} = \delta^4(p_1 + p_2 - p_3 - p_4)$$

$$\times \frac{e^4}{128\pi^2} \frac{1}{E^6} [E^2 + p'^2 \cos^2\theta + m_\mu^2] d^3p_3 d^3p_4, \qquad (9.44)$$

where we have used $E^2 = p^2$, since we are neglecting the electron mass. We can integrate this expression over $\vec{p}_4$ and $|\vec{p}_3|$ using the conservation of the four-momentum given by the delta function. Using $d^3p_3 = p'^2 dp' d\Omega$ we get

$$\frac{d\sigma}{d\Omega} = \int p'^2 dp' \delta(E_1 + E_2 - E_3 - E_4) \frac{e^4}{128\pi^2} \frac{1}{E^6} (E^2 + p'^2 \cos^2\theta + m_\mu^2)$$

$$= \left[ \frac{\partial(E_3 + E_4)}{\partial p'} \right]^{-1} \frac{e^4}{128\pi^2} \frac{p'^2}{E^6} (E^2 + p'^2 \cos^2\theta + m_\mu^2). \qquad (9.45)$$

The derivative can be evaluated by noticing that $E_3^2 = E_4^2 = m_\mu^2 + p'^2$

$$\frac{\partial(E_3 + E_4)}{\partial p'} = 2\frac{p'}{E}. \qquad (9.46)$$

Using $e^2 = 4\pi\alpha$ we get the differential cross-section

$$\frac{d\sigma}{d\Omega} = \frac{\alpha^2}{8} \frac{p'^2}{E^6} \frac{E}{2p'} (E^2 + p'^2 \cos^2\theta + m_\mu^2) = \frac{\alpha^2}{16E^4} \frac{p'}{E} (E^2 + p'^2 \cos^2\theta + m_\mu^2). \qquad (9.47)$$

In the extreme relativistic limit $E \gg m_\mu$ ($p' \approx E$) we get the expression

$$\frac{d\sigma}{d\Omega} = \frac{\alpha^2}{16E^2} (1 + \cos^2\theta), \qquad (9.48)$$

from which the total cross-section

$$\sigma = \frac{\alpha^2}{16E^2} \int d\Omega (1 + \cos^2\theta) = \frac{\alpha^2}{16E^2} 2\pi \int_{-1}^{1} dw (1 + w^2) = \frac{\alpha^2 \pi}{3E^2}. \qquad (9.49)$$

In the general case we get

$$\sigma = \frac{\alpha^2}{16E^4} \frac{p'}{E} \int d\Omega (E^2 + p'^2 \cos^2\theta + m_\mu^2)$$
$$= \frac{\pi\alpha^2}{4E^4} \frac{p'}{E} \left( E^2 + \frac{1}{3}p'^2 + m_\mu^2 \right). \tag{9.50}$$

In the high energy limit we can easily estimate the total cross-section. Recalling that

$$1 \text{ GeV}^{-2} = 0.389 \text{ mbarn}, \tag{9.51}$$

we get

$$\sigma \approx \frac{5.6 \cdot 10^{-5} (\text{mbarn})}{(E(\text{GeV}))^2} \cdot 0.389 \approx \frac{2.17 \cdot 10^{-5} (\text{mbarn})}{(E(\text{GeV}))^2}, \tag{9.52}$$

or in terms of nanobarns, 1 nbarn $= 10^{-6}$ mbarn,

$$\sigma \approx \frac{20 \text{ (nbarn)}}{(E(\text{GeV}))^2}. \tag{9.53}$$

The cross-section $e^+e^- \to \mu^+\mu^-$ is extremely useful in evaluating the ratio

$$R = \frac{\sigma(e^+e^- \to \text{hadrons})}{\sigma(e^+e^- \to \mu^+\mu^-)}. \tag{9.54}$$

In fact, it is possible to prove that this ratio, barring higher order corrections can be expressed in terms of the charges of the elementary constituents of hadrons, that is the quarks,

$$R = \sum_{\text{quarks with } m<E} Q_q^2 + \text{higher order corrections.} \tag{9.55}$$

The measurement of $R$ has been one of the milestones in showing the existence of quarks and that quarks besides having a fractional electric charge, have another degree of freedom, called **color**. Each quark exists in three different states of this new degree of freedom[3].

Let us comment about the case of massless fermions. As we have seen there is no problem in taking the limit $m_e \to 0$ in the evaluation of the cross-section. The same would be for $m_\mu$. The point is that for any factor $1/m$ arising from the projector there is a corresponding factor $m$ due to the wave function of the external fermions (see eq. (9.15)).

---

[3] For a history of the discovery of QCD and the formulation of the Quantum Chromodynamics (QCD), the theory of the strong interactions, see [Pickering (1984)] and [Peskin and Schroeder (1995)].

## 9.3   Coulomb scattering

Sometimes one can think to the electromagnetic field as an assigned quantity, in that case it will be described by a classical function rather than by an operator. This is the case for the scattering of electrons and positrons from an external field as the Coulomb field of a heavy nucleus. The full electromagnetic field will be, of course, the sum of the classical part and of the quantized part. The expansion of the $S$ matrix is still given by

$$S = 1 + \sum_{n=1}^{\infty} \frac{(i)^n}{n!} \int \cdots \int d^4x_1 \cdots d^4x_n T\left(\mathcal{L}_I(x_1)\cdots\mathcal{L}_I(x_n)\right), \quad (9.56)$$

with

$$\mathcal{L}_I(x) = -e : \bar{\psi}(x)\gamma^\mu\psi(x)[A_\mu(x) + A_\mu^{\text{ext}}(x)] : . \quad (9.57)$$

We will consider here the scattering of an electron off an infinitely heavy

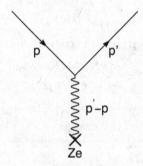

Fig. 9.4   The Feynman diagram for the Coulomb scattering of an electron. The external field is represented by a cross.

nucleus producing a static Coulomb potential. Let us introduce the Fourier transform of this field

$$A_\mu^{\text{ext}}(\vec{x}) = \int \frac{d^3q}{(2\pi)^3} e^{i\vec{q}\cdot\vec{x}} A_\mu^{\text{ext}}(\vec{q}). \quad (9.58)$$

At first order in the external field we have

$$S^{(1)} = -ie \int d^4x : \bar{\psi}(x)\gamma^\mu\psi(x) A_\mu^{\text{ext}}(\vec{x}) : \quad (9.59)$$

and the transition we consider is

$$|i\rangle = |e^-(p,r)\rangle \rightarrow |f\rangle = |e^-(p',s)\rangle. \quad (9.60)$$

This is described by the diagram of Fig. 9.4 with a contribution given by

$$\langle f|S^{(1)}|i\rangle = -ie \int d^4x \, \langle e^-(p',s)|\bar{\psi}^-(x)\gamma^\mu\psi^+(x)A_\mu^{\text{ext}}(\vec{x})|e^-(p,r)\rangle$$

$$= -ie \left(\frac{m}{E_pV}\right)^{1/2} \left(\frac{m}{E_{p'}V}\right)^{1/2} \int d^4x \, e^{ip'x}\bar{u}(p',s)$$

$$\times \int \frac{d^3q}{(2\pi)^3} e^{i\vec{q}\cdot\vec{x}} A_\mu^{\text{ext}}(\vec{q})\gamma^\mu u(p,s)e^{-ipx}$$

$$= -ie \left(\frac{m}{E_pV}\right)^{1/2} \left(\frac{m}{E_{p'}V}\right)^{1/2} (2\pi)\delta(E'-E)$$

$$\times \int \frac{d^3q}{(2\pi)^3} (2\pi)^3\delta^3(\vec{p}+\vec{q}-\vec{p}')\bar{u}(p',s)\gamma^\mu A_\mu^{\text{ext}}(\vec{q})u(p,s)$$

$$= (2\pi)\delta(E_{p'}-E_p)\left(\frac{m}{E_pV}\right)\mathcal{M}, \tag{9.61}$$

with

$$\mathcal{M} = \bar{u}(p',s)(-ie\gamma^\mu)A_\mu^{\text{ext}}(\vec{p}'-\vec{p})u(p,s). \tag{9.62}$$

Notice that in this case the spatial momentum is not conserved. In fact, the presence of the external field violates the translational invariance of the theory and, as a consequence, the nucleus absorbs the momentum $\vec{p}'-\vec{p}$ from the electron. From the previous expression we see also that when there are external fields the Feynman rules are modified, and we have to substitute the wave function of a photon

$$\left(\frac{1}{2E_qV}\right)^{1/2} \epsilon_\mu^{(\lambda)}(q), \tag{9.63}$$

with the Fourier transform of the external field

$$A_\mu^{\text{ext}}(\vec{q}). \tag{9.64}$$

The probability per unit time is given by

$$w = \frac{1}{T}\left|\langle f|S^{(1)}|i\rangle\right|^2 = 2\pi\delta(E'-E)\left(\frac{m}{EV}\right)^2 |\mathcal{M}|^2. \tag{9.65}$$

The expression for the density of final states is $(p' = |\vec{p}'|)$

$$dN_f = \frac{Vd^3p'}{(2\pi)^3} = \frac{Vp'^2dp'd\Omega}{(2\pi)^3}. \tag{9.66}$$

Since $E'^2 = E^2 = p'^2 + m^2 = p^2 + m^2$. we have $p = |\vec{p}| = p'$, and $E'dE' = p'dp'$. Therefore

$$dN_f = V\frac{p'E'dE'}{(2\pi)^3}d\Omega. \tag{9.67}$$

The incoming flux is $v/V = p/VE$, and we get

$$d\sigma = \frac{VwdN_f}{v} = \left(\frac{m}{2\pi}\right)^2 dE'\delta(E'-E)|\mathcal{M}|^2 d\Omega. \qquad (9.68)$$

The differential cross-section is obtained by integrating over the final energy of the electron

$$\frac{d\sigma}{d\Omega} = \left(\frac{m}{2\pi}\right)^2 |\mathcal{M}|^2 = \left(\frac{me}{2\pi}\right)^2 \left|\bar{u}(p',s)\gamma^\mu A_\mu^{\text{ext}}(\vec{q})u(p,r)\right|^2, \qquad (9.69)$$

where $\vec{q} = \vec{p}\,' - \vec{p}$. Averaging over the initial polarizations and summing over the final ones we obtain

$$\frac{d\sigma}{d\Omega} = \left(\frac{me}{2\pi}\right)^2 \frac{1}{2} A_\mu^{\text{ext}}(\vec{q}) A_\nu^{\text{ext}}(\vec{q}) Tr\left[\frac{\hat{p}'+m}{2m}\gamma^\mu\frac{\hat{p}+m}{2m}\gamma^\nu\right]. \qquad (9.70)$$

From the evaluation of the trace we get

$$Tr[...] = \frac{1}{m^2}\left[p'_\mu p_\nu - g_{\mu\nu}(p'\cdot p) + p'_\nu p_\mu + g_{\mu\nu}m^2\right]. \qquad (9.71)$$

Assuming now that the external field is of Coulomb type, we have

$$A_\mu^{\text{ext}}(\vec{x}) = \left(-\frac{Ze}{4\pi|\vec{x}|}, \vec{0}\right) \qquad (9.72)$$

and

$$A_0^{\text{ext}}(\vec{q}) = -\frac{Ze}{|\vec{q}|^2}. \qquad (9.73)$$

We see that we need only the terms with $\mu = \nu = 0$ from the trace

$$Tr[...]_{\mu=\nu=0} = \frac{1}{m^2}[E^2 + m^2 + \vec{p}\cdot\vec{p}\,'] \qquad (9.74)$$

and since $\vec{p}\cdot\vec{p}\,' = p^2\cos\theta$

$$Tr[...]_{\mu=\nu=0} = \frac{1}{m^2}[E^2 + m^2 + p^2\cos\theta]. \qquad (9.75)$$

Then, we get

$$\frac{d\sigma}{d\Omega} = 2\frac{(Z\alpha)^2}{|\vec{q}|^4}(E^2 + m^2 + p^2\cos\theta). \qquad (9.76)$$

Using

$$|\vec{q}|^2 = |\vec{p}\,' - \vec{p}|^2 = 4p^2\sin^2\frac{\theta}{2}, \qquad (9.77)$$

$m^2 = E^2 - |\vec{p}|^2$, and $v = p/E$, we finally obtain

$$\frac{d\sigma}{d\Omega} = 2\frac{(Z\alpha)^2}{(4p^2\sin^2(\theta/2))^2}(E^2 + m^2 + p^2\cos\theta)$$

$$= \frac{(Z\alpha)^2}{4E^2v^4\sin^4(\theta/2)}[1 - v^2\sin^2(\theta/2)]. \qquad (9.78)$$

In the non-relativistic limit $v \ll 1$, $E \approx m$, we find

$$\frac{d\sigma}{d\Omega} = \frac{(Z\alpha)^2}{4m^2\sin^4(\theta/2)}\frac{1}{v^4} = \frac{(Z\alpha)^2}{16T^2\sin^4(\theta/2)}, \qquad (9.79)$$

which is the classical Rutherford formula for the Coulomb scattering with $T = mv^2/2$.

## 9.4 Application to atomic systems

We consider here the interaction of an atomic system with the electromagnetic field. As we have shown, when we have quantized the electromagnetic field, the physical states can be taken as containing only transverse photons. The total system will be described by the following non-relativistic Hamiltonian

$$H = \sum_i \frac{(\vec{p}_i(t) - e\vec{A}(\vec{x}_i(t), t))^2}{2m_i} + \sum_i eA^0(\vec{x}_i(t), t) + H_{\text{em}} = H_0 + H_I, \quad (9.80)$$

where $H_{\text{em}}$ is the the Hamiltonian for the non-interacting electromagnetic field and

$$H_I = -e \sum_i \frac{\vec{p}_i(t) \cdot \vec{A}(\vec{x}_i(t), t)}{m_i} + e^2 \sum_i \frac{\vec{A}(\vec{x}_i(t), t)^2}{2m_i} \quad (9.81)$$

is the interaction term[4]. $H_0$ is the sum of the atomic Hamiltonian (describing the interaction of the electrons with the nucleus) with $H_{\text{em}}$. The sum is over all the electrons in the atom. Notice that the part in $A^0$ describes the interaction of the electrons with the nucleus and it has been included in $H_0$. The possible processes generated at the first order in perturbation theory are represented in Fig. 9.5. Here we will consider only the emission and absorption processes.

Let us start considering the emission process. At the lowest order consider the transition

$$|A, n_k\rangle \rightarrow |A', (n_k + 1)\rangle, \quad (9.82)$$

where $A$ and $A'$ are the labels for the atomic system in the energy eigenstates corresponding to the eigenvalues $E_A$ and $E_{A'}$ respectively. We assume also an initial radiation field with $n_k$ photons each with momentum $\vec{k}$. We want to evaluate the probability for the atom to emit a photon of energy $\omega_k = |\vec{k}|$. From our rules we can evaluate the relevant $S$-matrix element

$$\int_{-\infty}^{+\infty} dt \, ie \, \langle A', (n_k + 1)| \sum_i \vec{v}_i(t) \cdot \vec{A}^{(-)}(\vec{x}_i(t), t)|A, n_k\rangle. \quad (9.83)$$

By using the fact that the operators are in the interaction picture we can write the previous expression as follows

$$\int_{-\infty}^{+\infty} dt \, ie \, \langle A', (n_k+1)|e^{iH_{\text{at}}t} \sum_i \vec{v}_i(0) \cdot \vec{A}^{(-)}(\vec{x}_i(0), t)e^{-iH_{\text{at}}t}|A, n_k\rangle, \quad (9.84)$$

---

[4]We are neglecting the fact that $[\vec{p}, \vec{A}(\vec{x}, t)] \neq 0$. This is because in the approximation that we will use (dipole approximation, see later) $\vec{A}$ is evaluated at $\vec{x} = 0$.

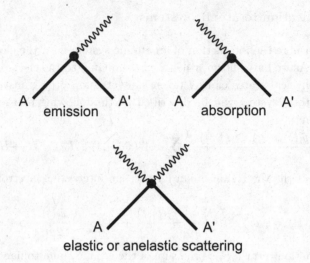

**Fig. 9.5** The possible first order processes for an atomic system interacting with a radiation field, as described by the Hamiltonian interaction of (9.81).

where $H_{at}$ is the part of $H_0$ relative to the atom and acts only on the position and the velocity operators. Inserting the expansion for the field $\vec{A}$ we get

$$\int_{-\infty}^{+\infty} dt\, ie \int \frac{d^3\vec{q}}{\sqrt{2\omega_q V}} \, e^{i(E_{A'} - E_A - \omega_q)t}$$
$$\times \langle A', (n_k + 1)| \sum_i \vec{v}_i(0) \cdot \sum_{\lambda=1,2} \vec{\epsilon}_\lambda^*(\vec{q}) e^{-i\vec{q}\cdot\vec{x}_i(0)} a_\lambda^\dagger(\vec{q})|A, n_k\rangle, \quad (9.85)$$

from which

$$2\pi\delta(E_{A'} - E_A - \omega_k)(ie)\sqrt{\frac{(n_k + 1)}{2\omega_k V}}\langle A'|\sum_i \vec{v}_i(0)\cdot\vec{\epsilon}^*(\vec{k})e^{-i\vec{k}\cdot\vec{x}_i(0)}|A\rangle. \quad (9.86)$$

Here we have used the following properties of the creation and annihilation operators

$$|n\rangle = \frac{a^{\dagger n}}{\sqrt{n!}}|0\rangle, \quad (9.87)$$

$$a^\dagger|n\rangle = \sqrt{n+1}|n+1\rangle, \quad a|n\rangle = \sqrt{n}|n-1\rangle. \quad (9.88)$$

By taking the modulus square of the matrix element, dividing by the time $T$, and multiplying by the final states density we get the differential probability for the emission of one photon

$$dw_e = 2\pi\delta(E_{A'} - E_A - \omega_k)\,e^2(n_k + 1)$$

$$\times |\langle A'| \sum_i \vec{v}_i(0) \cdot \vec{\epsilon}^*(\vec{k}) e^{-i\vec{k}\cdot\vec{x}_i(0)} |A\rangle|^2 \frac{d^3\vec{k}}{2\omega_k(2\pi)^3}. \qquad (9.89)$$

For low frequencies, that is when $|\vec{k}| \ll 1/a$, where $a$ is the typical atomic size, it is possible to make the so-called dipole-approximation,

$$e^{-i\vec{k}\cdot\vec{x}} \approx 1. \qquad (9.90)$$

Furthermore we can use the property

$$\dot{\vec{x}}_{A'A} = -i[H_{\text{at}}, \vec{x}]_{AA'} = -i(E_{A'} - E_A)\vec{x}_{A'A} = -i\omega_k x_{A'A}. \qquad (9.91)$$

Using also $\alpha = e^2/4\pi$ we get

$$\frac{dw_e}{d\Omega}\Big|_{A \to A'} = \frac{\alpha}{2\pi e}(n_k + 1)\omega_k^3 |\langle A'|\vec{d} \cdot \vec{\epsilon}^*(\vec{k})|A\rangle|^2, \qquad (9.92)$$

where we have introduced the dipole-operator

$$\vec{d} = e \sum_i \vec{x}_i(0). \qquad (9.93)$$

If we do not detect the polarization of the emitted photon we have to sum over the polarizations and this can be done by using

$$\sum_{\lambda=1,2} \epsilon_\lambda^i(\vec{k})\,\epsilon_\lambda^{j\,*}(\vec{k}) = \delta_{ij} - \frac{k_i k_j}{|\vec{k}|^2}, \qquad (9.94)$$

following from

$$\sum_{\lambda=0,1,2,3} \epsilon_\lambda^\mu(\vec{k})\,\epsilon_\lambda^{\nu\,*}(\vec{k})g^{\lambda\lambda} = g^{\mu\nu} \qquad (9.95)$$

and recalling that (choosing $n^\mu = (1, \vec{0})$)

$$\epsilon_3^i = \frac{k^i - (n \cdot k)n^i}{(n \cdot k)} = \frac{k^i}{|\vec{k}|}, \quad \epsilon_0^i = n^i = 0. \qquad (9.96)$$

Therefore, by introducing the angle $\theta$ between $\vec{k}$ and $\vec{d}$ we obtain

$$\frac{dw_e}{d\Omega}\Big|_{A \to A'} = \frac{\alpha}{2\pi}(n_k + 1)\omega_k^3 |\vec{d}_{AA'}|^2 \sin^2\theta. \qquad (9.97)$$

Integrating over the solid angle we find the emission probability

$$w_e = \frac{4}{3}\alpha\omega_k^3(n_k + 1)|\vec{d}_{AA'}|^2. \qquad (9.98)$$

In a completely analogous way we find the absorption probability

$$w_a = \frac{4}{3}\alpha\omega_k^3 n_k |\vec{d}_{AA'}|^2. \tag{9.99}$$

We can easily establish the relation between $n_k$ and the intensity per unit volume of a radiation consisting of $n_k$ photons of energy $\omega_k$

$$I(\omega_k)d\omega_k = \frac{n_k}{V}\omega_k 2 \int_{\text{sol. ang.}} dN_f = \frac{n_k}{V}\omega_k \frac{V\omega_k^2 d\omega_k}{\pi^2} = n_k \frac{\omega_k^3}{\pi^2}d\omega_k, \tag{9.100}$$

where we integrate over the solid angle. Here $dN_f$ is the number of states of momentum between $\vec{k}$ and $\vec{k}+d\vec{k}$ and the factor 2 comes from the possible polarizations. We have also used

$$\int_{\text{sol. ang.}} dN_f = \int_{\text{sol. ang.}} \frac{d^3\vec{k}V}{(2\pi)^3} = \frac{\omega_k^2 d\omega_k V}{2\pi^2}. \tag{9.101}$$

We are now in the position to determine the black body radiation law. In fact, inside the block body the thermal equilibrium will be established when

$$\frac{w_e}{w_a} = e^{\omega_k/kT}. \tag{9.102}$$

However we have

$$\frac{w_e}{w_a} = 1 + \frac{1}{n_k}. \tag{9.103}$$

It follows

$$n_k = \frac{1}{e^{\omega_k/kT} - 1}, \tag{9.104}$$

from which

$$I(\omega) = \frac{\omega^3}{\pi^2}\frac{1}{e^{\omega_k/kT} - 1}. \tag{9.105}$$

From the previous results we can also evaluate easily the life-time relative to the radiation emission as

$$\frac{1}{\tau} = w_e = \frac{4}{3}\alpha\omega^3(n+1)|\vec{d}_{AA'}|^2. \tag{9.106}$$

## 9.5    Exercises

(1) Consider the scattering $e^+e^- \to \mu^+\mu^-$ in the limit $E \gg m_e, m_\mu$. Show, at the same order of perturbation theory considered in the text, and using dimensional analysis, that the total cross-section must be of the form

$$\sigma \propto \frac{\alpha^2}{E^2}. \tag{9.107}$$

(2) In a first approximation the proton can be thought as a point-like particle. In this case it can be described in terms of a Dirac field, $\psi_p$. The interaction of the proton with the electromagnetic field is given by

$$\mathcal{L}_I = -e\bar{\psi}_p\gamma^\mu\psi_p A_\lambda. \tag{9.108}$$

Consider the process

$$e^-p \to e^-p \tag{9.109}$$

in the reference frame where the initial proton is at rest. Then, consider the following two cases:

(a) The electron energy $E \ll m$, where $m$ is the mass of the electron.

(b) The electron energy $E$ is of the same order or bigger than the mass of the proton, $M$ and therefore we can neglect $m$.

Evaluate the differential cross-section for the scattering $ep$ showing that in the first case it coincides with the differential cross-section obtained for the Coulomb scattering with a fixed potential.

(3) Consider the Compton scattering $e^-\gamma \to e^-\gamma$ for unpolarized initial photons and electrons. Assume that the final polarizations are undetected. Evaluate the analog of the quantity $X$ defined in eq. (9.20). The result obtained in eq. (8.147) of Section 8.6 will be useful. Neglecting the mass of the electron, express the result in terms of the Mandelstam variables:

$$s = (p_i + k_i)^2, \quad t = (p_f - p_i)^2, \quad u = (p_f - k_i)^2. \tag{9.110}$$

Notice also, that these variables are not independent, since

$$s + t + u = 0. \tag{9.111}$$

(4) Under the same conditions of the previous exercise evaluate the quantity $X$ for the pair annihilation process $e^+e^- \to 2\gamma$. The result is related to the one obtained in the exercise (3) via the so-called **crossing symmetry**, $p_i \to -p_f$ and $k_i \to -k_f$. Explain why.

# Chapter 10

# One-loop renormalization

## 10.1 Divergences of the Feynman integrals

Let us consider again the Coulomb scattering. We will expand the $S$ matrix up to the third order in the electric charge, assuming that the external field is weak enough so we can consider it at first order. Then, the relevant Feynman diagrams are the ones in Fig. 10.1. The Coulomb scattering can be used, in principle, to define the physical electric charge of the electron. This is done assuming that the amplitude is linear in $e_{\text{phys}}$, from which we get an expansion of the type

$$e_{\text{phys}} = e + a_2 e^3 + \cdots = e(1 + a_2 e^2 + \cdots), \qquad (10.1)$$

in terms of the parameter $e$ which appears in the original Lagrangian. The first problem we encounter is that we would like to express the results of our calculations in terms of measured quantities as, for instance, the electric charge, $e_{\text{phys}}$. This could be done by inverting the previous expansion but, and here comes the second problem, the coefficients of the expansion are divergent quantities. To show this, consider, for instance, the self-energy contribution to one of the external electrons (as the one in Fig. 10.1a). We have

$$\mathcal{M}_a = \bar{u}(p')(-ie\gamma^\mu)A_\mu^{\text{ext}}(\vec{p}\,' - \vec{p})\frac{i}{\hat{p} - m + i\epsilon}\left[ie^2\Sigma(p)\right]u(p), \qquad (10.2)$$

where (recall that we are using the Lorenz gauge for the electromagnetic field)

$$ie^2\Sigma(p) = \int \frac{d^4k}{(2\pi)^4}\,(-ie\gamma_\mu)\frac{-ig^{\mu\nu}}{k^2 + i\epsilon}\frac{i}{\hat{p} - \hat{k} - m + i\epsilon}(-ie\gamma_\nu)$$

$$= -e^2\int \frac{d^4k}{(2\pi)^4}\,\gamma_\mu\frac{1}{k^2 + i\epsilon}\frac{1}{\hat{p} - \hat{k} - m + i\epsilon}\gamma^\mu, \qquad (10.3)$$

223

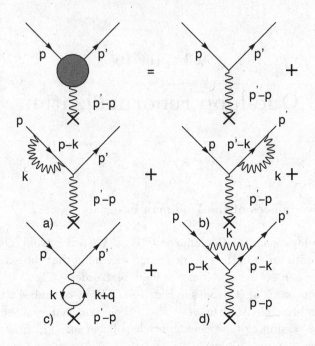

Fig. 10.1  The Feynman diagrams for the Coulomb scattering at the third order in the electric charge and at the first order in the external field, denoted by the cross.

that is

$$\Sigma(p) = i \int \frac{d^4k}{(2\pi)^4} \, \gamma_\mu \frac{\hat{p} - \hat{k} + m}{(p-k)^2 - m^2 + i\epsilon} \gamma^\mu \frac{1}{k^2 + i\epsilon}. \qquad (10.4)$$

For large momentum, $k$, the integrand behaves as $1/k^3$ and the integral diverges linearly. Analogously, one can check that all the other third order contributions are divergent. Let us write explicitly the amplitudes for the other diagrams

$$\mathcal{M}_b = \bar{u}(p')ie^2\Sigma(p')\frac{i}{\hat{p}' - m + i\epsilon}(-ie\gamma^\mu)A_\mu^{\text{ext}}(\vec{p}' - \vec{p})u(p), \qquad (10.5)$$

$$\mathcal{M}_c = \bar{u}(p')(-ie\gamma^\mu)\frac{-ig_{\mu\nu}}{q^2 + i\epsilon}ie^2\Pi^{\nu\rho}(q)A_\rho^{\text{ext}}(\vec{p}' - \vec{p})u(p), \quad q = p' - p, \quad (10.6)$$

where

$$ie^2\Pi^{\mu\nu}(q) = (-1)\int \frac{d^4k}{(2\pi)^4}$$

$$\times Tr\left[\frac{i}{\hat{k} + \hat{q} - m + i\epsilon}(-ie\gamma^\mu)\frac{i}{\hat{k} - m + i\epsilon}(-ie\gamma^\nu)\right], \quad (10.7)$$

(the minus sign originates from the fermion loop) and therefore

$$\Pi^{\mu\nu}(q) = i \int \frac{d^4k}{(2\pi)^4} \, Tr \left[ \frac{1}{\hat{k} + \hat{q} - m} \gamma^\mu \frac{1}{\hat{k} - m + i\epsilon} \gamma^\nu \right]. \tag{10.8}$$

The last contribution is

$$\mathcal{M}_d = \bar{u}(p')(-ie)e^2 \Lambda^\mu(p', p) u(p) A_\mu^{\text{ext}}(\vec{p}\,' - \vec{p}), \tag{10.9}$$

where

$$e^2 \Lambda^\mu(p', p) = \int \frac{d^4k}{(2\pi)^4} \, (-ie\gamma^\alpha) \frac{i}{\hat{p}\,' - \hat{k} - m + i\epsilon} \gamma^\mu$$
$$\times \frac{i}{\hat{p} - \hat{k} - m + i\epsilon} (-ie\gamma^\beta) \frac{-ig_{\alpha\beta}}{k^2 + i\epsilon}, \tag{10.10}$$

or

$$\Lambda^\mu(p', p) = -i \int \frac{d^4k}{(2\pi)^4} \, \gamma^\alpha \frac{1}{\hat{p}\,' - \hat{k} - m + i\epsilon} \gamma^\mu$$
$$\times \frac{1}{\hat{p} - \hat{k} - m + i\epsilon} \gamma_\alpha \frac{1}{k^2 + i\epsilon}. \tag{10.11}$$

The problem of the divergences is a serious one and in order to give some sense to the theory we have to find a way to define our integrals. This is what is called the regularization procedure of the Feynman integrals, consisting in formulating a prescription in order to make the integrals finite. This can be done in various ways, for instance, introducing an ultraviolet cut-off or, as we shall see later, by the more convenient method of dimensional regularization. However, the theory should not depend on the way in which we define the integrals, otherwise we would have to assign some physical meaning to the regularization procedure. This brings us to the other problem, the inversion of eq. (10.1). Since, after regularization, the coefficients are finite, we can proceed to perform the inversion, obtaining $e$ as a function of $e_{\text{phys}}$ and therefore all the other observables in terms of the physical electric charge (that is the one measured in the Coulomb scattering). By doing so we introduce in the observables a dependence on the renormalization procedure. When this dependence cancels out, we say that the theory is **renormalizable**. Thinking of the regularization in terms of a cut-off, this means that considering the observable quantities in terms of $e_{\text{phys}}$ and removing the cut-off, that is taking the limit for the cut-off going to the infinity, the result should be finite. Of course, this cancellation is not obvious at all, and in fact in most of the theories this does not happen. However there is a restricted class of renormalizable theories as, for instance, QED. We will not discuss the renormalization at all order, or prove which criteria a theory should satisfy in order to be renormalizable. We will give these criteria without proof, but we will try to justify them on a physical basis. As far as QED is concerned we will study in detail the renormalization at one-loop level.

The previous way of defining a renormalizable theory amounts to saying that the original parameters in the Lagrangian, as $e$, should be infinite and that their divergences should compensate the divergences of the Feynman diagrams. Then one can try to separate the infinite from the finite part of the parameters (however this separation is ambiguous, see later). The infinite contributions are called counterterms, and by definition they have the same operator structure of the original terms in the Lagrangian. Therefore, the procedure of regularization can be performed by adding to the original Lagrangian counterterms cooked in such a way that their contribution kills the divergent part of the Feynman integrals. This means that the coefficients of these counterterms have to be infinite when we remove the regularization. We see that the theory will be renormalizable if the counterterms we add to make the theory finite have the same structure of the original terms in the Lagrangian. In fact, if this is the case, they can be absorbed by the original parameters giving rise to finite values. However the final, finite parameters are arbitrary, since they have to be fixed by the experiments (renormalization conditions).

In the case of QED all the divergences can be brought back to the three functions $\Sigma(p)$, $\Pi_{\mu\nu}(q)$ and $\Lambda^{\mu}(p',p)$ (see later in Section 10.4). This does not mean that an arbitrary diagram is not divergent, but it can be made finite if the previous functions are finite. In such a case one has only to show that eliminating these three divergences (primitive divergences) the theory is automatically finite. In particular we will show that the divergent part of $\Sigma(p)$ can be absorbed into the definition of the mass of the electron and a redefinition of the electron field (wave function renormalization). The divergence in $\Pi_{\mu\nu}$, the photon self-energy, can be absorbed in the wave function renormalization of the photon (the mass of the photon is not renormalized due to the gauge invariance). And finally the divergence of $\Lambda_{\mu}(p',p)$ goes into a re-definition of the parameter $e$. To realize this program we divide the Lagrangian density into two parts, one written in terms of the physical parameters and the other containing the counterterms. We will call the original parameters and fields of the theory, bare parameters and bare fields. We will make use of an index $B$ in order to distinguish the bare quantities from the physical ones. Therefore the two pieces of the Lagrangian density should look as follows: the piece in terms of the physical parameters (from now on we will omit the subscript, phys, from the physical parameters) $\mathcal{L}_{\mathrm{p}}$

$$\mathcal{L}_{\mathrm{p}} = \bar{\psi}(i\hat{\partial} - m)\psi - e\bar{\psi}\gamma_\mu\psi A^\mu - \frac{1}{4}F_{\mu\nu}F^{\mu\nu} - \frac{1}{2}(\partial_\mu A^\mu)^2 \qquad (10.12)$$

and the counterterms part $\mathcal{L}_{\mathrm{c.t.}}$

$$\mathcal{L}_{\mathrm{c.t.}} = iB\bar{\psi}\hat{\partial}\psi - A\bar{\psi}\psi - \frac{C}{4}F_{\mu\nu}F^{\mu\nu} - \frac{E}{2}(\partial_\mu A^\mu)^2 - eD\bar{\psi}\hat{A}\psi. \qquad (10.13)$$

Of course, the sum of these two contributions should coincide with the original Lagrangian written in terms of the bare quantities. Adding together $\mathcal{L}_{\mathrm{p}}$ and $\mathcal{L}_{\mathrm{c.t.}}$, we get

$$\mathcal{L} = (1 + B)i\bar{\psi}\hat{\partial}\psi - (m + A)\bar{\psi}\psi - e(1 + D)\bar{\psi}\gamma_\mu\psi A^\mu - \frac{1 + C}{4}F_{\mu\nu}F^{\mu\nu}$$

$$+ \text{gauge-fixing}, \qquad (10.14)$$

where, for the sake of simplicity, we have omitted the gauge fixing term. Defining the renormalization constant of the fields

$$Z_1 = (1 + D), \qquad Z_2 = (1 + B), \qquad Z_3 = (1 + C), \qquad (10.15)$$

we write the bare fields as

$$\psi_B = Z_2^{1/2}\psi, \qquad A_B^\mu = Z_3^{1/2}A^\mu, \qquad (10.16)$$

obtaining

$$\mathcal{L} = i\bar{\psi}_B\hat{\partial}\psi_B - \frac{m + A}{Z_2}\bar{\psi}_B\psi_B - \frac{eZ_1}{Z_2 Z_3^{1/2}}\bar{\psi}_B\gamma_\mu\psi_B A_B^\mu - \frac{1}{4}F_{B,\mu\nu}F_B^{\mu\nu}$$

$$+ \text{gauge-fixing}. \qquad (10.17)$$

Defining

$$m_B = \frac{m + A}{Z_2}, \qquad e_B = \frac{eZ_1}{Z_2 Z_3^{1/2}}, \qquad (10.18)$$

we get the initial Lagrangian expressed in terms of the bare fields and bare parameters

$$\mathcal{L} = i\bar{\psi}_B\hat{\partial}\psi_B - m_B\bar{\psi}_B\psi_B - e_B\bar{\psi}_B\gamma_\mu\psi_B A_B^\mu - \frac{1}{4}F_{B,\mu\nu}F_B^{\mu\nu} + \text{gauge-fixing}. \qquad (10.19)$$

Since the finite piece can be fixed in terms of the observable quantities, the division of the parameters in physical and counterterm part is well defined. This way of fixing the finite part of the Lagrangian gives rise to the renormalization conditions. The counterterms $A$, $B$, ... are determined recursively at each perturbative order in such a way to eliminate the divergent parts and to respect the renormalization conditions. We will see later how this works in practice at one-loop level. Another observation is that $Z_1$ and $Z_2$ have to do with the self-energy of the electron, and as such they depend on the electron mass. Therefore, if we consider the theory for a different particle, as the muon, which has the same interactions as the electron and differs only for the value of the mass $(m_\mu \approx 200\, m_e)$, one would get a different electric charge. Or, phrased in a different way, one would have to tune the bare electric charges of $e$ and $\mu$ at different values, in order to get the same physical charges. This looks very unnatural, but the gauge invariance of the theory implies that at any order in perturbation theory $Z_1 = Z_2$. As a consequence $e_B = e/Z_3^{1/2}$, and since $Z_3$ comes from the photon self-energy, the relation between the bare and the physical electric charge is universal (that is, it does not depend on the kind of charged particle under consideration).

Summarizing, we start dividing up the bare Lagrangian in two pieces. Then we regularize the theory giving some prescription to get finite Feynman integrals. The part containing the counterterms is determined, order by order, by requiring that the divergences of the Feynman integrals, which come about when removing the regularization, be cancelled out by the counterterm contributions. Since the separation of an infinite quantity into an infinite plus a finite term is not well defined, we use the renormalization conditions to fix the finite part. After evaluating a physical quantity we remove the regularization. Notice that although the counterterms are divergent quantities, when we remove the cut-off, we can order them according to the power of the coupling in which we are doing the perturbative calculation. That is to say, we have to do with a double limit, one in the coupling and the other in some parameter, regulator, which defines the regularization. The order of the limits is first to work at some order in the coupling, at fixed regulator, and then remove the regularization.

Before going into the calculations for QED we want to illustrate some general result about the renormalization. If one considers a theory involving scalar, fermion and massless spin 1 (as the photon) fields, it is not difficult to construct an algorithm which allows to evaluate the ultraviolet (that is for large momenta) divergence of any Feynman diagram. For example, in the case of the electron self-energy (see Figs. 10.1a and 10.1b) one has an integration over the four-momentum $p$ and a behavior of the integrand, coming from the propagators, as $1/p^3$, giving rise to a linear divergence (as we shall see later the divergence is only logarithmic). From this counting one can see that *the only terms in the Lagrangian density containing monomials in the fields with mass dimension smaller than or equal to the number of space-time dimensions are renormalizable.* Theories satisfying this criterium are renormalizable (barring some small technicalities). The mass dimensions of the fields can be easily evaluated from the observation that the action is dimensionless in our units ($\hbar = 1$). Therefore, in $n$ space-time dimensions, the Lagrangian density, defined as

$$\int d^n x \, \mathcal{L} \tag{10.20}$$

has a mass dimension $n$. Looking at the kinetic terms of the bosonic fields (two derivatives) and of the fermionic fields (one derivative), we see that

$$\dim[\phi] = \dim[A_\mu] = \frac{n}{2} - 1, \quad \dim[\psi] = \frac{n-1}{2}. \tag{10.21}$$

In particular, in 4 dimensions the bosonic fields have dimension 1 and the fermionic 3/2. Then, we see that QED is renormalizable, since all the terms in the Lagrangian density have dimensions smaller than or equal to 4

$$\dim[\bar{\psi}\psi] = 3, \quad \dim[\bar{\psi}\gamma_\mu\psi A^\mu] = 4. \tag{10.22}$$

The condition on the dimensions of the operators appearing in the Lagrangian can be translated into a condition on the coupling constants. In fact, each monomial $\mathcal{O}_i$ will appear multiplied by a coupling $g_i$

$$\mathcal{L} = \sum_i g_i \mathcal{O}_i, \tag{10.23}$$

with

$$\dim[g_i] = n - \dim[\mathcal{O}_i]. \tag{10.24}$$

The renormalizability requires

$$\dim[\mathcal{O}_i] \leq n, \tag{10.25}$$

from which

$$\dim[g_i] \geq 0, \tag{10.26}$$

that is the couplings must have positive dimension in mass or be dimensionless. Therefore the renormalizability criterium can also be formulated in the following way: *only terms in the Lagrangian density containing couplings with positive mass dimension, $[g_i] \geq 0$, are renormalizable.*

In QED the only couplings are the mass of the electron and the electric charge which is dimensionless. As a further example consider a single scalar field. The most general renormalizable Lagrangian density is characterized by three parameters

$$\mathcal{L} = \frac{1}{2}\partial_\mu\phi\partial^\mu\phi - \frac{1}{2}m^2\phi^2 - \rho\phi^3 - \lambda\phi^4. \tag{10.27}$$

Here $m^2$ has dimension 2, $\rho$ has dimension 1 and $\lambda$ is dimensionless. In particular, linear $\sigma$-models are renormalizable theories. We will call theories with all the couplings such that $[g_i] \geq 0$ and all the possibly divergent operators present in the original Lagrangian, **strictly renormalizable**.

Giving these facts, let us try to understand what makes different renormalizable from non-renormalizable theories. In the renormalizable case the only divergent diagrams are the ones corresponding to the processes described by the operators appearing in the Lagrangian. Therefore adding to $\mathcal{L}$ the counterterm

$$\mathcal{L}_{\text{c.t.}} = \sum_i \delta g_i \mathcal{O}_i, \tag{10.28}$$

we can choose the $\delta g_i$ in such a way to cancel, order by order, the divergences. The theory depends on a finite number of arbitrary parameters equal to the number of parameters $g_i$. Therefore the theory is a predictive one. In the non-renormalizable case, the number of divergent diagrams increases with the perturbative order. At each order we have to introduce new counterterms having an operator structure different from the original one. Therefore the theory depends on an infinite number of arbitrary parameters. As an example, consider a fermionic theory with an interaction of the type $(\bar{\psi}\psi)^2$. Since this term has dimension 6, the relative coupling has dimension $-2$:

$$\mathcal{L}_{\text{int}} = -g_2(\bar{\psi}\psi)^2. \tag{10.29}$$

At one loop the theory gives rise to the divergent diagrams of Fig. 10.2. The divergence of the first diagram can be absorbed into a counterterm of the original type

$$-\delta g_2(\bar{\psi}\psi)^2. \tag{10.30}$$

The other two diagrams need counterterms of the type

$$\delta g_3(\bar{\psi}\psi)^3 + \delta g_4(\bar{\psi}\psi)^4. \tag{10.31}$$

These counterterms give rise to new one-loop divergent diagrams, as for instance the ones in Fig. 10.3. The first diagram modifies the already introduced counterterm $(\bar{\psi}\psi)^3$, but the second one needs a new counterterm

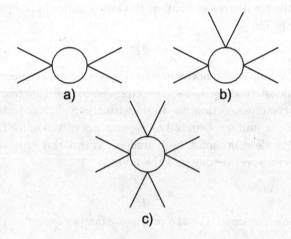

Fig. 10.2   Divergent diagrams coming from the interaction $(\bar\psi\psi)^2$.

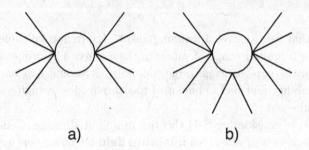

Fig. 10.3   Divergent diagrams coming from the interactions $(\bar\psi\psi)^2$ and $(\bar\psi\psi)^3$.

$$\delta g_5(\bar\psi\psi)^5. \tag{10.32}$$

This process never ends.

The renormalization requirement restricts in a fantastic way the possible field theories. However, one could think that this requirement is too technical and one could imagine other ways of giving a meaning to Lagrangians which do not satisfy this condition. This can be done but at the expense of limiting the validity of the theory within a well defined range of energy. Consider again a theory with a four-fermion interaction. Since dim $g_2 = -2$, if we consider the scattering $\psi + \psi \to \psi + \psi$ in the high energy limit (where we can neglect all the masses), on dimensional ground we get that the total cross-section, at the lowest order in the perturbation theory, behaves like

$$\sigma \approx g_2^2 E^2. \tag{10.33}$$

Analogously, in any non-renormalizable theory, due to the presence of couplings with negative dimensions, the cross-section will increase with the energy. But the cross-section has to do with the $S$ matrix which must be unitary. Since a unitary matrix has eigenvalues of modulus 1, it follows that its matrix elements must be bounded. Translating this argument in terms of the cross-section one gets the bound

$$\sigma \leq \frac{c^2}{E^2}, \tag{10.34}$$

where $c$ is some constant. For the previous example we get

$$g_2 E^2 \leq c. \tag{10.35}$$

This implies a violation of the $S$ matrix unitarity at energies such that

$$E \geq \sqrt{\frac{c}{g_2}}. \tag{10.36}$$

It follows that we can give a meaning also to non-renormalizable theories, but only for a limited range of values of the energy. This range is fixed by the value of the non-renormalizable couplings. It should be clear that non renormalizability and bad behavior of the amplitudes at high energies are strictly connected.

The idea of considering field theories in a limited range of energy leads to the modern view of **effective quantum field theories** (see for example, [Weinberg (1995b)]).

## 10.2 Dimensional regularization of the Feynman integrals

As we have discussed in the previous Section we need a procedure to give sense to the otherwise divergent Feynman integrals. The simplest of these procedures is to introduce a cut-off $\Lambda$ and define a divergent integral as

$$\int_0^\infty \rightarrow \lim_{\Lambda \to \infty} \int_0^\Lambda. \tag{10.37}$$

Of course, in the spirit of renormalization, we have first to perform the perturbative expansion and then take the limit over the cut-off. Although this procedure is very simple, it is inadequate in situations like gauge theories.

In fact, one can show that the cut-off regularization breaks the translational invariance and this creates problems in gauge theories. One kind of regularization, which nowadays is very popular, is the dimensional regularization. This consists in considering the integration in an arbitrary number of space-time dimensions and taking the limit of four dimensions at the end. Dimensional regularizing is very convenient because it respects all the usual symmetries. In fact, except for a very limited number of cases, the symmetries do not depend on the number of space-time dimensions. Let us now illustrate dimensional regularization. We want to evaluate integrals of the type

$$I_4(k) = \int d^4p \, F(p,k), \tag{10.38}$$

with $F(p,k) \approx p^{-2}$ or $p^{-4}$. The idea is that integrating on a lower number of dimensions improves the convergence properties of the integral in the ultraviolet. For instance, if $F(p,k) \approx p^{-4}$, the integral is convergent in 2 and in 3 dimensions. Therefore, we would like to introduce a quantity

$$I(\omega, k) = \int d^{2\omega}p \, F(p,k), \tag{10.39}$$

to be regarded as a function of the complex variable $\omega$. Then, if we can define a complex function $I'(\omega, k)$ on the entire complex plane, with definite singularities, and such to coincide with $I$ on some common domain, then by analytic continuation, $I$ and $I'$ define the same analytic function. A simple example of this procedure is given by Euler's Gamma function. This complex function is defined for $Re \, z > 0$ by the integral representation

$$\Gamma(z) = \int_0^\infty dt \, e^{-t}t^{z-1}. \tag{10.40}$$

If Re $z \leq 0$, the integral diverges as

$$\frac{dt}{t^{1+|Re \, z|}}, \tag{10.41}$$

in the limit $t \to 0$. However it is easy to get a representation valid also for $Re \, z \leq 0$. Let us divide the integration region in two parts defined by a parameter $\alpha$

$$\Gamma(z) = \int_0^\alpha dt \, e^{-t}t^{z-1} + \int_\alpha^\infty dt \, e^{-t}t^{z-1}. \tag{10.42}$$

Expanding the exponential in the first integral and integrating term by term we get

$$\Gamma(z) = \sum_{n=0}^\infty \frac{(-1)^n}{n!} \int_0^\alpha dt \, t^{n+z-1} + \int_\alpha^\infty dt \, e^{-t}t^{z-1}$$

$$= \sum_{n=0}^\infty \frac{(-1)^n}{n!} \frac{\alpha^{n+z}}{n+z} + \int_\alpha^\infty dt \, e^{-t}t^{z-1}. \tag{10.43}$$

The second integral converges for any $z$ since $\alpha > 0$. This expression coincides, for $Re\, z > 0$, with the representation (10.40) for the $\Gamma$ function, but it is defined also for $Re\, z < 0$ where it has simple poles located at $z = -n$. Therefore it is a meaningful expression on all the complex $z$-plane. Notice that in order to isolate the divergences we need to introduce an arbitrary parameter $\alpha$. However the result does not depend on the particular value of this parameter. This is the Weierstrass representation of the Euler $\Gamma(z)$. From this example we see that we need to perform the following three steps

- Find a domain where $I(\omega, k)$ is convergent. Typically this will be for $Re\, \omega < 2$.
- Construct an analytic function identical to $I(\omega, k)$ in the domain of convergence, but defined on a larger domain including the point $\omega = 2$.
- At the end of the calculation take the limit $\omega \to 2$.

## 10.3    Integration in arbitrary dimensions

Let us consider the integral

$$I_N = \int d^N p F(p^2),\tag{10.44}$$

with $N$ an integer number and $p$ a vector in a Euclidean $N$-dimensional space. Since the integrand is invariant under rotations of the $N$-dimensional vector $p$, we can perform the angular integration by means of

$$d^N p = d\Omega_N p^{N-1} dp,\tag{10.45}$$

where $d\Omega_N$ is the element of solid angle in $N$ dimensions. Therefore $\int d\Omega_N = S_N$, with $S_N$ the surface of the unit sphere in $N$ dimensions. Then

$$I_N = S_N \int_0^\infty p^{N-1} F(p^2) dp.\tag{10.46}$$

The value of the surface of the sphere, $S_N$, can be evaluated by the following trick. Consider

$$I = \int_{-\infty}^{+\infty} e^{-x^2} dx = \sqrt{\pi}.\tag{10.47}$$

By taking $N$ of these factors we get

$$I^N = \int dx_1 \cdots dx_N e^{-(x_1^2 + \cdots + x_N^2)} = \pi^{N/2}.\tag{10.48}$$

The same integral can be evaluated in polar coordinates

$$\pi^{N/2} = S_N \int_0^\infty \rho^{N-1} e^{-\rho^2} d\rho. \tag{10.49}$$

By putting $x = \rho^2$ we have

$$\pi^{N/2} = \frac{1}{2} S_N \int_0^\infty x^{N/2-1} e^{-x} dx = \frac{1}{2} S_N \Gamma\left(\frac{N}{2}\right), \tag{10.50}$$

where we have used the representation of the Euler $\Gamma$ function given in the previous Section. Therefore

$$S_N = \frac{2\pi^{N/2}}{\Gamma\left(\frac{N}{2}\right)} \tag{10.51}$$

and

$$I_N = \frac{\pi^{N/2}}{\Gamma\left(\frac{N}{2}\right)} \int_0^\infty x^{N/2-1} F(x) dx, \tag{10.52}$$

with $x = p^2$.

The integrals we will be interested in are of the type

$$I_N^{(M)} = \int \frac{d^N p}{(p^2 - a^2 + i\epsilon)^A}, \tag{10.53}$$

with $p$ a vector in an $N$-dimensional Minkowski space. We have also used the index $(M)$ to distinguish explicitly this integral, defined in Minkowski space, from an analogous integral in the Euclidean space (see later). We can perform an anti-clockwise rotation of 90 degrees (Wick's rotation) in the complex plane of $p_0$ without hitting any singularity. Then we perform a change of variable $p_0 \to ip_0$ obtaining

$$I_N^{(M)} = i \int \frac{d^N p}{(-p^2 - a^2)^A} = i(-1)^A I_N, \tag{10.54}$$

where $I_N$ is a Euclidean integral of the kind discussed at the beginning of this Section with $F(x)$ given by

$$F(x) = (x + a^2)^{-A}. \tag{10.55}$$

Therefore we have to evaluate

$$I_N = \frac{\pi^{N/2}}{\Gamma\left(\frac{N}{2}\right)} \int_0^\infty \frac{x^{N/2-1}}{(x + a^2)^A} dx. \tag{10.56}$$

By putting $x = a^2 y$ we get

$$I_N = (a^2)^{N/2-A} \frac{\pi^{N/2}}{\Gamma\left(\frac{N}{2}\right)} \int_0^\infty y^{N/2-1} (1 + y)^{-A} dy \tag{10.57}$$

and recalling the integral representation for the Euler $B(x,y)$ function (valid for $Re\, x, y > 0$)

$$B(x,y) = \frac{\Gamma(x)\Gamma(y)}{\Gamma(x+y)} = \int_0^\infty t^{x-1}(1+t)^{-(x+y)}dt, \qquad (10.58)$$

it follows

$$I_N = \pi^{N/2}\frac{\Gamma(A-N/2)}{\Gamma(A)}\frac{1}{(a^2)^{A-N/2}}. \qquad (10.59)$$

We have obtained this representation for $N/2 > 0$ and $Re(A - N/2) > 0$. But we know how to extend the Euler Gamma function to the entire complex plane, and therefore we can extend this formula to complex dimensions $N = 2\omega$

$$I_{2\omega} = \pi^\omega\frac{\Gamma(A-\omega)}{\Gamma(A)}\frac{1}{(a^2)^{A-\omega}}. \qquad (10.60)$$

This shows that $I_{2\omega}$ has simple poles located at

$$\omega = A, A+1, \cdots. \qquad (10.61)$$

Therefore our integral is defined in all the $\omega$ complex plane excluding the points (10.61). At the end we will have to consider the limit $\omega \to 2$. The original integral in Minkowski space is then

$$\int \frac{d^{2\omega}p}{(p^2 - a^2)^A} = i\pi^\omega(-1)^A\frac{\Gamma(A-\omega)}{\Gamma(A)}\frac{1}{(a^2)^{A-\omega}}. \qquad (10.62)$$

For the following it will be useful to derive another formula. Let us put in the previous equation $p = p' + k$

$$\int \frac{d^{2\omega}p'}{(p'^2 + 2p'\cdot k + k^2 - a^2)^A} = i\pi^\omega(-1)^A\frac{\Gamma(A-\omega)}{\Gamma(A)}\frac{1}{(a^2)^{A-\omega}}. \qquad (10.63)$$

Defining $b^2 = -a^2 + k^2$ we find

$$\int \frac{d^{2\omega}p}{(p^2 + 2p\cdot k + b^2)^A} = i\pi^\omega(-1)^A\frac{\Gamma(A-\omega)}{\Gamma(A)}\frac{1}{(k^2 - b^2)^{A-\omega}}. \qquad (10.64)$$

Differentiating this expression with respect to $k_\mu$ we get various useful relations as

$$\int d^{2\omega}p\frac{p_\mu}{(p^2 + 2p\cdot k + b^2)^A} = i\pi^\omega(-1)^A\frac{\Gamma(A-\omega)}{\Gamma(A)}\frac{-k_\mu}{(k^2 - b^2)^{A-\omega}} \qquad (10.65)$$

and

$$\int d^{2\omega}p\frac{p_\mu p_\nu}{(p^2 + 2p\cdot k + b^2)^A} = i\frac{\pi^\omega(-1)^A}{\Gamma(A)(k^2 - b^2)^{A-\omega}}$$
$$\times \left[\Gamma(A-\omega)k_\mu k_\nu - \frac{1}{2}g_{\mu\nu}(k^2 - b^2)\Gamma(A-\omega-1)\right]. \qquad (10.66)$$

Since at the end of our calculations we will have to take the limit $\omega \to 2$, it will be useful to recall the expansion of the Gamma function around its poles

$$\Gamma(\epsilon) = \frac{1}{\epsilon} - \gamma + \mathcal{O}(\epsilon), \tag{10.67}$$

where

$$\gamma = 0.5772... \tag{10.68}$$

is the Euler-Mascheroni constant, and for $(n \geq 1)$:

$$\Gamma(-n + \epsilon) = \frac{(-1)^n}{n!} \left[ \frac{1}{\epsilon} + \psi(n+1) + \mathcal{O}(\epsilon) \right], \tag{10.69}$$

with

$$\psi(n+1) = 1 + \frac{1}{2} + \cdots + \frac{1}{n} - \gamma. \tag{10.70}$$

## 10.4 One-loop regularization of QED

We will show now that at one-loop the divergent contributions come only from the three diagrams discussed before. In fact, let us define $D$, the superficial degree of divergence of a one-loop diagram, as the difference between 4 (the number of integration over the momentum) and the number of powers of momentum coming from the propagators (we assume here that no powers of momenta are coming from the vertices, which is true in QED, but not for scalar electrodynamics. We get

$$D = 4 - I_F - 2I_B \tag{10.71}$$

where $I_F$ and $I_B$ are the number of internal fermionic and bosonic lines. Going through the loop one has to have a number of vertices, $V$ equal to the number of internal lines

$$V = I_F + I_B. \tag{10.72}$$

Always for the one-loop diagrams we have

$$2V = 2I_F + E_F, \quad V = 2I_B + E_B \tag{10.73}$$

with $E_F$ and $E_B$ the number of external fermionic and bosonic lines. From these equations we can get $V$, $I_F$ and $I_B$ in terms of $E_F$ and $E_B$

$$I_F = \frac{1}{2}E_F + E_B, \quad I_B = \frac{1}{2}E_F, \quad V = E_F + E_B \tag{10.74}$$

from which

$$D = 4 - \frac{3}{2}E_F - E_B. \qquad (10.75)$$

It follows that the only one-loop superficially divergent diagrams in QED are the ones in Fig. 10.4. In fact, all the diagrams with an odd number of external photons vanish. This is trivial at one-loop, since the trace of an odd number of $\gamma$ matrices is zero. In general (Furry's theorem) it follows from the conservation of charge conjugation, since the photon is an eigenstate of $C$ with eigenvalue equal to $-1$.

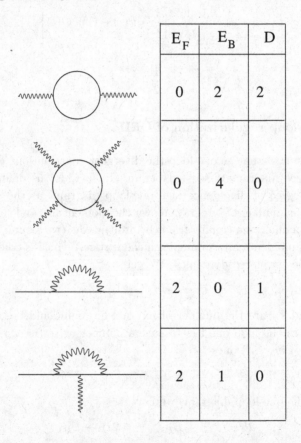

| $E_F$ | $E_B$ | D |
|-------|-------|---|
| 0 | 2 | 2 |
| 0 | 4 | 0 |
| 2 | 0 | 1 |
| 2 | 1 | 0 |

Fig. 10.4   The superficially divergent one-loop diagrams in spinor QED.

Furthermore, the effective degree of divergence can be less than the superficial degree. In fact, gauge invariance often lowers the divergence.

For instance the diagram of Fig. 10.4, with four external photons (light-light scattering), has $D = 0$ (logarithmic divergence), but gauge invariance makes it finite. Also the vacuum polarization diagram has an effective degree of divergence equal to 0. Therefore, there are only three divergent one-loop diagrams and all the divergences can be brought back to the three functions $\Sigma(p)$, $\Pi_{\mu\nu}(q)$ and $\Lambda^{\mu}(p', p)$. This does not mean that an arbitrary diagram is not divergent, but that it can be made finite if the divergences are eliminated in the previous three functions. In particular we will show that the divergent part of $\Sigma(p)$ can be absorbed into the definition of the mass of the electron and a redefinition of the electron field (wave function renormalization). The divergence in $\Pi_{\mu\nu}$, the photon self-energy, can be absorbed in the wave function renormalization of the photon (the mass of the photon is not renormalized due to gauge invariance). And finally the divergence of $\Lambda_{\mu}(p', p)$ goes into the definition of the electric charge.

We will now use dimensional regularization in order to regularize the relevant divergent quantities in QED, that is $\Sigma(p)$, $\Pi_{\mu\nu}(q)$ and $\Lambda^{\mu}(p, p')$. Furthermore, in order to define the counterterms we will determine the divergent parts of these expressions in the limit $\omega \to 2$. It will also be convenient to introduce a parameter $\mu$ such that these quantities have the same dimensions in the space with $d = 2\omega$ as in $d = 4$. The algebra of the Dirac matrices is also easily extended to arbitrary dimensions $d$. For instance, starting from

$$[\gamma_{\mu}, \gamma_{\nu}]_{+} = 2g_{\mu\nu}, \tag{10.76}$$

we get

$$\gamma^{\mu}\gamma_{\mu} = d \tag{10.77}$$

and

$$\gamma_{\mu}\gamma_{\nu}\gamma^{\mu} = (2 - d)\gamma_{\nu}. \tag{10.78}$$

Other relations can be obtained by starting from the algebraic properties of the $\gamma$-matrices. Let us begin with the electron self-energy which we will require to have dimension 1 as in $d = 4$. From eq. (10.4) we have

$$\Sigma(p) = i\mu^{4-2\omega} \int \frac{d^{2\omega}k}{(2\pi)^{2\omega}} \gamma_{\mu} \frac{\hat{p} - \hat{k} + m}{(p - k)^2 - m^2 + i\epsilon} \gamma^{\mu} \frac{1}{k^2 + i\epsilon}. \tag{10.79}$$

In order to use the equations of the previous Section it is convenient to combine together the denominators of this expression into a single one. This is done by using a formula due to Feynman

$$\frac{1}{ab} = \int_0^1 \frac{dz}{[az + b(1 - z)]^2}, \tag{10.80}$$

which is proven using

$$\frac{1}{ab} = \frac{1}{b-a}\left[\frac{1}{a} - \frac{1}{b}\right] = \frac{1}{b-a}\int_a^b \frac{dx}{x^2}. \qquad (10.81)$$

Performing the change of variable

$$x = az + b(1-z). \qquad (10.82)$$

We get

$$\Sigma(p) = i\mu^{4-2\omega}\int_0^1 dz \int \frac{d^{2\omega}k}{(2\pi)^{2\omega}}\gamma_\mu \frac{\hat{p} - \hat{k} + m}{[(p-k)^2 z - m^2 z + k^2(1-z)]^2}\gamma^\mu. \qquad (10.83)$$

The denominator can be written in the following way

$$[...] = p^2 z - m^2 z + k^2 - 2p \cdot kz. \qquad (10.84)$$

The term $p \cdot k$ can be eliminated through the following change of variable $k = k' + pz$. We find

$$[...] = (p^2 - m^2)z + (k' + pz)^2 - 2p\cdot(k' + pz)z = k'^2 - m^2 z + p^2 z(1-z). \qquad (10.85)$$

It follows (we put $k' = k$)

$$\Sigma(p) = i\mu^{4-2\omega}\int_0^1 dz \int \frac{d^{2\omega}k}{(2\pi)^{2\omega}}\gamma_\mu \frac{\hat{p}(1-z) - \hat{k} + m}{[k^2 - m^2 z + p^2 z(1-z)]^2}\gamma^\mu. \qquad (10.86)$$

The linear term in $k$ has vanishing integral, therefore

$$\Sigma(p) = i\mu^{4-2\omega}\int_0^1 dz \int \frac{d^{2\omega}k}{(2\pi)^{2\omega}}\gamma_\mu \frac{\hat{p}(1-z) + m}{[k^2 - m^2 z + p^2 z(1-z)]^2}\gamma^\mu \qquad (10.87)$$

and integrating over $k$

$$\Sigma(p) = i\mu^{4-2\omega}\int_0^1 dz \frac{1}{(2\pi)^{2\omega}}i\pi^\omega \frac{\Gamma(2-\omega)}{\Gamma(2)}\gamma_\mu \frac{\hat{p}(1-z) + m}{[m^2 z - p^2 z(1-z)]^{2-\omega}}\gamma^\mu. \qquad (10.88)$$

By defining $\epsilon = 4 - 2\omega$ we get

$$\Sigma(p) = -\mu^\epsilon \int_0^1 dz \frac{1}{(2\pi)^{4-\epsilon}}\pi^{(4-\epsilon)/2}\Gamma(\epsilon/2)\gamma_\mu \frac{\hat{p}(1-z) + m}{[m^2 z - p^2 z(1-z)]^{\epsilon/2}}\gamma^\mu. \qquad (10.89)$$

Contracting the $\gamma$ matrices

$$\Sigma(p) = -\frac{1}{16\pi^2}\Gamma(\epsilon/2)\int_0^1 dz(4\pi\mu^2)^{\epsilon/2}\frac{(\epsilon-2)\hat{p}(1-z) + (4-\epsilon)m}{[m^2 z - p^2 z(1-z)]^{\epsilon/2}}, \qquad (10.90)$$

we obtain

$$\Sigma(p) = \frac{1}{16\pi^2}\Gamma(\epsilon/2)\int_0^1 dz[2\hat{p}(1-z) - 4m - \epsilon(\hat{p}(1-z) - m)]$$
$$\times \left[\frac{m^2 z - p^2 z(1-z)}{4\pi\mu^2}\right]^{-\epsilon/2}. \qquad (10.91)$$

Defining

$$A = 2\hat{p}(1-z) - 4m, \quad B = -\hat{p}(1-z) + m, \quad C = \frac{m^2 z - p^2 z(1-z)}{4\pi\mu^2} \quad (10.92)$$

and expanding for $\epsilon \to 0$

$$\Sigma(p) = \frac{1}{16\pi^2}\int_0^1 dz\left[\frac{2A}{\epsilon} + 2B - \gamma A\right]\left[1 - \frac{\epsilon}{2}\log C\right]$$
$$\approx \frac{1}{16\pi^2}\int_0^1 dz\left[\frac{2A}{\epsilon} - A\log C + 2B - \gamma A\right]$$
$$= \frac{1}{8\pi^2\epsilon}(\hat{p} - 4m) - \frac{1}{16\pi^2}[\hat{p} - 2m + \gamma(\hat{p} - 4m)]$$
$$- \frac{1}{8\pi^2}\int_0^1 dz[\hat{p}(1-z) - 2m]\log\frac{m^2 z - p^2 z(1-z)}{4\pi\mu^2}$$
$$= \frac{1}{8\pi^2\epsilon}(\hat{p} - 4m) + \text{finite terms.} \qquad (10.93)$$

Consider now the self-energy of the photon (vacuum polarization) (see eq. (10.8))

$$\Pi^{\mu\nu}(q) = i\mu^{4-2\omega}\int\frac{d^{2\omega}k}{(2\pi)^{2\omega}}Tr\left[\frac{1}{\hat{k} + \hat{q} - m}\gamma^\mu\frac{1}{\hat{k} - m}\gamma^\nu\right]$$
$$= i\mu^{4-2\omega}\int\frac{d^{2\omega}k}{(2\pi)^{2\omega}}\frac{Tr[\gamma_\mu(\hat{k} + m)\gamma_\nu(\hat{k} + \hat{q} + m)]}{(k^2 - m^2)((k+q)^2 - m^2)}. \quad (10.94)$$

We use again the Feynman representation

$$\Pi^{\mu\nu}(q) = i\mu^{4-2\omega}\int_0^1 dz\int\frac{d^{2\omega}k}{(2\pi)^{2\omega}}\frac{Tr[\gamma_\mu(\hat{k} + m)\gamma_\nu(\hat{k} + \hat{q} + m)]}{[(k^2 - m^2)(1-z) + ((k+q)^2 - m^2)z]^2}. \qquad (10.95)$$

Then we write the denominator as

$$[...] = k^2 + q^2 z - m^2 + 2k \cdot qz. \qquad (10.96)$$

We then perform the change of variable $k = k' - qz$ in order to cancel the mixed term $2k \cdot qz$. In this way we get

$$[...] = (k' - qz)^2 + 2(k' - qz)\cdot qz + q^2 z - m^2 = k'^2 + q^2 z(1-z) - m^2 \quad (10.97)$$

and therefore (we put again $k' = k$)

$$\Pi^{\mu\nu}(q) = i\mu^{4-2\omega} \int_0^1 dz \int \frac{d^{2\omega}k}{(2\pi)^{2\omega}}$$

$$\times \frac{Tr[\gamma_\mu(\hat{k} - \hat{q}z + m)\gamma_\nu(\hat{k} + \hat{q}(1-z) + m)]}{[k^2 + q^2z(1-z) - m^2]^2}. \qquad (10.98)$$

Since the integral of the odd terms in $k$ is zero, it is enough to evaluate the contribution of the even terms to the trace

$$Tr[...]_{\text{even}} = Tr[\gamma_\mu\hat{k}\gamma_\nu\hat{k}] - Tr[\gamma_\mu\hat{q}\gamma_\nu\hat{q}]z(1-z) + m^2 Tr[\gamma_\mu\gamma_\nu]. \qquad (10.99)$$

If we define the $\gamma$'s as matrices of dimension $2^\omega \times 2^\omega$ we can repeat the calculation of Section 9.2 obtaining a factor $2^\omega$ instead of 4. Therefore

$$Tr[...]_{\text{even}} = 2^\omega[2k_\mu k_\nu - g_{\mu\nu}k^2$$
$$- (2q_\mu q_\nu - g_{\mu\nu}q^2)z(1-z) + m^2 g_{\mu\nu}]$$
$$= 2^\omega[2k_\mu k_\nu - 2z(1-z)(q_\mu q_\nu - g_{\mu\nu}q^2) - g_{\mu\nu}(k^2 - m^2$$
$$+ q^2z(1-z)] \qquad (10.100)$$

and we find

$$\Pi^{\mu\nu}(q) = i\mu^{4-2\omega}2^\omega \int_0^1 dz \int \frac{d^{2\omega}k}{(2\pi)^{2\omega}} \left[ \frac{2k_\mu k_\nu}{[k^2 + q^2z(1-z) - m^2]^2} \right.$$

$$- \frac{2z(1-z)(q_\mu q_\nu - g_{\mu\nu}q^2)}{[k^2 + q^2z(1-z) - m^2]^2}$$

$$\left. - \frac{g_{\mu\nu}}{[k^2 + q^2z(1-z) - m^2]} \right]. \qquad (10.101)$$

From the relations of the previous Section we get

$$\int d^{2\omega}p \frac{2p_\mu p_\nu}{[p^2 - a^2]^2} = -ig_{\mu\nu}\pi^\omega \frac{\Gamma(1-\omega)}{(a^2)^{1-\omega}}, \qquad (10.102)$$

whereas

$$\int d^{2\omega}p \frac{1}{[p^2 - a^2]} = -i\pi^\omega \frac{\Gamma(1-\omega)}{(a^2)^{1-\omega}}. \qquad (10.103)$$

Therefore the first and the third contributions to the vacuum polarization cancel out and we are left with

$$\Pi^{\mu\nu}(q) = -i\mu^{4-2\omega}2^\omega(q_\mu q_\nu - g_{\mu\nu}q^2) \int_0^1 dz\, 2z(1-z)$$

$$\times \int \frac{d^{2\omega}k}{(2\pi)^{2\omega}} \frac{1}{[k^2 + q^2z(1-z) - m^2]^2}. \qquad (10.104)$$

Notice that the original integral was quadratically divergent, but in the final result there is only a logarithmic divergence. The reason is gauge invariance. In fact, the photon self-energy is coupled to the photon field. Gauge invariance (in momentum space) implies that the theory should be invariant under the replacement $A_\mu(q) \to A_\mu(q) + \lambda q_\mu$. But this means that we must have $q^\mu \Pi_{\mu\nu}(q) = q^\nu \Pi_{\mu\nu} = 0$. In order to satisfy this condition the tensor $\Pi_{\mu\nu}$ must necessarily be of the form

$$\Pi_{\mu\nu} = (q_\mu q_\nu - g_{\mu\nu} q^2)\Pi(q). \tag{10.105}$$

Since the photon self-energy has dimension 2, it follows that $\Pi(q)$ has dimension zero. This quantity is an integral over $d^4k$, therefore the integrand must have dimension $-4$ and it is logarithmically divergent. With the same argument one shows that the one-loop diagram with 4 external photons, which is logarithmically divergent by the power counting made in Section 10.4, is finite.

Performing the momentum integration we have

$$\Pi^{\mu\nu}(q) = -i2\mu^{4-2\omega}2^\omega(q_\mu q_\nu - g_{\mu\nu} q^2)$$
$$\times \int_0^1 dz\, z(1-z)\frac{i\pi^\omega}{(2\pi)^{2\omega}}\frac{\Gamma(2-\omega)}{[m^2 - q^2 z(1-z)]^{2-\omega}}. \tag{10.106}$$

By putting again $\epsilon = 4 - 2\omega$ and expanding the previous expression

$$\Pi^{\mu\nu}(q) = 2\mu^\epsilon 2^{2-\epsilon/2}(q_\mu q_\nu - g_{\mu\nu} q^2)$$
$$\times \int_0^1 dz\, z(1-z)\frac{\pi^{2-\epsilon/2}}{(2\pi)^{4-\epsilon}}\frac{\Gamma(\epsilon/2)}{[m^2 - q^2 z(1-z)]^{\epsilon/2}}$$
$$= \frac{2}{16\pi^2}2^{2-\epsilon/2}(q_\mu q_\nu - g_{\mu\nu} q^2)$$
$$\times \int_0^1 dz\, z(1-z)\Gamma(\epsilon/2)\left[\frac{m^2 - q^2 z(1-z)}{4\pi\mu^2}\right]^{-\epsilon/2}$$
$$\approx \frac{2}{16\pi^2}(4 - 2\epsilon\log 2)(q_\mu q_\nu - g_{\mu\nu} q^2)$$
$$\times \int_0^1 dz\, z(1-z)(\frac{2}{\epsilon} - \gamma)\left[1 - \frac{\epsilon}{2}\log C\right], \tag{10.107}$$

where $C$ is definite in eq. (10.92). Finally

$$\Pi_{\mu\nu}(q) = \frac{1}{8\pi^2}(q_\mu q_\nu - g_{\mu\nu} q^2)\int_0^1 dz\, z(1-z)\left[\frac{8}{\epsilon} - 4\gamma - 4\log 2\right]\left[1 - \frac{\epsilon}{2}\log C\right] \tag{10.108}$$

and

$$\Pi_{\mu\nu}(q) \approx \frac{1}{2\pi^2}(q_\mu q_\nu - g_{\mu\nu}q^2)\left[\frac{1}{3\epsilon} - \frac{\gamma}{6}\right.$$
$$\left. - \int_0^1 dz \; z(1-z)\log\left[\frac{m^2 - q^2 z(1-z)}{2\pi\mu^2}\right]\right], \quad (10.109)$$

from which

$$\Pi_{\mu\nu}(q) = \frac{1}{6\pi^2}(q_\mu q_\nu - g_{\mu\nu}q^2)\frac{1}{\epsilon} + \text{finite terms.} \quad (10.110)$$

Now we have to evaluate $\Lambda_\mu(p',p)$. From eq. (10.11) we have

$$\Lambda^\mu(p',p) = -i\mu^{4-2\omega}\int\frac{d^{2\omega}k}{(2\pi)^{2\omega}}\gamma^\alpha\frac{\hat{p}' - \hat{k} + m}{(p'-k)^2 - m^2}\gamma^\mu\frac{\hat{p} - \hat{k} + m}{(p-k)^2 - m^2}\gamma_\alpha\frac{1}{k^2}. \quad (10.111)$$

The general formula to reduce $n$ denominators to a single one is

$$\prod_{i=1}^n\frac{1}{a_i} = (n-1)!\int_0^1\prod_{i=1}^n d\beta_i\frac{\delta(1 - \sum_{i=1}^n\beta_i)}{[\sum_{i=1}^n\beta_i a_i]^n}. \quad (10.112)$$

To show this equation notice that

$$\prod_{i=1}^n\frac{1}{a_i} = \int_0^\infty\prod_{i=1}^n d\alpha_i e^{-\sum_{i=1}^n\alpha_i a_i}. \quad (10.113)$$

Introducing the identity

$$1 = \int_0^\infty d\lambda\delta(\lambda - \sum_{i=1}^n\alpha_i) \quad (10.114)$$

and changing variables $\alpha_i = \lambda\beta_i$

$$\prod_{i=1}^n\frac{1}{a_i} = \int_0^\infty\prod_{i=1}^n d\alpha_i d\lambda\delta(\lambda - \sum_{i=1}^n\alpha_i)e^{-\sum_{i=1}^n\alpha_i a_i}$$
$$= \int_0^\infty\prod_{i=1}^n d\beta_i\lambda^n\frac{d\lambda}{\lambda}e^{-\lambda\sum_{i=1}^n\beta_i a_i}\delta(1 - \sum_{i=1}^n\beta_i). \quad (10.115)$$

The integration over $\beta_i$ can be restricted to the interval $[0,1]$ due to the delta function. We get

$$\int_0^\infty d\lambda\lambda^{n-1}e^{-\rho\lambda} = \frac{(n-1)!}{\rho^n}. \quad (10.116)$$

proving eq. (10.112). In the case of the vertex function we get

$$\Lambda^\mu(p',p) = -2i\mu^{4-2\omega}\int\frac{d^{2\omega}k}{(2\pi)^{2\omega}}\int_0^1 dx\int_0^{1-x} dy$$
$$\times\frac{\gamma^\alpha(\hat{p}' - \hat{k} + m)\gamma^\mu(\hat{p} - \hat{k} + m\gamma_\alpha}{[k^2(1-x-y) + (p-k)^2 x - m^2 x + (p'-k)^2 y - m^2 y]^3}. \quad (10.117)$$

The denominator is

$$[...] = k^2 - m^2(x+y) + p^2x + p'^2y - 2k \cdot (px + p'y). \tag{10.118}$$

Changing variable, $k = k' + px + py$

$$\begin{aligned}[...] &= (k' + px + p'y)^2 - m^2(x+y) + p^2x \\ &\quad + p'^2y - 2(k' + px + p'y) \cdot (px + p'y) \\ &= k'^2 - m^2(x+y) + p^2x(1-x) + p'^2y(1-y) \\ &\quad - 2p \cdot p'xy. \end{aligned} \tag{10.119}$$

Letting again $k' \to k$

$$\Lambda^\mu(p',p) = -2i\mu^{4-2\omega} \int_0^1 dx \int_0^{1-x} dy \int \frac{d^{2\omega}k}{(2\pi)^{2\omega}}$$
$$\times \frac{\gamma^\alpha(\hat{p}'(1-y) - \hat{p}x - \hat{k} + m)\gamma^\mu(\hat{p}(1-x) - \hat{p}'y - \hat{k} + m)\gamma_\alpha}{[k^2 - m^2(x+y) + p^2x(1-x) + p'^2y(1-y) - 2p \cdot p'xy]^3}. \tag{10.120}$$

The odd term in $k$ is zero after integration. The term in $k^2$ is logarithmically divergent, whereas the remaining part is convergent. Separating the divergent piece, $\Lambda_\mu^{(1)}$, from the convergent one, $\Lambda_\mu^{(2)}$,

$$\Lambda_\mu = \Lambda_\mu^{(1)} + \Lambda_\mu^{(2)}, \tag{10.121}$$

we get for the first term

$$\Lambda_\mu^{(1)}(p',p) = -2i\mu^{4-2\omega} \int_0^1 dx \int_0^{1-x} dy \int \frac{d^{2\omega}k}{(2\pi)^{2\omega}}$$
$$\times \frac{\gamma^\alpha \gamma_\lambda \gamma_\mu \gamma_\sigma \gamma_\alpha k^\lambda k^\sigma}{[k^2 - m^2(x+y) + p^2x(1-x) + p'^2y(1-y) - 2p \cdot p'xy]^3}$$
$$= -2i\mu^{4-2\omega} \int_0^1 dx \int_0^{1-x} dy \frac{i\pi^\omega(-1)^3}{(2\pi)^{2\omega}\Gamma(3)} \left(-\frac{1}{2}\right) \Gamma(2-\omega)$$
$$\times \frac{\gamma^\alpha \gamma_\lambda \gamma^\mu \gamma^\lambda \gamma_\alpha}{[m^2(x+y) - p^2x(1-x) - p'^2y(1-y) + 2p \cdot p'xy]^{2-\omega}}$$
$$= \frac{1}{2}\mu^{4-2\omega} \left(\frac{1}{4\pi}\right)^\omega \Gamma(2-\omega) \int_0^1 dx \int_0^{1-x} dy$$
$$\times \frac{\gamma^\alpha \gamma_\lambda \gamma^\mu \gamma^\lambda \gamma_\alpha}{[m^2(x+y) - p^2x(1-x) - p'^2y(1-y) + 2p \cdot p'xy]^{2-\omega}}. \tag{10.122}$$

Using the relation

$$\gamma^\alpha \gamma_\lambda \gamma^\mu \gamma^\lambda \gamma_\alpha = (2-d)^2 \gamma^\mu, \tag{10.123}$$

we obtain

$$\Lambda_\mu^{(1)}(p',p) = \frac{1}{2}\mu^\epsilon \left(\frac{1}{4\pi}\right)^{2-\epsilon/2} \Gamma(\epsilon/2)(\epsilon-2)^2\gamma_\mu \int_0^1 dx \int_0^{1-x} dy$$

$$\times \frac{1}{[m^2(x+y) - p^2x(1-x) - p'^2y(1-y) + 2p\cdot p'xy]^{\epsilon/2}}$$

$$\approx \left(\frac{1}{32\pi^2}\right)\left[\frac{2}{\epsilon} - \gamma\right][4 - 4\epsilon]\gamma_\mu \int_0^1 dx \int_0^{1-x} dy$$

$$\times \left[\frac{m^2(x+y) - p^2x(1-x) - p'^2y(1-y) + 2p\cdot p'xy}{4\pi\mu^2}\right]^{-\epsilon/2}$$

$$\approx \frac{1}{8\pi^2\epsilon}\gamma_\mu - \frac{1}{16\pi^2}(\gamma+2)\gamma_\mu - \frac{1}{8\pi^2}\gamma_\mu \int_0^1 dx \int_0^{1-x} dy$$

$$\times \log\left[\frac{m^2(x+y) - p^2x(1-x) - p'^2y(1-y) + 2p\cdot p'xy}{4\pi\mu^2}\right] \tag{10.124}$$

and finally

$$\Lambda_\mu^{(1)}(p',p) = \frac{1}{8\pi^2\epsilon}\gamma_\mu + \text{finite terms.} \tag{10.125}$$

In the convergent part we can put directly $\omega = 2$

$$\Lambda_\mu^{(2)}(p',p) = -\frac{i}{8\pi^4}\int_0^1 dx \int_0^{1-x} dy \frac{i\pi^2(-1)^3}{\Gamma(3)}$$

$$\times \frac{\gamma^\alpha(\hat{p}'(1-y) - \hat{p}x + m)\gamma^\mu(\hat{p}(1-x) - \hat{p}'y + m)\gamma_\alpha}{m^2(x+y) - p^2x(1-x) - p'^2y(1-y) + 2p\cdot p'xy}$$

$$= -\frac{1}{16\pi^2}\int_0^1 dx \int_0^{1-x} dy$$

$$\times \frac{\gamma^\alpha(\hat{p}'(1-y) - \hat{p}x + m)\gamma^\mu(\hat{p}(1-x) - \hat{p}'y + m)\gamma_\alpha}{m^2(x+y) - p^2x(1-x) - p'^2y(1-y) + 2p\cdot p'xy}. \tag{10.126}$$

## 10.5    One-loop renormalization

We summarize here the results of the previous Section

$$\Sigma(p) = \frac{1}{8\pi^2\epsilon}(\hat{p} - 4m) + \Sigma^f(p). \tag{10.127}$$

$$\Pi_{\mu\nu}(q) = (q_\mu q_\nu - g_{\mu\nu}q^2)\left[\frac{1}{6\pi^2\epsilon} + \Pi^f(q)\right] \equiv (q_\mu q_\nu - g_{\mu\nu}q^2)\Pi(q^2). \tag{10.128}$$

$$\Lambda_\mu(p',p) = \frac{1}{8\pi^2\epsilon}\gamma_\mu + \Lambda_\mu^f(p',p), \qquad (10.129)$$

where the functions with the superscript $f$ represent the finite contributions. Let us start discussing the electron self-energy. As shown in eq. (10.2), the effect of $\Sigma(p)$ is to correct the electron propagator. In fact we have (see Fig. 10.5):

$$S_F(p) = \frac{i}{\hat{p}-m} + \frac{i}{\hat{p}-m}ie^2\Sigma(p)\frac{i}{\hat{p}-m} + \cdots, \qquad (10.130)$$

from which, at the same order in the perturbative expansion

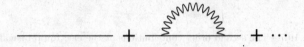

Fig. 10.5 The loop expansion for the electron propagator.

$$S_F(p) = \frac{i}{\hat{p}-m}\left(1 + \frac{e^2\Sigma(p)}{\hat{p}-m}\right)^{-1} = \frac{i}{\hat{p}-m+e^2\Sigma(p)}. \qquad (10.131)$$

Therefore the effect of the divergent terms is to modify the coefficients of $\hat{p}$ and $m$:

$$iS_F^{-1}(p) = \hat{p}-m+e^2\Sigma(p) = \hat{p}\left(1 + \frac{e^2}{8\pi^2\epsilon}\right) - m\left(1 + \frac{e^2}{2\pi^2\epsilon}\right) + \text{finite terms.} \qquad (10.132)$$

This allows us to define the counterterms to be added to the Lagrangian expressed in terms of the physical parameters in such a way to cancel these divergences

$$(\mathcal{L}_1)_{ct} = iB\bar{\psi}\hat{\partial}\psi - A\bar{\psi}\psi, \qquad (10.133)$$

$(\mathcal{L}_1)_{ct}$ modifies the Feynman rules adding two operators parametrized by $A$ and $B$. These coefficients can be evaluated noticing that the expression of the propagator, taking into account $(\mathcal{L}_1)_{ct}$, is

$$\frac{i}{(1+B)\hat{p}-(m+A)} \approx \frac{i}{\hat{p}-m}\left(1 - \frac{B\hat{p}-A}{\hat{p}-m}\right)$$

$$\approx \frac{i}{\hat{p}-m} + \frac{i}{\hat{p}-m}(iB\hat{p}-iA)\frac{i}{\hat{p}-m}, (10.134)$$

where, consistently with our expansion we have taken only the first order terms in $A$ and $B$. We can associate to these two terms the diagrams of Fig. 10.6, with contributions $-iA$ to the mass term, and $iB\hat{p}$ to $\hat{p}$.

Fig. 10.6   The counterterms for the self-energy ($\not{p}$ in the figure should be read as $\hat{p}$).

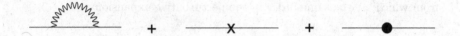

Fig. 10.7   The second order contributions at the electron propagator.

The propagator at the second order in the coupling constant is then obtained by adding the diagrams of Fig. 10.7 to the free part. We get

$$S_F(p) = \frac{i}{\hat{p} - m} + \frac{i}{\hat{p} - m} \left( ie^2 \Sigma(p) + iB\hat{p} - iA \right) \frac{i}{\hat{p} - m} \qquad (10.135)$$

and the correction is given by

$$e^2 \Sigma(p) + B\hat{p} - A = \left( \frac{e^2}{8\pi^2 \epsilon} + B \right) \hat{p} - \left( \frac{e^2}{2\pi^2 \epsilon} m + A \right) + \text{finite terms.} \quad (10.136)$$

We can now fix the counterterms by choosing

$$B = -\frac{e^2}{8\pi^2} \left( \frac{1}{\epsilon} + F_2 \left( \frac{m}{\mu} \right) \right), \qquad (10.137)$$

$$A = -\frac{me^2}{2\pi^2} \left( \frac{1}{\epsilon} + F_m \left( \frac{m}{\mu} \right) \right), \qquad (10.138)$$

with $F_2$ and $F_m$ finite for $\epsilon \to 0$. Notice that these two functions are dimensionless and for the moment being completely arbitrary. However they can be determined by the renormalization conditions, that is by fixing the arbitrary constants appearing in the Lagrangian. In fact, given

$$iS_F^{-1}(p) \equiv \Gamma^{(2)}(p) = \hat{p} - m + B\hat{p} - A + e^2 \Sigma(p)$$

$$= \left( 1 - \frac{e^2}{8\pi^2} F_2 \right) \hat{p} - m + \frac{me^2}{2\pi^2} F_m + e^2 \Sigma^f, \qquad (10.139)$$

we can require that at the physical pole, $\hat{p} = m$, the propagator coincides with the free propagator

$$S_F(p) \approx \frac{i}{\hat{p} - m}, \quad \text{for } \hat{p} = m. \qquad (10.140)$$

From here we get two conditions. The first one is

$$\Gamma^{(2)}(\hat{p} = m) = 0 \tag{10.141}$$

from which

$$e^2 \Sigma^f(\hat{p} = m) - \frac{me^2}{8\pi^2} F_2 + \frac{me^2}{2\pi^2} F_m = 0. \tag{10.142}$$

The second condition is

$$\frac{\partial \Gamma^{(2)}(p)}{\partial p^\mu}\bigg|_{\hat{p}=m} = \gamma_\mu, \tag{10.143}$$

giving

$$e^2 \frac{\partial \Sigma^f(p)}{\partial p^\mu}\bigg|_{\hat{p}=m} - \frac{e^2}{8\pi^2} F_2 \gamma_\mu = 0. \tag{10.144}$$

One should be careful because these particular conditions of renormalization give some problem related to the zero mass of the photon. In fact, one finds some ill-defined integral in the infrared region. However these are harmless divergences, and can be eliminated giving a small mass to the photon and sending this mass to zero at the end of the calculations. Notice that these conditions of renormalization have the advantage of being expressed directly in terms of the measured parameters, as the electron mass. However, one could renormalize at an arbitrary mass scale, $M$. In this case the parameters comparing in $\mathcal{L}_p$ are not the directly measured parameters, but they can be correlated to the actual parameters by evaluating some observable quantity. From this point of view one could avoid the problems mentioned above by choosing a different point of renormalization.

As far as the vacuum polarization is concerned, $\Pi_{\mu\nu}$ gives rise to the following correction to the photon propagator (illustrated in Fig. 10.8)

$$D'_{\mu\nu}(q) = \frac{-ig_{\mu\nu}}{q^2} + \frac{-ig_{\mu\lambda}}{q^2} ie^2 \Pi^{\lambda\rho}(q) \frac{-ig_{\rho\nu}}{q^2} + \cdots, \tag{10.145}$$

from which

$$D'_{\mu\nu}(q) = \frac{-ig_{\mu\nu}}{q^2} + \frac{-ig_{\mu\lambda}}{q^2} \left[ (ie^2)(q^\lambda q^\rho - g^{\lambda\rho} q^2) \Pi(q^2) \right] \frac{-ig_{\rho\nu}}{q^2} + \cdots$$

$$= \frac{-ig_{\mu\nu}}{q^2} \left[ 1 - e^2 \Pi(q^2) \right] - i \frac{q_\mu q_\nu}{q^4} e^2 \Pi(q^2) + \cdots. \tag{10.146}$$

We see that the one-loop propagator has a divergent part in $g_{\mu\nu}$, as well in the term proportional to the momenta. Therefore the propagator does not correspond any more to the Lorenz gauge and we need to add to the following terms in $\mathcal{L}_p$

$$\mathcal{L}_2 = -\frac{1}{4} F_{\mu\nu} F^{\mu\nu} - \frac{1}{2} (\partial_\mu A^\mu)^2, \tag{10.147}$$

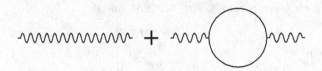

Fig. 10.8   The loop expansion for the photon propagator.

the two counterterms

$$(\mathcal{L}_2)_{\text{ct}} = -\frac{C}{4}F_{\mu\nu}F^{\mu\nu} - \frac{E}{2}(\partial_\mu A^\mu)^2$$

$$= -C\left(\frac{1}{4}F_{\mu\nu}F^{\mu\nu} + \frac{1}{2}(\partial_\mu A^\mu)^2\right) - \frac{E-C}{2}(\partial_\mu A^\mu)^2. \quad (10.148)$$

As for the electron propagator, we can look at these two contributions as perturbations to the free Lagrangian, and evaluate the corresponding Feynman rules, or evaluate the effect on the propagator. The modification in the equation defining the propagator due to these two terms is

$$[(1+C)q^2 g_{\mu\nu} - (C-E)q_\mu q_\nu]D^{\nu\lambda}(q) = -ig_\mu^\lambda. \quad (10.149)$$

We solve this equation by putting

$$D_{\mu\nu}(q) = \alpha(q^2)g_{\mu\nu} + \beta(q^2)q_\mu q_\nu. \quad (10.150)$$

Substituting in the previous equation we determine $\alpha$ and $\beta$. The result is

$$\alpha = -\frac{i}{q^2(1+C)}, \qquad \beta = -\frac{i}{q^4}\frac{C-E}{(1+C)(1+E)}. \quad (10.151)$$

The free propagator, including the corrections at the first order in $C$ and $E$ is

$$D_{\mu\nu}(q) = \frac{-ig_{\mu\nu}}{q^2}(1-C) - i\frac{q_\mu q_\nu}{q^4}(C-E) \quad (10.152)$$

and the total propagator

$$D'_{\mu\nu}(q) = \frac{-ig_{\mu\nu}}{q^2}[1 - e^2\Pi(q^2) - C] - i\frac{q_\mu q_\nu}{q^4}[e^2\Pi(q^2) + C - E]. \quad (10.153)$$

We can now choose $E = 0$ and cancel the divergence by the choice

$$C = -\frac{e^2}{6\pi^2\epsilon} + F_3\left(\frac{m}{\mu}\right). \quad (10.154)$$

In fact, we are free to choose the finite term in the gauge fixing, since this choice does not change the physics. This is because the terms proportional to $q_\mu q_\nu$, as it follows from gauge invariance and the consequent conservation of the electromagnetic current, do not contribute to the physical amplitude. For instance, if we have a vertex with a virtual photon (that is a vertex connected to an internal photon line) and two external electrons, the term proportional to $q_\mu q_\nu$ is saturated with

$$\bar{u}(p')\gamma_\mu u(p), \qquad q = p' - p. \qquad (10.155)$$

The result is zero, by taking into account the Dirac equation. Let us now consider the mass of the photon. We have

$$D'_{\mu\nu}(q) = \frac{-ig_{\mu\nu}}{q^2}\left[1 - F_3 - e^2\Pi^f\right] - i\frac{q_\mu q_\nu}{q^4}\left[F_3 + e^2\Pi^f\right]$$

$$\equiv \frac{-ig_{\mu\nu}}{q^2}[1 - \tilde{\Pi}] - i\frac{q_\mu q_\nu}{q^4}\tilde{\Pi}, \qquad (10.156)$$

where

$$\tilde{\Pi} = e^2\Pi^f + F_3. \qquad (10.157)$$

At the second order in the electric charge we can write the propagator in the form

$$D'_{\mu\nu}(q) = \frac{-i}{q^2(1 + \tilde{\Pi}(q))}\left[g_{\mu\nu} + \frac{q_\mu q_\nu}{q^2}\tilde{\Pi}(q)\right] \qquad (10.158)$$

and we see that the propagator has a pole at $q^2 = 0$, since $\tilde{\Pi}(q^2)$ is finite for $q^2 \to 0$. Therefore the photon remains massless after renormalization. This is part of a rather general aspect of renormalization which says that, if the regularization procedure respects the symmetries of the original Lagrangian, the symmetries are preserved at any perturbative order. However, there are cases where it is not possible to devise a regularization procedure such as to preserve a given symmetry. This is the case of the **anomalous symmetries** which are symmetries only at the classical level but broken by quantum corrections.

Also for the photon we will require that at the physical pole, $q^2 \to 0$, the propagator has the free form, that is

$$\tilde{\Pi}(0) = 0, \qquad (10.159)$$

from which

$$F_3 = -e^2\Pi^f(0). \qquad (10.160)$$

## 10.6  Lamb shift and $g - 2$

In this Section we will use the technology developed in this Chapter to evaluate two important physical quantities. The first one is a contribution to the Lamb shift. The levels $2S_{1/2}$ and $2P_{1/2}$ of the hydrogen atom are degenerate except for corrections coming from QED. Among these corrections there is one coming from the radiative correction to the photon propagator and this is the one that we will evaluate here. The second quantity that we will evaluate comes from the vertex corrections and it is a contribution to the gyromagnetic factor of the electron which is equal to 2 in the Dirac theory. Let us start with the vacuum polarization.

We will consider the expression (10.157) for small momenta:

$$\tilde{\Pi}(q) \approx e^2 q^2 \frac{d\Pi^f(q)}{dq^2}\Big|_{q^2=0}, \tag{10.161}$$

where use has been made of $F_3$ being constant. Using the expression for $\Pi^f$ from the previous Section, we get

$$\Pi^f(q) \approx -\frac{1}{2\pi^2}\left(\frac{\gamma}{6} + \frac{1}{6}\log\left(\frac{m^2}{2\pi\mu^2}\right)\right) + \frac{1}{2\pi^2}\int_0^1 dz\, z^2(1-z)^2\frac{q^2}{m^2} + \cdots \tag{10.162}$$

and

$$\tilde{\Pi}(q) \approx \frac{e^2 q^2}{60\pi^2 m^2}. \tag{10.163}$$

It follows

$$D'_{\mu\nu} = -i\frac{g_{\mu\nu}}{q^2}\left[1 - \frac{e^2 q^2}{60\pi^2 m^2}\right] + \text{gauge terms}. \tag{10.164}$$

The first term, $1/q^2$, gives rise to the Coulomb potential, $e^2/4\pi r$. The second gives a correction by a term proportional to a delta function in the real space (we are using the same notations of Section 7.2)

$$\Delta_{12} = e^2 \int_{-\infty}^{+\infty} dt \int \frac{d^4q}{(2\pi)^4} e^{-iq(x_1-x_2)}\left[\frac{1}{q^2} - \frac{e^2}{60\pi^2 m^2}\right]$$

$$= -\frac{e^2}{4\pi r} - \frac{e^4}{60\pi^2 m^2}\delta^3(\vec{r}). \tag{10.165}$$

This modification of the Coulomb potential changes the energy levels of the hydrogen atom, and it is one of the contributions to the Lamb shift, which produces a splitting of the levels $2S_{1/2}$ and $2P_{1/2}$. The total Lamb shift is the sum of all the self-energy and vertex corrections, and turns out to be about 1057.9 MHz. The contribution we have just evaluated is only

$-27.1$ MHz, but it is important since the agreement between experiment and theory is at the level of 0.1 MHz.

Now we have to discuss the vertex corrections. We have seen that the divergent contribution is $\Lambda_\mu^{(1)}(p',p)$, and this is proportional to $\gamma_\mu$. The counterterm to add to the interacting part of $\mathcal{L}_{\mathrm{p}}$,

$$\mathcal{L}_{\mathrm{p}}^{\mathrm{int}} = -e\bar{\psi}\gamma_\mu\psi A^\mu, \tag{10.166}$$

is

$$(\mathcal{L}_3)_{\mathrm{ct}} = -eD\bar{\psi}\gamma_\mu\psi A^\mu. \tag{10.167}$$

The complete vertex (omitting the factor $-ie$) is given by (see Fig. 10.9)

$$\Lambda_\mu^T = \gamma_\mu(1+D) + e^2\Lambda^\mu. \tag{10.168}$$

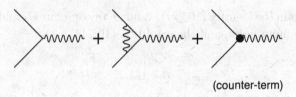

(counter-term)

Fig. 10.9 The one-loop vertex corrections.

Due to the conservation of the electromagnetic current, it is possible to prove the validity of the Ward-Takahashi identity [Ward (1950); Takahashi (1957)] between the total self-energy and the total vertex correction

$$\frac{\partial \Sigma^T(p)}{\partial p^\mu} = \Lambda_\mu^T(p,p), \tag{10.169}$$

where $\Sigma^T$ and $\Lambda_\mu^T$ are defined as

$$\Gamma^{(2)} = \hat{p} - m + \Sigma^T, \qquad \Gamma_\mu = \gamma_\mu + \Lambda_\mu^T. \tag{10.170}$$

This identity applied to the expressions (10.139) and (10.168) gives

$$B\gamma_\mu + e^2\frac{\partial \Sigma(p)}{\partial p^\mu} = D\gamma_\mu + e^2\Lambda_\mu(p,p), \tag{10.171}$$

where $\Sigma$ and $\Lambda$ are the self-energy and vertex corrections evaluated from the Feynman diagrams associated to $\mathcal{L}_p$. Therefore, from eq. (10.201) in the first exercise of this Chapter, it follows

$$B = D, \tag{10.172}$$

at this order in perturbation theory. This equality is certainly satisfied by the divergent parts of these counterterms, since they are devised in order to eliminate the terms in $1/\epsilon$ in $\Sigma$ and $\Lambda_\mu$ which are equal (see eqs. (10.127) and (10.129)). However, in order to satisfy (10.172) also the finite parts of $B$ and $D$ should be equal. As a consequence we fix the counterterm $D$ by

$$D = -\frac{e^2}{8\pi^2}\left[\frac{1}{\epsilon} + F_2\right], \tag{10.173}$$

with $F_2$ the same as in eq. (10.144). Notice that in this way the wave function renormalization terms $Z_1$ and $Z_2$ are made equal, and the charge renormalization depends only on $Z_3$, that is from the vacuum polarization.

The complete one-loop vertex is then

$$\Lambda_\mu^T = \left[\gamma_\mu + e^2(\Lambda_\mu^f - \frac{F_2}{8\pi^2}\gamma_\mu)\right]. \tag{10.174}$$

By using again the identity (10.201), valid at any order in $1/\epsilon$ and, therefore, also for the finite parts, and the condition (10.144)

$$\Lambda_\mu^f\Big|_{\hat{p}=m} = \frac{\partial\Sigma^f}{\partial p^\mu}\Big|_{\hat{p}=m} = \frac{F_2}{8\pi^2}\gamma_\mu, \tag{10.175}$$

we see that for spinors on shell we have

$$\bar{u}(p)\Lambda_\mu^T u(p)|_{\hat{p}=m} = \bar{u}(p)\gamma_\mu u(p). \tag{10.176}$$

It is interesting to notice that we have been able to satisfy the Ward-Takahashi identity since we had the freedom to choose the finite part of the counterterm $D$. There are situations where this is not possible and this is the case of the **anomalies**, that is transformations that are symmetries at the classical level but that are broken at quantum level. A celebrated example is the one of the axial-vector anomaly. As we know from the exercises in Chapter 6, the Lagrangian for a massless Dirac field is invariant under the chiral transformation

$$\psi(x) \to e^{i\alpha\gamma_5}\psi(x), \tag{10.177}$$

at the classical level, besides being invariant under a phase transformation. Both symmetries are preserved by making the Dirac field to interact with the electromagnetic field. However, it turns out that it is impossible to define a renormalization procedure (choice of the counterterms) in order to satisfy both the Ward-Takahashi identities following from the conservations of the two currents associated to the two symmetries. Usually one makes the choice of satisfying the Ward-Takahashi identity associated to the phase symmetry and the chiral symmetry is broken at the quantum level (axial anomaly). This point will be discussed in the final Chapter of the book.

We will now evaluate the radiative corrections to the $g-2$ of the electron. Here $g$ is the gyromagnetic ratio, which is predicted to be equal to 2 by the Dirac equation (see Section 4.7). To this end we first need to prove the Gordon identity for the current of a Dirac particle

$$\bar{u}(p')\gamma_\mu u(p) = \bar{u}(p') \left[ \frac{p^\mu + p^{\mu'}}{2m} + \frac{i}{2m}\sigma_{\mu\nu}q^\nu \right] u(p), \tag{10.178}$$

with $q = p' - p$. For the proof we start from

$$\hat{p}\gamma_\mu u(p) = (-m\gamma_\mu + 2p_\mu)u(p) \tag{10.179}$$

and

$$\gamma_\mu \hat{p} u(p) = m\gamma_\mu u(p). \tag{10.180}$$

Subtracting these two expressions we obtain

$$\gamma_\mu u(p) = \left( \frac{p_\mu}{m} - \frac{i}{m}\sigma_{\mu\nu}p^\nu \right) u(p). \tag{10.181}$$

An analogous operation on the barred spinor leads to the result. We observe also that the Gordon identity shows immediately that the value of the gyromagnetic ratio is 2, because it implies that the coupling with the electromagnetic field is just

$$\frac{e}{2m}\sigma_{\mu\nu}F^{\mu\nu}(q). \tag{10.182}$$

To evaluate the correction to this term from the one-loop diagrams, it is enough to evaluate the matrix element

$$e^2\bar{u}(p')\Lambda_\mu^{(2)}(p',p)u(p) \tag{10.183}$$

in the limit $p' \to p$ and for on-shell momenta. In fact $\Lambda_\mu^{(1)}$ contributes only to the terms in $\gamma_\mu$ and, in the previous limit, they have to build up the free vertex, as implied by the renormalization condition. Therefore we will ignore all the terms proportional to $\gamma_\mu$ and we will take the first order in the momentum $q$. For momenta on shell, the denominator of $\Lambda_\mu^{(2)}$ is given by

$$[...] = m^2(x+y) - m^2x(1-x) - m^2y(1-y) + 2m^2xy = m^2(x+y)^2. \tag{10.184}$$

In order to evaluate the numerator, let us define

$$V_\mu = \bar{u}(p')\gamma^\alpha \left[ \hat{p}'(1-y) - \hat{p}x + m \right] \gamma^\mu \left[ \hat{p}(1-x) - \hat{p}'y + m \right] \gamma_\alpha u(p). \tag{10.185}$$

Using $\hat{p}\gamma_\alpha = -\gamma_\alpha\hat{p} + 2p_\alpha$, and an analogous equation for $\hat{p}'$, we can bring $\hat{p}$ to act on the spinor to the right of the expression, and $\hat{p}'$ on the spinor to the left, obtaining

$$V_\mu = \bar{u}(p') \left[ my\gamma^\alpha + 2(1-y)p'^\alpha - \gamma^\alpha \hat{p}x \right] \gamma^\mu$$
$$\times \left[ mx\gamma_\alpha + 2(1-x)p_\alpha - \hat{p}'y\gamma_\alpha \right] u(p). \tag{10.186}$$

Making use of

$$\gamma_\alpha \gamma^\mu \gamma_\alpha = -2\gamma^\mu, \tag{10.187}$$

$$\gamma_\alpha \gamma_\mu \gamma_\nu \gamma^\alpha = 4g_{\mu\nu}, \tag{10.188}$$

$$\gamma_\alpha \hat{p} \gamma^\mu \hat{p}' \gamma^\alpha = -2\hat{p}' \gamma^\mu \hat{p}, \tag{10.189}$$

we get

$$V_\mu = \bar{u}(p') \Big[ -2m^2 xy\gamma^\mu + 2my(1-x)(-m\gamma^\mu + 2p^\mu)$$
$$-4my^2 p'^\mu + 2mx(1-y)(-m\gamma^\mu + 2p'^\mu) + 4(1-x)(1-y)m^2\gamma^\mu$$
$$-2y(1-y)m^2\gamma^\mu - 4mx^2 p^\mu$$
$$-2m^2 x(1-x)\gamma^\mu - 2xym^2\gamma^\mu \Big] u(p) \tag{10.190}$$

and for the piece which does not contain $\gamma_\mu$

$$V_\mu = 4m\bar{u}(p') \left[ p^\mu (y - xy - x^2) + p'^\mu (x - xy - y^2) \right] u(p). \tag{10.191}$$

Therefore the relevant part of the vertex contribution is

$$e^2 \bar{u}(p')\Lambda_\mu^{(2)}(p',p)u(p) \rightarrow -\frac{e^2}{16\pi^2} \int_0^1 dx \int_0^{1-x} dy \, \frac{4}{m} \bar{u}(p') \Big[ p^\mu \frac{(y - xy - x^2)}{(x+y)^2}$$
$$+ p'^\mu \frac{(x - xy - y^2)}{(x+y)^2} \Big] u(p). \tag{10.192}$$

By changing variable, $z = x + y$, we obtain

$$\int_0^1 dx \int_0^{1-x} dy \frac{y - xy - x^2}{(x+y)^2} = \int_0^1 dx \int_x^1 dz \left[ \frac{1-x}{z} - \frac{x}{z^2} \right]$$
$$= \int_0^1 \left[ -(1-x)\log x + x - 1 \right]$$
$$= \left[ -x\log x + x + \frac{x^2}{2}\log x - \frac{x^2}{4} + \frac{x^2}{2} - x \right]_0^1 = \frac{1}{4}, \tag{10.193}$$

from which

$$e^2 \bar{u}(p')\Lambda_\mu^{(2)}(p',p)u(p) \rightarrow -\frac{e^2}{16\pi^2 m} \bar{u}(p')[p_\mu + p'_\mu]u(p). \tag{10.194}$$

Using the Gordon identity in this expression, and eliminating the further contribution in $\gamma_\mu$, we obtain the correction to the magnetic moment

$$e^2 \bar{u}(p') \Lambda_\mu^{(2)}(p', p) u(p)|_{\text{magn. mom.}} \to \frac{ie^2}{16\pi^2 m} \bar{u}(p') \sigma_{\mu\nu} q^\nu u(p). \quad (10.195)$$

Finally we have to add this correction to the vertex part taken at $p' = p$, which coincides with the free vertex

$$\bar{u}(p')[\gamma_\mu + \frac{ie^2}{16\pi^2 m} \sigma_{\mu\nu} q^\nu] u(p)\bigg|_{p' \approx p}$$

$$\approx \bar{u}(p') \left[ \frac{p_\mu + p'_\mu}{2m} + \frac{i}{2m} \left( 1 + \frac{e^2}{8\pi^2} \right) \sigma_{\mu\nu} q^\nu \right] u(p). \quad (10.196)$$

Therefore the correction is

$$\frac{e}{2m} \to \frac{e}{2m} \left( 1 + \frac{\alpha}{2\pi} \right). \quad (10.197)$$

Recalling that $g$ is the ratio between $\vec{S} \cdot \vec{B}$ and $e/2m$, we get

$$\frac{g}{2} = 1 + \frac{\alpha}{2\pi} + \mathcal{O}(\alpha^2). \quad (10.198)$$

This correction was evaluated by Schwinger in 1948. Actually we know the first three terms of the expansion

$$a_{\text{th}} = \frac{1}{2}(g - 2) = \frac{1}{2}\frac{\alpha}{\pi} - 0.32848 \left( \frac{\alpha}{\pi} \right)^2 + 1.49(\frac{\alpha}{\pi})^3 + \cdots$$

$$= (1159652.4 \pm 0.4) \times 10^{-9}, \quad (10.199)$$

to be compared with the experimental value

$$a_{\text{exp}} = (1159652.4 \pm 0.2) \times 10^{-9}. \quad (10.200)$$

## 10.7  Exercises

(1) Using the two regularized expressions (10.79) and (10.111) prove the identity

$$\frac{\partial \Sigma(p)}{\partial p} = \Lambda_\mu(p, p). \quad (10.201)$$

This relation must hold order by order in the expansion in $\epsilon = 4 - 2\omega$. Check that this is true for the terms of order $1/\epsilon$.

(2) Shows that the Lagrangian of a classical massless Dirac field interacting with an electromagnetic field is invariant under the chiral transformation

$$\psi \to e^{i\alpha\gamma_5} \psi. \quad (10.202)$$

(3) Verify the relations (10.187), (10.188) and (10.189) satisfied by the $\gamma$-matrices.

(4) Another regularization used in QED is the Pauli-Villars, consisting in changing the photon propagator

$$\frac{1}{k^2} \to \frac{1}{k^2} - \frac{1}{k^2 - M^2}. \tag{10.203}$$

where $M$ is a large mass. Consider the divergent integral

$$I = \int d^4k \frac{1}{k^2[(k+p)^2 - m^2]}. \tag{10.204}$$

Show that this integral is finite using Pauli-Villars regularization, and evaluate it. Show that the result diverges logarithmically for $M \to \infty$.

(5) Given a scalar and a Dirac field, assume that they are interacting via the interaction

$$\mathcal{L}_I = g\bar{\psi}\gamma_5\psi\phi. \tag{10.205}$$

Although this theory is similar to spinor QED, and its couplings satisfy $[g_i] \geq 0$, it is not strictly renormalizable. Explain why and which kind of modification is necessary in order to make it strictly renormalizable.

# Chapter 11

# Path integral formulation of quantum mechanics

## 11.1 Feynman's formulation of quantum mechanics

In 1948 Feynman (see, for instance, [Feynman (1948a)] and [Feynman and Hibbs (2010)]) gave a formulation of quantum mechanics quite different from the standard one. This formulation did not rely upon the standard tools of quantum mechanics as Hilbert spaces, linear operators, Schrödinger equation etc., but it gave an explicit expression for the quantum-mechanical amplitudes. This approach was based on a mathematical formalism far from being well defined. In fact, the basis was rather the physical idea that the probability amplitude for going from one state to another could be obtained by summing together all the possible paths connecting the two states, weighted with an appropriate factor. This way of facing the problem was very close to the use of composition law in probability theory. However, Feynman was applying the composition law to probability amplitudes rather than to probabilities. The key observation to get the appropriate weight was due to Dirac (see, for example [Dirac (2001)]), who proved that, for one degree of freedom, the probability amplitude $\langle q', t + \Delta t | q, t \rangle$, in the case of an infinitesimal $\Delta t$, was given by $e^{iS}$, with $S$ the classical action to go from the point $q$ at the time $t$ to the point $q'$ at the time $t + \Delta t$. In fact we will show that Feynman's formulation can be derived from the usual formalism of quantum mechanics. The path integral formulation (as Feynman's approach is usually called) looks as just another approach to quantum mechanics. However, at the end of the '60s it turned out to be a crucial tool for the quantization of gauge field theories [Faddeev and Popov (1967)]. Furthermore, it is one of the few methods which can be used for studying situations outside the perturbative approach. In order to derive Feynman's formulation, we will start studying a quantum system consisting in one degree of freedom. We will consider later on the extension to infinite degrees of freedom, as needed in quantum field theory.

We recall that the quantum mechanical problem is essentially the evaluation of the time evolution of a dynamical system, that is the calculation of the propagator. This can be evaluated in any basis, for instance, in configuration space. Therefore, we need to evaluate the expression

$$\langle q', t' | q, t \rangle, \quad t' \geq t. \tag{11.1}$$

It will be convenient to work in the Heisenberg representation, where the operators are time dependent. Then, the states in the previous equation are defined as eigenstates of the operator $\mathbf{q}(t)$ at the time $t$, that is

$$\mathbf{q}(t) | q, t \rangle = | q, t \rangle q. \tag{11.2}$$

In the case of a time-independent Hamiltonian (but the approach can be extended to the more general case of a time-dependent Hamiltonian), the following relation holds:

$$|q, t\rangle = e^{iHt} |q, 0\rangle. \tag{11.3}$$

The matrix element (11.1) can be evaluated by slicing the time interval $t' - t$ in infinitesimal pieces and using the completeness relation. We get

$$\langle q', t' | q, t \rangle = \lim_{n \to \infty} \int dq_1 \cdots dq_{n-1} \langle q', t' | q_{n-1}, t_{n-1} \rangle \langle q_{n-1}, t_{n-1} | q_{n-2}, t_{n-2} \rangle \cdot$$
$$\cdots \langle q_1, t_1 | q, t \rangle, \tag{11.4}$$

where

$$t_k = t + k\epsilon, \quad t_0 \equiv t, \quad t_n \equiv t'. \tag{11.5}$$

It follows

$$\langle q', t' | q, t \rangle = \lim_{n \to \infty} \int \left( \prod_{k=1}^{n-1} dq_k \right) \left( \prod_{k=0}^{n-1} \langle q_{k+1}, t_{k+1} | q_k, t_k \rangle \right), \tag{11.6}$$

where, $q_1 = q$ and $q_n = q'$. We have reduced the problem of evaluating (11.1) to the case of an infinitesimal time interval $\epsilon$. Using the completeness in momentum space, with states taken at a time $\tilde{t}$ in between $t_k$ and $t_{k+1}$:

$$\langle q_{k+1}, t_k + \epsilon | q_k, t_k \rangle = \int dp \langle q_{k+1}, t_k + \epsilon | p, \tilde{t} \rangle \langle p, \tilde{t} | q_k, t_k \rangle. \tag{11.7}$$

The mixed matrix elements can be written as

$$\langle q_{k+1}, t_k + \epsilon | p, \tilde{t} \rangle = \langle q_{k+1}, 0 | e^{-iH(\mathbf{p}, \mathbf{q})(t_k + \epsilon - \tilde{t})} | p, 0 \rangle. \tag{11.8}$$

Making use of the canonical commutation relations for $\mathbf{p}$ and $\mathbf{q}$, the Hamiltonian can be written in a form where all the operators $\mathbf{q}$ are at the left of all the operators $\mathbf{p}$. The result will be denoted by the symbol $H_+(\mathbf{q}, \mathbf{p})$. We get:

$$\langle q_{k+1}, t_k + \epsilon | p, \tilde{t} \rangle \simeq \langle q_{k+1}, 0 | p, 0 \rangle \left( 1 - iH_+(q_{k+1}, p)(t_k + \epsilon - \tilde{t}) \right)$$
$$\simeq \langle q_{k+1}, 0 | p, 0 \rangle e^{-iH_+(q_{k+1}, p)(t_k + \epsilon - \tilde{t})}. \tag{11.9}$$

The function $H_+(q, p)$ is obtained by substituting in $H_+(\mathbf{q}, \mathbf{p})$ the operators $\mathbf{q}$ and $\mathbf{p}$ with their eigenvalues. Notice that, although $H(\mathbf{q}, \mathbf{p})$ and $H_+(\mathbf{q}, \mathbf{p})$ are the same operator, $H(q, p)$ and $H_+(q, p)$ are, in general, different functions. In this way we get

$$\langle q_{k+1}, t_k + \epsilon | p, \tilde{t} \rangle \simeq \frac{1}{\sqrt{2\pi}} e^{i(pq_{k+1} - H_+(q_{k+1}, p)(t_k + \epsilon - \tilde{t}))}, \tag{11.10}$$

with an analogous relation for

$$\langle p, \tilde{t} | q_k, t_k \rangle \simeq \frac{1}{\sqrt{2\pi}} e^{i(-pq_k - H_-(q_k, p)(\tilde{t} - t_k))}. \tag{11.11}$$

In the previous equation, $H_-$ is the eigenvalue of the operator $H_-(\mathbf{p}, \mathbf{q})$ defined by moving the operators $\mathbf{q}$ to the right of $\mathbf{p}$ by using the canonical commutation relations. By choosing $\tilde{t} = (t_k + t_{k+1})/2$ in the previous two equations, we find

$$\langle q_{k+1}, t_k + \epsilon | q_k, t \rangle \simeq \int \frac{dp}{2\pi} e^{i\epsilon(p(q_{k+1} - q_k)/\epsilon - H_c(q_{k+1}, q_k, p))}, \tag{11.12}$$

where

$$H_c(q_{k+1}, q_k, p) = \frac{1}{2}(H_+(q_{k+1}, p) + H_-(q_k, p)). \tag{11.13}$$

Since in the limit $\epsilon \to 0$ we have

$$\langle q_{k+1}, t_k + \epsilon | q_k, t_k \rangle \to \delta(q_{k+1} - q_k), \tag{11.14}$$

it is reasonable to define a **velocity** variable:

$$\dot{q}_k = \frac{q_{k+1} - q_k}{\epsilon}, \tag{11.15}$$

and

$$H_c(q_{k+1}, q_k, p) \equiv H(q_k, p). \tag{11.16}$$

In the same limit, $H(q_k, p)$ is the Hamiltonian of the system evaluated in the phase space, except for some reordering term. Therefore

$$\langle q_{k+1}, t_k + \epsilon | q_k, t_k \rangle \simeq \int \frac{dp}{2\pi} e^{i\epsilon L(q_k, p)}, \tag{11.17}$$

where

$$L(q_k, p) = p\dot{q}_k - H(q_k, p) \tag{11.18}$$

is the Lagrangian in phase space. Coming back to (11.6) and using (11.17), we obtain

$$\langle q', t' | q, t \rangle = \lim_{n \to \infty} \int \left( \prod_{k=1}^{n-1} dq_k \right) \left( \prod_{k=0}^{n-1} \frac{dp_k}{2\pi} \right) e^{i \sum_{k=0}^{n-1} \epsilon L(q_k, p_k)}. \tag{11.19}$$

In the limit $n \to \infty$, the argument of the exponential gives the integral between $t$ and $t'$ of the Lagrangian in the phase space, that is the canonical action

$$S = \int_t^{t'} dt\, L(q, p). \tag{11.20}$$

In the limit, the multiple integral in eq. (11.19) is called the **functional integral**, and we will denote it symbolically as

$$\langle q', t' | q, t \rangle = \int_{q,t}^{q',t'} d\mu(q(t)) d\mu(p(t)) e^{i \int_t^{t'} dt L(q,p)}, \tag{11.21}$$

where

$$d\mu(q(t)) = \prod_{t < t'' < t'} dq(t''), \quad d\mu(p(t)) = \prod_{t \le t'' \le t'} \frac{dp(t'')}{2\pi}. \tag{11.22}$$

The asymmetry between the integration in $q(t)$ and $p(t)$ implies that the integral is not invariant under a generic canonical transformation. Notice that the integration is made over all the functions $q(t'')$ such that $q(t) = q$ and $q(t') = q'$. On the contrary, there are no limitations on $p(t)$ and $p(t')$. This is related to the uncertainty principle. The expression (11.21) is the **path integral in phase space**.

## 11.2   Path integral in configuration space

The path integral in eq. (11.19) can be easily written as a functional integral in configuration space, if the Hamiltonian of the system is of the type

$$H(\mathbf{q}, \mathbf{p}) = \frac{\mathbf{p}^2}{2m} + V(\mathbf{q}). \tag{11.23}$$

In such a case

$$H_+(q, p) = H_-(q, p) = H(q, p) = \frac{p^2}{2m} + V(q), \tag{11.24}$$

and the momentum integration can be performed explicitly

$$\langle q', t'|q, t\rangle = \lim_{n\to\infty} \int \left(\prod_{k=1}^{n-1} dq_k\right)$$

$$\times \prod_{k=0}^{n-1} \frac{dp_k}{2\pi} e^{i[p_k(q_{k+1}-q_k)-\epsilon p_k^2/2m-\epsilon(V(q_{k+1})+V(q_k))/2]}$$

$$= \lim_{n\to\infty} \int \left(\prod_{k=1}^{n-1} dq_k\right) \left(\frac{1}{2\pi}\sqrt{\frac{2m\pi}{i\epsilon}}\right)^n \prod_{k=0}^{n-1} e^{(-2m/i\epsilon)(q_{k+1}-q_k)^2/4}$$

$$\times e^{-i\epsilon(V(q_{k+1})+V(q_k))/2}$$

$$= \lim_{n\to\infty} \int \left(\sqrt{\frac{m}{2i\pi\epsilon}}\right)^n \left(\prod_{k=1}^{n-1} dq_k\right)$$

$$\times \prod_{k=0}^{n-1} e^{i\epsilon[m((q_{k+1}-q_k)/\epsilon)^2/2-(V(q_{k+1})+V(q_k))/2]}, \tag{11.25}$$

with the result

$$\langle q', t'|q, t\rangle = \lim_{n\to\infty} \int \left(\sqrt{\frac{m}{2i\pi\epsilon}}\right)^n \left(\prod_{k=1}^{n-1} dq_k\right)$$

$$\times e^{i\epsilon \sum_{k=0}^{n-1}[m((q_{k+1}-q_k)/\epsilon)^2/2-(V(q_{k+1})+V(q_k))/2]}. \tag{11.26}$$

We also have

$$\lim_{n\to\infty} \sum_{k=0}^{n-1} \epsilon \left[\frac{m}{2}\left(\frac{q_{k+1}-q_k}{\epsilon}\right)^2 - \frac{V(q_{k+1})+V(q_k)}{2}\right] = \int_t^{t'} dt L(q, \dot{q}), \tag{11.27}$$

where $L(q, \dot{q})$ is the Lagrangian in configuration space. Therefore, we will write

$$\langle q', t'|q, t\rangle = \int_{q,t}^{q',t'} \mathcal{D}(q(t)) e^{i\int_t^{t'} dt L(q,\dot{q})}, \tag{11.28}$$

with the functional measure given by

$$\mathcal{D}(q(t)) = \lim_{n\to\infty} \left(\sqrt{\frac{m}{2i\pi\epsilon}}\right)^n \prod_{k=1}^{n-1} dq_k. \tag{11.29}$$

Let us notice that for $N$ degrees of freedom, with Hamiltonian of the type

$$H(\mathbf{q}_i, \mathbf{p}_i) = \sum_{i=1}^{N} \frac{\mathbf{p}_i^2}{2m} + V(\mathbf{q}_i), \quad i = 1, \ldots, N, \tag{11.30}$$

the expression (11.28) can be easily generalized. However, for arbitrary dynamical systems, one has to start with the formulation in the phase space given in the previous Section, which can be easily generalized to an arbitrary number of degrees of freedom.

The expression (11.28) (really defined by (11.26)), formulated originally by Feynman, has a simple physical interpretation. Recalling the boundary conditions

$$q(t) = q, \qquad q(t') = q', \tag{11.31}$$

we see that the expression in the exponent of (11.28) is nothing but the action evaluated for a continuous path starting from $q$ at time $t$ and ending at $q'$ at time $t'$. This path is approximated in eq. (11.26) by many infinitesimal straight paths as shown in Fig. 11.1.

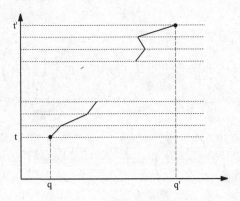

Fig. 11.1 The Feynman definition for the path integral.

According to Feynman, the probability amplitude to go from $(q, t)$ to $(q', t')$ is obtained by multiplying together the amplitudes corresponding to the infinitesimal straight paths. Each of these amplitudes (neglecting the normalization factor $\sqrt{m/2i\pi\epsilon}$) is given by the exponential of $i$ times the action evaluated along the path. Furthermore, we need to sum (or integrate functionally) over all the possible paths joining the two points. In a more concise way, one says that the amplitude to go from one point to another is obtained by summing over all the possible paths joining the two points with a weight proportional to the exponential of $iS$, where $S$ is the classical action evaluated along the path.

## 11.3  The physical interpretation of the path integral

Feynman arrived at the path integral formulation in configuration space using physical arguments, but he also knew, from Dirac's work, that the amplitude, for an infinitesimal time difference, is proportional to $\exp(iS)$, with $S$ the classical action evaluated along the infinitesimal path. From that, Feynman could derive directly the final result (see [Feynman and Hibbs (2010)]). Let us consider the classical double slit experiment, as illustrated in Fig. 11.2, where S is the electron source. The electrons are focused by the slit A, go through one of the two holes in B, and arrive at the screen C, where they are detected by the detector D. By counting the electrons arriving at the detector, one can determine the number of electrons arriving at C as a function of the coordinate $x$. This number, divided by the total number of electrons emitted by the source, is proportional to the probability distribution of the electrons on the screen C. Under the conditions of Fig. 11.2, that is with both slits 1 and 2 open, one gets the distribution given in Fig. 11.3. In order to explain this distribution, it would seem reasonable to make the following assumptions:

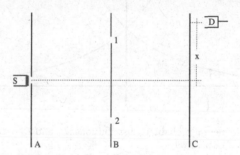

Fig. 11.2   The double slit experiment.

- Each electron must go through the slit 1 or through the slit 2.
- As a consequence, the probability for the electron to reach the screen at the point $x$ must be given by the sum of the probability to reach $x$ through 1 with the probability to go through 2.

These hypotheses can be easily checked by closing one of the two slits and doing again the experiment. The results obtained with this option are given in Fig. 11.4.

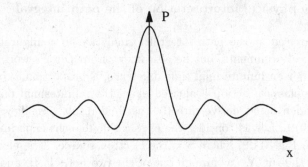

Fig. 11.3   The experimental probability distribution, $P$, with both slits open.

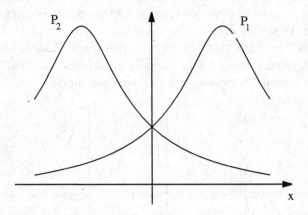

Fig. 11.4   The experimental probability distributions, $P_1$ and $P_2$ obtained keeping open the slit 1 or the slit 2 respectively.

Since $P \neq P_1 + P_2$, one (or both) of our hypotheses must be wrong. In fact, quantum mechanics tells us that we can say that an electron goes through one of the two slits, say 1, with absolute certainty, only if we detect it going through the slit 1. Only in this case, one would get $P = P_1 + P_2$. In fact, this is the result that it is obtained when doing the experiment with a further detector telling us which of the holes have been crossed by the electron. This means that, with this experimental setting, the interference gets destroyed. On the contrary, the experiment tells us that, when we do not detect the hole crossed by the electron, the total amplitude for the process is given by the sum of the amplitudes corresponding to the two different possibilities, that is

$$P = |A|^2 = |A_1 + A_2|^2 = |A_1|^2 + |A_2|^2 + A_1^* A_2 + A_1 A_2^*$$
$$\neq |A_1|^2 + |A_2|^2 = P_1 + P_2. \tag{11.32}$$

The two expressions differ for a term which is not positive definite, and which is the direct responsible for the interference pattern. Therefore, quantum mechanics tells us that the usual rules for combining together probabilities are not right, when we do not know the path followed by the particle. According to the usual interpretation of quantum mechanics, in this case, the concept of the path followed by a particle is misleading.

Feynman's interpretation is different, though the theories turn out to be equivalent. According to him, it is still possible to speak about trajectories followed by the electron, if one gives up to the usual rules for combining probabilities. This point of view requires taking into account different types of alternatives according to different experimental settings. In particular, Feynman defines the following alternatives:

(1) **Exclusive alternatives** - He speaks about exclusive alternatives, when use is made of an apparatus to determine the alternatives followed by the system. For instance, if one makes use of a detector to determine the hole crossed by the electron. In this case, the total probability is given by the sum of the probabilities corresponding to the different alternatives:

$$P = \sum_i P_i. \tag{11.33}$$

(2) **Interfering alternatives** - This is the situation occurring when the possible alternatives for the system are not detected by an experimental apparatus. In this case, we associate to each alternative an amplitude $A_i$ and assume that the total amplitude is given by the sum of all the amplitudes $A_i$

$$A = \sum_i A_i. \tag{11.34}$$

The total probability is given by

$$P = |A|^2 = |\sum_i A_i|^2. \tag{11.35}$$

Therefore, the probability rules are strictly related to the experimental setting. There are situations in which both types of alternatives occur. For instance, suppose we want to know the total probability of having an electron in C, within the interval $(a, b)$, having a detector in C and not looking at the slit crossed by the electrons. In this case, the total probability will be given by

$$P = \int_a^b dx |A_1(x) + A_2(x)|^2. \tag{11.36}$$

Since the alternatives of getting the electron at different points in C are exclusive ones.

From this point of view, it is possible to speak about trajectories of the electron, but we have to associate to each possible path an amplitude and then sum over all the paths. For instance, suppose that we want to evaluate the amplitude to go from S to the screen C when the screen B is eliminated. This can be done by increasing the number of holes in screen B. Then, if we denote by $A(x, y)$ the amplitude for the electron to reach $x$ going through the hole placed at a distance $y$ from the center of B, the total amplitude will be given by

$$A(x) = \int dy \, A(x, y). \tag{11.37}$$

We can continue by using more and more screens of type B with more and more holes. That is, we can construct a lattice of points between A and C and consider all the possible paths from A to the point $x$ on C, joining the sites of the lattice, as shown in Fig. 11.5. By denoting with $\mathcal{C}$ the generic path, the total amplitude will be given by

$$A(x) = \sum_{\mathcal{C}} A(x, \mathcal{C}). \tag{11.38}$$

At this point, it is easy to get the path integral expression for the total amplitude, since each path is obtained by joining together many infinitesimal paths, and for each of these paths the amplitude is proportional to $\exp(iS)$, where $S$ is the action for the infinitesimal path. Therefore, for the single path $\mathcal{C}$ the amplitude will be given by $\exp(iS(\mathcal{C}))$. This is the final result, except for a proportionality factor, that can be obtained from the phase space formulation. This is important, because in several cases the simple result outlined above is not the correct one (see exercise (1) in this Chapter).

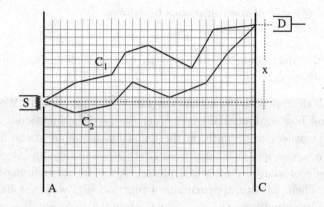

Fig. 11.5 The lattice for the evaluation of equation (11.38).

## 11.4 Functional formalism

The most natural mathematical setting to deal with path integration, as defined in Section 11.1, is the theory of functionals. The concept of functional is nothing but an extension of the concept of function. Consider, for instance, a real function. This is a mapping from a manifold $\mathcal{M}$ to $R^1$, that is a correspondence which associates to a point of the manifold $\mathcal{M}$ a real number. A real functional is defined as an application from a space of functions (and therefore an $\infty$-dimensional space) to $R^1$. If we denote the space of real functions by $\mathcal{L}$, a real functional is the mapping

$$F: \quad \mathcal{L} \to R^1. \tag{11.39}$$

We will make use of the notations

$$F[\eta], \quad F, \quad F[\cdot], \qquad \eta \in \mathcal{L} \tag{11.40}$$

to denote the functional defined in eq. (11.39). Examples of functionals are

- $\mathcal{L} = L_w^1[-\infty, +\infty]$ = the space of the integrable functions in $R^1$ with respect to the measure $w(x)dx$:

$$F_1[\eta] = \int_{-\infty}^{+\infty} dx\, w(x)\eta(x). \tag{11.41}$$

- $\mathcal{L} = L^2[-\infty, +\infty]$ = space of the square integrable functions in $R^1$:

$$F_2[\eta] = e^{-\frac{1}{2}\int_{-\infty}^{+\infty} dx\, \eta^2(x)}. \tag{11.42}$$

- $\mathcal{L} = \mathcal{C}$ = space of the continuous functions:

$$F_3[\eta] = \eta(x_0). \tag{11.43}$$

$F_3$ is the functional that associates to each continuous function its value at a given point $x_0$.

The next step is to define the derivative of a functional. We need to understand how a functional varies when we vary the functions. A simple way is to consider some path on the definition manifold of the functions. For instance, consider functions from $R^1$ to $R^1$, and take an interval $(t, t')$ of $R^1$ made of many infinitesimal pieces, as we did for defining the path integral. Then, we can approximate a function $\eta(t)$ with the limit of its discrete approximations $(\eta_1, \eta_2, \cdots, \eta_n)$ obtained by evaluating the functions at the discrete points in which the interval has been divided. In this approximation, we can think of the functional $F$ as an application from $R^n \to R^1$. That is

$$F[\eta] \to F(\eta_1, \cdots, \eta_n) \equiv F(\eta_i), \quad i = 1, \cdots, n. \tag{11.44}$$

Then, we consider the variation of $F(\eta_i)$ when we vary the quantities $\eta_i$:

$$F(\eta_i + \delta\eta_i) - F(\eta_i) = \sum_{j=1}^{n} \frac{\partial F(\eta_i)}{\partial \eta_j} \delta\eta_j. \tag{11.45}$$

Let us put

$$K_i = \frac{1}{\epsilon} \frac{\partial F}{\partial \eta_i}, \tag{11.46}$$

where $\epsilon$ is the infinitesimal interval where we evaluate the discrete apoproximation to the function $\eta$. Then, for $\epsilon \to 0$

$$F(\eta_i + \delta\eta_i) - F(\eta_i) = \sum_{j=1}^{n} K_j \delta\eta_j \epsilon \to \int K \, \delta\eta \, dx, \tag{11.47}$$

that is

$$\delta F[\eta] = \int K(x) \, \delta\eta(x) \, dx. \tag{11.48}$$

Since $\delta\eta(x)$ is the variation of the function evaluated at $x$, we will define the *functional derivative* as

$$\delta F[\eta] = \int \frac{\delta F[\eta]}{\delta \eta(x)} \delta\eta(x) \, dx. \tag{11.49}$$

In other words, to evaluate the functional derivative of a functional, we first evaluate its infinitesimal variation and then we extract the functional derivative by comparison with the previous formula (the coefficient of $\delta\eta$). Looking at eq. (11.46) we get

$$\frac{\delta F[\eta]}{\delta\eta(x)} = \lim_{\epsilon\to 0}\frac{1}{\epsilon}\frac{\partial F(\eta_i)}{\partial\eta_i}, \qquad i = 1,\cdots,n,$$

$$\eta_i = \eta(x_i), \qquad x_i = x_1 + (i-1)\epsilon, \qquad \epsilon = \frac{x_n - x_1}{n-1}. \qquad (11.50)$$

Let us now evaluate the functional derivatives of the previously defined functionals:

$$\delta F_1[\eta] = \int_{-\infty}^{+\infty} w(x)\delta\eta(x)dx, \qquad (11.51)$$

giving

$$\frac{\delta F_1[\eta]}{\delta\eta(x)} = w(x). \qquad (11.52)$$

Then,

$$\delta F_2[\eta] = e^{-\frac{1}{2}\int dx\eta^2(x)}\left(-\int \eta(x)\,\delta\eta(x)\,dx\right), \qquad (11.53)$$

and, therefore

$$\frac{\delta F_2[\eta]}{\delta\eta(x)} = -\eta(x)F_2[\eta]. \qquad (11.54)$$

In order to evaluate the derivative of $F_3$ let us write

$$F_3[\eta] = \int \eta(x)\delta(x-x_0)dx. \qquad (11.55)$$

In this way, $F_3$ looks formally as a functional of type $F_1$, and

$$\frac{\delta F_3[\eta]}{\delta\eta(x)} = \delta(x-x_0). \qquad (11.56)$$

As a final example, consider

$$F_4[\eta] = e^{-\frac{1}{2}\int dxdyK(x,y)\eta(x)\eta(y)}. \qquad (11.57)$$

We find

$$\delta F_4[\eta] = e^{-\frac{1}{2}\int dxdyK(x,y)\eta(x)\eta(y)}\left(-\int dxdyK(x,y)\delta\eta(x)\eta(y)\right), \quad (11.58)$$

where we have used the symmetry of the integral with respect to the exchange $x \leftrightarrow y$. We obtain

$$\frac{\delta F_4[\eta]}{\delta\eta(x)} = -\int dyK(x,y)\eta(y)F_4[\eta]. \qquad (11.59)$$

Starting from the functional derivative, it is possible to generalize the Taylor expansion to the functional case. This generalization is called the Volterra expansion of a functional:

$$F[\eta_0 + \eta_1] = \sum_{k=0}^{\infty}\frac{1}{k!}\int dx_1\cdots dx_k\frac{\delta^k F[\eta]}{\delta\eta(x_1)\cdots\delta\eta(x_k)}\bigg|_{\eta=\eta_0}\eta_1(x_1)\cdots\eta_1(x_k).$$

$$(11.60)$$

All these definitions are easily generalized to the case of $\eta$ being a function from $R^m \to R^n$, or from a manifold $M$ to a manifold $N$.

## 11.5　General properties of the path integral

In this Section, we would like to point out some important properties of the path integral, namely (remember that $t$ and $t'$ define the integration limits of the action):

- **Invariance of the functional measure under translations** - If the translation is defined as follows

$$q(\tau) \to q(\tau) + \eta(\tau), \qquad \eta(t) = \eta(t') = 0,$$
$$p(\tau) \to p(\tau) + \pi(\tau), \qquad \text{for any } \pi(t),\ \pi(t'), \qquad (11.61)$$

  the property follows immediately from the definition of the path integral in phase space.

- **Factorization of the path integral**. From the very definition of the path integral, it follows that in order to evaluate the amplitude for going from $q$ at time $t$ to $q'$ at time $t'$ we can first go to $q''$ at a time $t \leq t'' \leq t'$, then, sum over all the paths from $q$ to $q''$ and from $q''$ to $q'$, and eventually integrate over all the possible intermediate points $q''$ (see Fig. 11.6). In equations, we get

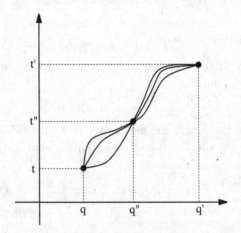

Fig. 11.6　The factorization of the path integral.

$$\langle q', t' | q, t \rangle = \int_{q,t}^{q't'} d\mu(q(\tau)) d\mu(p(\tau)) e^{i \int_t^{t''} L d\tau + i \int_{t''}^{t'} L d\tau}$$

$$= \int dq'' \int_{q,t}^{q'',t''} d\mu(q(\tau)) d\mu(p(\tau)) e^{i \int_t^{t''} L d\tau}$$

$$\times \int_{q'',t''}^{q',t'} d\mu(q(\tau)) d\mu(p(\tau)) e^{i \int_{t''}^{t'} L d\tau}$$

$$= \int dq'' \langle q', t' | q'', t'' \rangle \langle q'', t'' | q, t \rangle, \qquad (11.62)$$

as it follows from the factorization properties of the measure

$$d\mu(q(\tau)) = \prod_{t < \tau < t'} dq(\tau) = dq'' \prod_{t < \tau < t''} dq(\tau) \prod_{t'' < \tau < t'} dq(\tau),$$

$$d\mu(p(\tau)) = \prod_{t \leq \tau \leq t''} dp(\tau) = \prod_{t \leq \tau \leq t''} dp(\tau) \prod_{t'' \leq \tau \leq t'} dp(\tau). \qquad (11.63)$$

Therefore, the factorization property is equivalent to the completeness relation.

The path integral allows us to evaluate expectation values. To this end, we will consider expressions of the type (called the average value of $F$)

$$\int_{q,t}^{q',t'} d\mu(q(\tau)) d\mu(p(\tau)) F[q(\tau), p(\tau)] e^{iS}, \qquad (11.64)$$

where $F$ is an arbitrary functional of $q$, $p$ and of their time derivatives. Let us start with

$$\int_{q,t}^{q',t'} d\mu(q(\tau)) d\mu(p(\tau)) e^{iS} q(t''), \qquad t \leq t'' \leq t'. \qquad (11.65)$$

We first notice that

$$\int_{q,t}^{q',t'} d\mu(q(\tau)) d\mu(p(\tau)) e^{iS} q(t') = q' \int_{q,t}^{q',t'} d\mu(q(\tau)) d\mu(p(\tau)) e^{iS}$$

$$= q' \langle q', t' | q, t \rangle, \qquad (11.66)$$

since $q(t)$ satisfies the boundary condition $q(t') = q'$. By using the factorization, we get

$$\int_{q,t}^{q',t'} d\mu(q(\tau)) d\mu(p(\tau)) e^{iS} q(t'') = \int dq'' \int_{q,t}^{q'',t''} d\mu(q(\tau)) d\mu(p(\tau)) e^{iS} q(t'')$$

$$\times \int_{q'',t''}^{q',t'} d\mu(q(\tau)) d\mu(p(\tau)) e^{iS} = \int dq'' \langle q', t' | q'', t'' \rangle q'' \langle q'', t'' | q, t \rangle. \qquad (11.67)$$

On the other hand, $|q, t\rangle$ is an eigenstate of $\mathbf{q}(t)$, therefore

$$\int_{q,t}^{q',t'} d\mu(q(\tau))d\mu(p(\tau))e^{iS}q(t'') = \langle q',t'|\mathbf{q}(t'')|q,t\rangle. \tag{11.68}$$

The average value of $q(t'')$ coincides with the expectation value of the operator $\mathbf{q}(t'')$.

Then, let us study the average value of $q(t_1)q(t_2)$, with $t \le t_1, t_2 \le t'$. By proceeding as before, we will get the expectation value of $\mathbf{q}(t_1)\mathbf{q}(t_2)$ if $t_1 \ge t_2$, or $\mathbf{q}(t_2)\mathbf{q}(t_1)$ if $t_2 \ge t_1$. In terms of the $T$-product we obtain

$$\int_{q,t}^{q',t'} d\mu(q(\tau))d\mu(p(\tau))e^{iS}q(t_1)q(t_2) = \langle q',t'|T(\mathbf{q}(t_1)\mathbf{q}(t_2))|q,t\rangle. \tag{11.69}$$

Analogously,

$$\int_{q,t}^{q',t'} d\mu(q(\tau))d\mu(p(\tau))e^{iS}q(t_1)\cdots q(t_n) = \langle q',t'|T(\mathbf{q}(t_1)\cdots\mathbf{q}(t_n))|q,t\rangle. \tag{11.70}$$

Let us now consider the Volterra expansion of a functional of $q(t)$:

$$\int_{q,t}^{q',t'} d\mu(q(\tau))d\mu(p(\tau))F[q]e^{iS}$$

$$= \sum_{k=0}^{\infty} \frac{1}{k!} \int dt_1 \cdots dt_k \frac{\delta^k F[q]}{\delta q(t_1)\cdot \delta q(t_k)}\bigg|_{q=0}$$

$$\times \int_{q,t}^{q',t'} d\mu(q(\tau))d\mu(p(\tau))e^{iS}q(t_1)\cdots q(t_k)$$

$$= \sum_{k=0}^{\infty} \frac{1}{k!} \int dt_1 \cdots dt_k \frac{\delta^k F[q]}{\delta q(t_1)\cdots \delta q(t_k)}\bigg|_{q=0}$$

$$\times \langle q',t'|T(\mathbf{q}(t_1)\cdots\mathbf{q}(t_k))|q,t\rangle \equiv \langle q',t'|T(F[\mathbf{q}])|q,t\rangle, \tag{11.71}$$

where

$$T(F[\mathbf{q}]) \equiv \sum_{k=0}^{\infty} \frac{1}{k!} \int dt_1 \cdots dt_k \frac{\delta^k F[q]}{\delta q(t_1)\cdots \delta q(t_k)}\bigg|_{q=0} T(\mathbf{q}(t_1)\cdots\mathbf{q}(t_k)). \tag{11.72}$$

If the functional depends also on the time derivatives, a certain caution is necessary. For instance, let us study the average value of $q(t_1)\dot{q}(t_2)$

$(t \leq t_1, t_2 \leq t')$:

$$\int_{q,t}^{q',t'} d\mu(q(\tau))d\mu(p(\tau))e^{iS}q(t_1)\dot{q}(t_2)$$

$$= \lim_{\epsilon \to 0} \int_{q,t}^{q',t'} d\mu(q(\tau))d\mu(p(\tau))e^{iS}q(t_1)\frac{q(t_2+\epsilon) - q(t_2)}{\epsilon}$$

$$= \lim_{\epsilon \to 0} \frac{1}{\epsilon}\langle q',t'|T\Big(\mathbf{q}(t_1)\mathbf{q}(t_2+\epsilon)\Big) - T\Big(\mathbf{q}(t_1)\mathbf{q}(t_2)\Big)|q,t\rangle$$

$$= \frac{d}{dt_2}\langle q',t'|T\Big(\mathbf{q}(t_1)\mathbf{q}(t_2)\Big)|q,t\rangle. \tag{11.73}$$

Let us define an ordered $T^\star$-product (or covariant T-product) as

$$T^\star\Big(\mathbf{q}(t_1)\dot{\mathbf{q}}(t_2)\Big) = \frac{d}{dt_2}T\Big(\mathbf{q}(t_1)\mathbf{q}(t_2)\Big). \tag{11.74}$$

Then, it is not difficult to show that, for an arbitrary functional of $q(t)$ and $\dot{q}(t)$, one has

$$\int_{q,t}^{q',t'} d\mu(q(\tau))d\mu(p(\tau))F[q,\dot{q}]e^{iS} = \langle q',t'|T^\star(F[\mathbf{q},\dot{\mathbf{q}}])|q,t\rangle. \tag{11.75}$$

The proof can be done by expanding $F[q,\dot{q}]$ in a Volterra series of $q$ and $\dot{q}$.

The power of this formulation of quantum mechanics comes about when we want to get the fundamental properties of quantum mechanics, as the equations of motion for the operators, the commutation relations, etc. All these properties follow from Schwinger's action principle [Schwinger (1951c)], that, in this context, is a simple consequence of the relation:

$$\int_{q,t}^{q',t'} d\mu(q(\tau))d\mu(p(\tau))\frac{\delta}{\delta q(\tau)}\left[F[q(\tau),p(\tau)]e^{iS}\right] = 0. \tag{11.76}$$

This follows from the invariance under translations of the functional measure. In fact, let us introduce the following notation

$$d\mu_{q,p} = d\mu(q(\tau))d\mu(p(\tau)). \tag{11.77}$$

It follows that for an infinitesimal function $\eta$, such that $\eta(t) = \eta(t') = 0$

$$\int_{q,t}^{q',t'} d\mu_{q,p}\, A[q] = \int_{q,t}^{q',t'} d\mu_{q,p}\, A[q+\eta]$$

$$= \int_{q,t}^{q',t'} d\mu_{q,p}\, A[q] + \int_{q,t}^{q',t'} d\mu_{q,p} \int \eta(t)\frac{\delta A[q]}{\delta q(t)}dt. \tag{11.78}$$

Since this relation holds for an arbitrary $\eta$, the validity of eq. (11.76) follows, By performing explicitly the functional derivative in eq. (11.76), we get a relation which is nothing but the Schwinger formulation of quantum mechanics, that is

$$i\langle q',t'|T^\star\left(F[\mathbf{q}]\frac{\delta S}{\delta\mathbf{q}(\tau)}\right)|q,t\rangle = -\langle q',t'|T^\star\left(\frac{\delta F[\mathbf{q}]}{\delta\mathbf{q}(\tau)}\right)|q,t\rangle. \qquad (11.79)$$

By selecting the functional $F$ in a convenient way, it is possible to derive all the important properties of quantum mechanics. For instance, by choosing $F[q] = 1$ we get

$$\langle q',t'|T^\star\left(\frac{\delta S}{\delta\mathbf{q}(t_1)}\right)|q,t\rangle = 0, \qquad (11.80)$$

that is the quantum equations of motion. With the choice $F[q] = q$ we get

$$\langle q',t'|T^\star\left(\mathbf{q}(t_1)\frac{\delta S}{\delta\mathbf{q}(t_2)}\right)|q,t\rangle = i\langle q',t'|q,t\rangle\delta(t_1 - t_2). \qquad (11.81)$$

Since this relation holds for arbitrary states, we get

$$T^\star\left(\mathbf{q}(t_1)\frac{\delta S}{\delta\mathbf{q}(t_2)}\right) = i\delta(t_1 - t_2). \qquad (11.82)$$

For simplicity, let us consider the following action

$$S = \int_t^{t'}\left(\frac{1}{2}m\dot{q}^2 - V(q)\right)dt, \qquad (11.83)$$

then

$$\frac{\delta S}{\delta q(t_2)} = -\left(m\ddot{q}(t_2) + \frac{\partial V(q(t_2))}{\partial q(t_2)}\right), \qquad (11.84)$$

from which

$$T^\star\left[\mathbf{q}(t_1)\left(m\ddot{\mathbf{q}}(t_2) + \frac{\partial V(\mathbf{q})}{\partial\mathbf{q}(t_2)}\right)\right] = -i\delta(t_1 - t_2). \qquad (11.85)$$

By using the definition of the $T^\star$-product we get

$$T^\star(\mathbf{q}(t_1)\ddot{\mathbf{q}}(t_2) = \frac{d^2}{dt_2^2}T(\mathbf{q}(t_1)\mathbf{q}(t_2))$$
$$= \delta(t_1 - t_2)[\mathbf{q}(t_1),\dot{\mathbf{q}}(t_1)] + T(\mathbf{q}(t_1)\ddot{\mathbf{q}}(t_2)), \qquad (11.86)$$

and, therefore

$$-m\delta(t_1 - t_2)[\mathbf{q}(t_1),\dot{\mathbf{q}}(t_2)] + T\left[\mathbf{q}(t_1)\left(m\ddot{\mathbf{q}}(t_2) + \frac{\partial V}{\partial\mathbf{q}(t_2)}\right)\right] = -i\delta(t_1 - t_2). \qquad (11.87)$$

Using the equations of motion, we get

$$[\mathbf{q}(t), m\dot{\mathbf{q}}(t)] = i, \qquad (11.88)$$

which is nothing but the canonical commutation relation

$$[\mathbf{q}(t), \mathbf{p}(t)] = i. \qquad (11.89)$$

## 11.6  The generating functional of Green's functions

The relevant amplitudes in field theories are the ones evaluated between $t = -\infty$ and $t = +\infty$. Furthermore, as we will see in the next Chapter, we will be interested in evaluating the vacuum expectation value of $T$-products of local operators. Therefore, we will consider matrix elements of the type (limiting ourselves, for the moment being, to a single degree of freedom):

$$\langle 0|T(\mathbf{q}(t_1)\cdots\mathbf{q}(t_n))|0\rangle. \tag{11.90}$$

Here $|0\rangle$ is the ground state of the system. Notice that, more generally, one can consider the matrix element

$$\langle q',t'|T(\mathbf{q}(t_1)\cdots\mathbf{q}(t_n))|q,t\rangle$$
$$= \int_{q,t}^{q',t'} d\mu(q(\tau))d\mu(p(\tau))e^{iS}q(t_1)\cdots q(t_n). \tag{11.91}$$

These matrix elements can be derived in a more compact form, by introducing the generating functional

$$\langle q',t'|q,t\rangle_J = \int_{q,t}^{q',t'} d\mu(q(\tau))d\mu(p(\tau))e^{iS_J}, \tag{11.92}$$

where

$$S_J = \int_t^{t'} d\tau\,[p\dot{q} - H + Jq], \tag{11.93}$$

and $J(\tau)$ is an arbitrary external source. By differentiating this expression with respect to the external source and evaluating the derivatives at $J = 0$ we get

$$\langle q',t'|T(\mathbf{q}(t_1)\cdots\mathbf{q}(t_n))|q,t\rangle = \left(\frac{1}{i}\right)^n \frac{\delta^n}{\delta J(t_1)\cdots\delta J(t_n)}\langle q',t'|q,t\rangle_J\bigg|_{J=0}. \tag{11.94}$$

We want to show that it is possible to evaluate the matrix element in (11.90) using the generating functional. Let us introduce the wave function of the ground state

$$\Phi_0(q,t) = e^{-iE_0 t}\langle q|0\rangle = \langle q|e^{-iHt}|0\rangle = \langle q,t|0\rangle, \tag{11.95}$$

then

$$\langle 0|T(\mathbf{q}(t_1)\cdots\mathbf{q}(t_n))|0\rangle =$$
$$= \int dq'dq\langle 0|q',t'\rangle\langle q',t'|T(\mathbf{q}(t_1)\cdots\mathbf{q}(t_n))|q,t\rangle\langle q,t|0\rangle$$
$$= \int dq'dq\Phi_0^\star(q',t')\langle q',t'|T(\mathbf{q}(t_1)\cdots\mathbf{q}(t_n))|q,t\rangle\Phi_0(q,t). \tag{11.96}$$

Next, we define the functional $Z[J]$, the generating functional of the vacuum amplitudes, that is

$$Z[J] = \int dq' dq \Phi_0^\star(q', t') \langle q', t'|q, t\rangle_J \Phi_0(q, t) = \langle 0|0\rangle_J. \qquad (11.97)$$

It follows

$$\langle 0|T(\mathbf{q}(t_1)\cdots\mathbf{q}(t_n))|0\rangle = \left(\frac{1}{i}\right)^n \frac{\delta^n}{\delta J(t_1)\cdots\delta J(t_n)} Z[J]\Big|_{J=0}. \qquad (11.98)$$

We want to show that it is possible to evaluate $Z[J]$ as a path integral with arbitrary boundary conditions on $q$ and $q'$. More precisely, we want to prove the following relation

$$Z[J] = \lim_{t\to+i\infty,\ t'\to-i\infty} \frac{e^{iE_0(t'-t)}}{\Phi_0^\star(q)\Phi_0(q')} \langle q', t'|q, t\rangle_J, \qquad (11.99)$$

with $q$ and $q'$ arbitrary and with the ground state wave functions taken at $t = 0$. To this end, let us consider an external source $J(t)$ vanishing outside the interval $(t'', t''')$ with $t' > t''' > t'' > t$. Then, we can write

$$\langle q', t'|q, t\rangle_J = \int dq'' dq''' \langle q', t'|q''', t'''\rangle \langle q''', t'''|q'', t''\rangle_J \langle q'', t''|q, t\rangle, \qquad (11.100)$$

with

$$\langle q', t'|q''', t'''\rangle = \langle q'|e^{-iH(t'-t''')}|q'''\rangle. \qquad (11.101)$$

This can be evaluated inserting a complete sets of energy eigenstates

$$\langle q', t'|q''', t'''\rangle = \sum_n \Phi_n(q')\Phi_n^\star(q''')e^{-iE_n(t'-t''')}, \quad \Phi_n(q) = \langle q|n\rangle. \qquad (11.102)$$

Since $E_0$ is the lowest eigenvalue, we get

$$\lim_{t'\to-i\infty} e^{iE_0 t'} \langle q', t'|q''', t'''\rangle$$

$$= \lim_{t'\to-i\infty} \sum_n \Phi_n(q')\Phi_n^\star(q''')e^{iE_n t'''} e^{-i(E_n - E_0)t'}$$

$$= \Phi_0(q')\Phi_0^\star(q''')e^{iE_0 t'''} = \Phi_0^\star(q''', t''')\Phi_0(q'). \qquad (11.103)$$

In analogous way we find

$$\lim_{t\to+i\infty} e^{-iE_0 t} \langle q'', t''|q, t\rangle$$

$$= \lim_{t\to+i\infty} \sum_n \Phi_n(q'')\Phi_n^\star(q)e^{-iE_n t''} e^{i(E_n - E_0)t}$$

$$= \Phi_0(q'')\Phi_0^\star(q)e^{-iE_0 t''} = \Phi_0(q'', t'')\Phi_0^\star(q). \qquad (11.104)$$

Finally

$$\lim_{t\to+i\infty,t'\to-\infty} e^{iE_0(t'-t)} \langle q',t'|q,t\rangle_J$$

$$= \int dq''dq''' \Phi_0^\star(q)\Phi_0(q')\Phi_0^\star(q''',t''')\langle q''',t'''|q'',t''\rangle_J\Phi_0(q'',t'')$$

$$= \Phi_0^\star(q)\Phi_0(q')Z[J], \qquad (11.105)$$

from which, eq. (11.99) follows. Let us notice again that the values $q$ and $q'$ are completely arbitrary, and furthermore that the coefficient in front of the matrix element is $J$-independent. Therefore, it is physically irrelevant. In fact, the relevant physical quantities are the ratios

$$\frac{\langle 0|T(\mathbf{q}(t_1)\cdots\mathbf{q}(t_n))|0\rangle}{\langle 0|0\rangle} = \left(\frac{1}{i}\right)^n \frac{1}{Z[0]} \frac{\delta^n}{\delta J(t_1)\cdots\delta J(t_n)} Z[J]\Big|_{J=0}, \quad (11.106)$$

where the $J$-independent factor cancels out. Then, from (11.105), we get

$$Z[J] = \lim_{t\to+i\infty,t'\to-i\infty} N \int_{q,t}^{q',t'} \mathcal{D}(q(\tau))e^{i\int_t^{t'}(L+Jq)d\tau}, \qquad (11.107)$$

where $N$ is a $J$-independent normalization factor. This expression suggests the definition of a **Euclidean** generating functional $Z_E[J]$. This is obtained by introducing a Euclidean time $\tau_E = i\tau$ (Wick's rotation)

$$Z_E[J] = \lim_{\tau'\to+\infty,\tau\to-\infty} N \int_{q,\tau}^{q',\tau'} \mathcal{D}(q(\tau))e^{\int_\tau^{\tau'}(L(q,i\frac{dq}{d\tau})+Jq)d\tau_E}. \qquad (11.108)$$

Consider, for instance, the case

$$L = \frac{1}{2}m(\frac{dq}{dt})^2 - V(q), \qquad (11.109)$$

from which

$$L(q,i\frac{dq}{d\tau}) = -\frac{1}{2}m\left(\frac{dq}{d\tau_E}\right)^2 - V(q). \qquad (11.110)$$

It is then convenient to define a Euclidean Lagrangian and a Euclidean action:

$$L_E = \frac{1}{2}m\left(\frac{dq}{d\tau_E}\right)^2 + V(q), \qquad (11.111)$$

$$S_E = \int_\tau^{\tau'} L_E d\tau_E. \qquad (11.112)$$

Notice that the Euclidean Lagrangian coincides formally with the Hamiltonian. More generally

$$L_E = -L(q,i\frac{dq}{d\tau}). \qquad (11.113)$$

Therefore,

$$Z_E[J] = \lim_{\tau'_E \to +\infty, \tau_E \to -\infty} N \int_{q,\tau_E}^{q',\tau'_E} \mathcal{D}(q(\tau)) e^{-S_E + \int_\tau^{\tau'} J q d\tau_E}. \qquad (11.114)$$

Since $e^{-S_E}$ is a positive definite functional, the path integral turns out to be well defined and convergent. Then, the vacuum expectation value of the $T$-products can be evaluated by using $Z_E$ and continuing the result for real times

$$\frac{1}{Z[0]} \frac{\delta^n Z[J]}{\delta J(t_1) \cdots \delta J(t_n)}\bigg|_{J=0} = (i)^n \frac{1}{Z_E[0]} \frac{\delta^n Z_E[J]}{\delta J(\tau_1) \cdots \delta J(\tau_n)}\bigg|_{J=0, \tau_i = i t_i}. \qquad (11.115)$$

Since the values of $q$ and $q'$ are arbitrary, a standard choice is to set them to zero.

## 11.7   The Green function for the harmonic oscillator

Many dynamical systems, in a first approximation, can be described in terms of harmonic oscillators. Therefore, it is useful to evaluate the corresponding generating functional. Let us start from eq. (11.97), where we do not need to specify the boundary conditions

$$Z[J] = \int dq dq' \Phi_0^*(q', t') e^{iS_J} \Phi_0(q, t), \qquad (11.116)$$

with

$$\Phi_0(q, t) = \left(\frac{m\omega}{\pi}\right)^{\frac{1}{4}} e^{-\frac{1}{2}m\omega q^2} e^{-\frac{i}{2}\omega t}, \qquad (11.117)$$

$$\Phi_0^*(q', t') = \left(\frac{m\omega}{\pi}\right)^{\frac{1}{4}} e^{-\frac{1}{2}m\omega q'^2} e^{+\frac{i}{2}\omega t'}, \qquad (11.118)$$

and

$$S_J = \int_t^{t'} \left[\frac{1}{2}m\dot{q}^2 - \frac{1}{2}m\omega^2 q^2 + Jq\right] d\tau. \qquad (11.119)$$

In this calculation we are not interested in the normalization factors, but only on the $J$ dependence. Then, let us consider the argument of the exponential in eq. (11.116) and, for the moment being, let us neglect the time dependent terms from the oscillator wave functions,

$$\text{arg} = -\frac{1}{2}m\omega(q^2 + q'^2) + i \int_t^{t'} \left[\frac{1}{2}m\dot{q}^2 - \frac{1}{2}m\omega^2 q^2 + Jq\right] d\tau. \qquad (11.120)$$

We perform the change of variable

$$q(\tau) = x(\tau) + x_0(\tau), \tag{11.121}$$

where, we will choose $x_0(\tau)$ in such a way to extract all the dependence from the source $J$ out of the integral. We have

$$\arg = -\frac{1}{2}m\omega[x(t)^2 + x^2(t')] - \frac{1}{2}m\omega[x_0(t)^2 + x_0^2(t')]$$

$$-m\omega[x(t)x_0(t) + x(t')x_0(t')]$$

$$+i\int_t^{t'} \left[\frac{1}{2}m\dot{x}^2 - \frac{1}{2}m\omega^2 x^2\right] d\tau + i\int_t^{t'} \left[\frac{1}{2}m\dot{x}_0^2 - \frac{1}{2}m\omega^2 x_0^2 + Jx_0\right] d\tau$$

$$+i\int_t^{t'} \left[m\dot{x}\dot{x}_0 - m\omega^2 xx_0 + Jx\right] d\tau. \tag{11.122}$$

After integrating by parts, the last integral can be written as

$$im[x\dot{x}_0]_t^{t'} - i\int_t^{t'} [m\ddot{x}_0 + m\omega^2 x_0 - J]x d\tau. \tag{11.123}$$

Choosing $x_0$ as a solution of the oscillator equation of motion in the presence of the source $J$:

$$\ddot{x}_0 + \omega^2 x_0 = \frac{1}{m}J, \tag{11.124}$$

we get

$$im[x(t')\dot{x}_0(t') - x(t)\dot{x}_0(t)]. \tag{11.125}$$

Let us take the following boundary conditions on $x_0$:

$$i\dot{x}_0(t') = \omega x_0(t'), \quad i\dot{x}_0(t) = -\omega x_0(t). \tag{11.126}$$

In this way, the term (11.125) cancels the mixed terms in the second line of eq. (11.122). We are left with:

$$\arg = -\frac{1}{2}m\omega[x^2(t) + x^2(t')] - \frac{1}{2}m\omega[x_0^2(t) + x_0^2(t')]$$

$$+i\int_t^{t'} [\frac{1}{2}m\dot{x}^2 - \frac{1}{2}m\omega^2 x^2] d\tau$$

$$+i\int_t^{t'} [\frac{1}{2}m\dot{x}_0^2 - \frac{1}{2}m\omega^2 x_0^2 + Jx_0] d\tau. \tag{11.127}$$

Integrating by parts the last term, and using again the equation of motion for $x_0$, we get

$$i\int_t^{t'} [\frac{1}{2}m\dot{x}_0^2 - \frac{1}{2}m\omega^2 x_0^2 + Jx_0] d\tau = \frac{1}{2}m\omega[x_0^2(t') + x_0^2(t)] + \frac{i}{2}\int_t^{t'} Jx_0 d\tau. \tag{11.128}$$

Keeping in mind the boundary conditions (11.126), the term coming from the integration by parts cancels the second term in eq. (11.127), and therefore

$$\arg = -\frac{1}{2}m\omega[x^2(t) + x^2(t')] + i\int_t^{t'} [\frac{1}{2}m\dot{x}^2 - \frac{1}{2}m\omega^2 x^2 + \frac{i}{2}Jx_0]d\tau. \quad (11.129)$$

The $J$ dependence is now in the last term, and we can extract it from the integral. The first two terms give rise to the functional at $J = 0$. We obtain

$$Z[J] = e^{\frac{i}{2}\int_t^{t'} J(\tau)x_0(\tau)d\tau} Z[0], \quad (11.130)$$

where $x_0(\tau)$ is the solution of the equations of motion (11.124) with boundary conditions (11.126). For $t < \tau < t'$ this solution can be written as

$$x_0(\tau) = \int_t^{t'} \Delta(\tau - s)J(s)ds, \quad (11.131)$$

with

$$\left[\frac{d^2}{d\tau^2} + \omega^2\right]\Delta(\tau - s) = \frac{1}{m}\delta(\tau - s) \quad (11.132)$$

and

$$i\dot{\Delta}(t' - s) = \omega\Delta(t' - s), \quad i\dot{\Delta}(t - s) = -\omega\Delta(t - s). \quad (11.133)$$

The last equations are easily solved for $t' > s$ and $t < s$

$$\Delta(\tau) = Ae^{-i\omega\tau}, \qquad \tau > 0,$$
$$\Delta(\tau) = Be^{+i\omega\tau}, \qquad \tau < 0. \quad (11.134)$$

Therefore, we have

$$\Delta(\tau) = A\theta(\tau)e^{-i\omega\tau} + B\theta(-\tau)e^{+i\omega\tau}. \quad (11.135)$$

The quantities $A$ and $B$ are fixed by the requirement that $\Delta(\tau)$ satisfies the equation of motion eq. (11.132). We get

$$A = B = \frac{i}{2\omega m}, \quad (11.136)$$

and finally

$$\Delta(\tau) = \frac{i}{2\omega m}\left[\theta(\tau)e^{-i\omega\tau} + \theta(-\tau)e^{+i\omega\tau}\right]. \quad (11.137)$$

The functional, we were looking for, is

$$Z[J] = e^{\frac{i}{2}\int_t^{t'} dsds' J(s)\Delta(s-s')J(s')} Z[0]. \quad (11.138)$$

We see that $\Delta(s - s')$ is the vacuum expectation value of the $T$-product of the position operators at the times $s$ and $s'$:

$$\Delta(s - s') = -\frac{i}{Z[0]}\frac{\delta^2 Z[J]}{\delta J(s)\delta J(s')}\bigg|_{J=0} = i\frac{\langle 0|T(\mathbf{q}(s)\mathbf{q}(s'))|0\rangle}{\langle 0|0\rangle}. \quad (11.139)$$

In the following, we will need the Fourier transform of $\Delta(t)$:

$$\Delta(t) = \int d\nu\, e^{-i\nu t}\Delta(\nu). \quad (11.140)$$

Since it satisfies eq. (11.132) we obtain (here and in the following we will take the mass of the oscillator equal to 1):

$$(-\nu^2 + \omega^2)\Delta(\nu) = \frac{1}{2\pi}, \quad (11.141)$$

that is

$$\Delta(\nu) = -\frac{1}{2\pi}\frac{1}{\nu^2 - \omega^2}. \quad (11.142)$$

However, this expression is singular at $\nu = \pm\omega$, therefore, we need to specify how the integral (11.140) is defined. The proper way of doing it is illustrated in Fig. 11.7. For instance, for $t > 0$ by closing the integral in the lower plane around $\nu = \omega$, we get

$$(-2\pi i)\frac{-1}{2\pi}\frac{1}{2\omega}e^{-i\omega t} = \frac{i}{2\omega}e^{-i\omega t}. \quad (11.143)$$

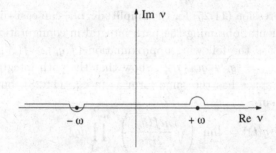

Fig. 11.7   The integration path for equation (11.140).

We would get the same result by integrating along the real axis, but translating the denominator of $+i\epsilon$, $\epsilon > 0$ (this is called the Feynman prescription), in such a way to move the pole at $\omega$ in the lower plane and the one at $-\omega$ in the upper plane, as follows

$$\Delta(t) = -\lim_{\epsilon\to 0^+}\frac{1}{2\pi}\int_{-\infty}^{+\infty}\frac{e^{-i\nu t}}{\nu^2 - \omega^2 + i\epsilon}d\nu. \quad (11.144)$$

One could arrive at this result also by noticing that the Feynman prescription is equivalent to substitute in the path integral the term

$$e^{-\frac{1}{2}\omega^2 \int q^2(t)dt},$$ (11.145)

with

$$e^{-\frac{i}{2}(\omega^2 - i\epsilon) \int q^2(t)dt}.$$ (11.146)

Therefore, the Feynman prescription is there to make the integral convergent:

$$e^{-\frac{1}{2}\epsilon \int q^2(t)dt}.$$ (11.147)

The same convergence is ensured by the Euclidean formulation.

## 11.8 Exercises

(1) Consider the following Lagrangian

$$L = \frac{m}{2}\dot{q}^2 f(q) - V(q),$$ (11.148)

where $f(q)$ is an arbitrary function of $q$ (a situation like this occurs for a particle in a gravitational field). Evaluate the corresponding Hamiltonian:

$$H = \frac{1}{2mf(q)}p^2 + V(q), \qquad p = \frac{\partial L}{\partial \dot{q}}.$$ (11.149)

Using the expression (11.25) for the amplitude, one can easily integrate over the momenta, obtaining the path integral in configuration space. To this end, use the following approximation $(f(q_k) + f(q_{k+1}))/2 \approx f(\bar{q}_k)$, where $\bar{q}_k = (q_k + q_{k+1})/2$. Show that the path integral in the configuration space has the same form as in eq. (11.28), but with a modified measure

$$\bar{\mathcal{D}}(q(t)) = \lim_{n \to \infty} \left(\sqrt{\frac{mf(\bar{q}_k)}{2i\pi\epsilon}}\right)^n \prod_{k=1}^{n-1} dq_k.$$ (11.150)

Alternatively, it is possible to write the path integral modifying the Lagrangian instead of the measure. Introducing an effective Lagrangian $L_{\text{eff}}$:

$$\int \mathcal{D}(q(t))e^{i \int L_{\text{eff}}dt}.$$ (11.151)

Evaluate $I_{\text{eff}}$ using

$$\lim_{n \to \infty} \sum_{k=0}^{n-1} \epsilon = \int_t^{t'} dt, \qquad \lim_{n \to \infty} \frac{\delta_{hk}}{\epsilon} = \delta(t_h - t_k).$$ (11.152)

(2) Consider the case of one-dimensional free motion. Starting from the expression (11.26) evaluate the probability amplitude by going from $(q, t)$ to $(q', t')$. The integrations can be done using the Gaussian integral

$$\int_{-\infty}^{\infty} dx\, e^{ax^2 + bx} = \sqrt{-\frac{\pi}{a}}\, e^{-b^2/4a}, \quad \mathrm{Re}\ a \le 0. \tag{11.153}$$

Show that the final result is given by

$$\langle q', t' | q, t \rangle = \left[\frac{m}{2\pi i (t' - t)}\right]^{\frac{1}{2}} e^{im(q'-q)^2/(2(t'-t))}. \tag{11.154}$$

# Chapter 12

# The path integral in field theory

In the previous Section we have shown that the path integral formalism allows the evaluation of the vacuum expectation values of $T$-products. In the first Section of this Chapter, we will prove that the matrix elements of the S-matrix can be reduced to the evaluation of the vacuum expectation value of $T$-products. This fact will allow us to apply directly the Feynman formulation to the field theory.

## 12.1   In- and out-states

In this Section, we will show that any $S$-matrix element can be reduced to the evaluation of the vacuum expectation value of a $T$-product of field operators. Such a formulation is called LSZ (from the authors, Lehman, Symanzik and Zimmermann [Lehman et al. (1955)]). This formalism is based on a series of very general properties as translation and Lorentz invariance. As in the previous Chapter, we will make use of the Heisenberg representation. Our assumptions will be the following ones:

(1) The eigenvalues of the four-momentum of physical staes lie within the forward light cone:

$$p^2 = p^\mu p_\mu \geq 0, \qquad p^0 \geq 0. \tag{12.1}$$

(2) There exists a nondegenerate Lorentz invariant state of minimum energy. This is called the **vacuum state**

$$\Phi_0 \equiv |0\rangle. \tag{12.2}$$

By convention, we assume that the corresponding energy is zero

$$\mathbf{P}_0|0\rangle = 0. \tag{12.3}$$

From eq. (12.1) we get also

$$\vec{\mathbf{P}}|0\rangle = 0 \to \mathbf{P}_\mu|0\rangle = 0. \qquad (12.4)$$

Since $p_\mu = 0$ is a Lorentz invariant condition, it follows that $\Phi_0$ appears as the vacuum state in all the Lorentz frames.

(3) For each stable particle of mass $m$, there exists a single particle stable state

$$\Phi_1 \equiv |p\rangle, \qquad (12.5)$$

where $\Phi_1$ is a momentum eigenstate with eigenvalue $p_\mu$ such that $p^2 = m^2$.

(4) Neglecting the complications from the zero mass states, we can add a further requirement: the vacuum and the single particle states form a discrete spectrum in $p_\mu$. For instance, in the case of particles of mass $m$, the spectrum is given in Fig. 12.1.

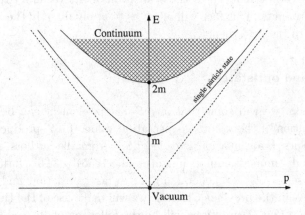

Fig. 12.1   Energy-momentum spectrum for particles of mass $m$.

As an example, consider the pion, $\pi^-$. This is not a stable particle and decays into $\mu + \bar{\nu}_\mu$, with a lifetime of the order of $\approx 2 \cdot 10^{-8}$ sec., but this number is very large with respect to the natural decay frequency $1/m_\pi \approx 5 \cdot 10^{-24}$ sec. ($m_\pi \approx 140$ MeV). Therefore, by neglecting the weak interactions (responsible for the decay), we can assume that the pion is a stable particle. The same argument can be applied to other particles, including composite particles, at least in the appropriate range of energy. For instance, pions are composite states of quark-antiquark pairs. Notice that this assumption is very much in the spirit of perturbation theory. In fact, we will introduce a field for each particle, assuming that the interactions do not modify the spectrum in a too violent way.

As we have discussed in Section 8.2, the main experiments realized in particle physics, are the scattering ones. Here, we would like to formalize better the concepts presented in that Section. We will describe the particles at $t = \pm\infty$ in terms of free fields, except that they will be subject to the self-interactions which, as we know, cannot be neglected. The free fields that we are going to use will be denoted by $\phi_{\text{in}}(x)$ for $t \to -\infty$ and by $\phi_{\text{out}}(x)$ for $t \to +\infty$. The interacting field $\phi(x)$ can be thought of as constructed in terms of these free field operators, such that at $t \to \pm\infty$ it reduces to $\phi_{\text{out}}(x)$ or $\phi_{\text{in}}(x)$. We will require a certain number of properties to the in- and out-fields. Namely:

(1) $\phi_{\text{in}}(x)$ must transform as the corresponding $\phi(x)$ with respect to the symmetries of the theory. In particular, with respect to translations, it must satisfy

$$[\mathbf{P}_\mu, \phi_{\text{in}}(x)] = -i\frac{\partial \phi_{\text{in}}(x)}{\partial x^\mu}, \tag{12.6}$$

equivalent to require, under a finite translation $x \to x + a$:

$$\phi_{\text{in}}(x + a) = e^{iP \cdot a}\phi_{\text{in}}(x)e^{-iP \cdot a}. \tag{12.7}$$

(2) $\phi_{\text{in}}(x)$ must satisfy the free Klein-Gordon equation (for a scalar field, or the Dirac free equation for a spinor field, etc.) corresponding to the **physical mass** $m$

$$(\Box + m^2)\phi_{\text{in}}(x) = 0. \tag{12.8}$$

Notice that this requirement implies that we have taken into account the self-interactions of the field responsible for the mass renormalization.

We will consider here the case of a hermitian scalar field. This approach can be easily extended to all the other cases, charged, spinor, vector fields, etc. The same properties are required for the out-fields. By using these two properties, it follows that the in-field creates from the vacuum the physical one particle state. In fact, let $|p\rangle$ be an eigenstate of $\mathbf{P}_\mu$ with eigenvalue $p_\mu$. We have

$$-i\frac{\partial}{\partial x^\mu}\langle p|\phi_{\text{in}}(x)|0\rangle = \langle p|[\mathbf{P}_\mu, \phi_{\text{in}}(x)]|0\rangle = p_\mu\langle p|\phi_{\text{in}}(x)|0\rangle. \tag{12.9}$$

By applying $-i\partial/\partial x_\mu$ once more, we find

$$-\Box\langle p|\phi_{\text{in}}(x)|0\rangle = p^2\langle p|\phi_{\text{in}}(x)|0\rangle, \tag{12.10}$$

and using the Klein-Gordon equation

$$(p^2 - m^2)\langle p|\phi_{\text{in}}(x)|0\rangle = 0. \tag{12.11}$$

We see that the only state that $\phi_{\text{in}}(x)$ can create out of the vacuum is the one with $p_\mu$ such that $p^2 = m^2$. From our hypotheses, it follows that this is the single particle state with mass $m$. Since the in- and out-fields are free fields, we can apply all the corresponding formalism. In particular they can be Fourier expanded

$$\phi_{\text{in}}(x) = \int d^3\vec{k}\left[a_{\text{in}}(\vec{k})f_{\vec{k}}(x) + a_{\text{in}}^\dagger(\vec{k})f_{\vec{k}}^*(x)\right], \tag{12.12}$$

where we make use of the same notations of Chapter 3 (see, in particular eq. (3.57)). Inverting this relation (see eq. (3.55)), we get

$$a_{\text{in}}(\vec{k}) = i\int d^3x f_{\vec{k}}^*(x)\partial_t^{(-)}\phi_{\text{in}}(x), \tag{12.13}$$

and, after a simple calculation, we find

$$[\mathbf{P}_i, a_{\text{in}}(\vec{k})] = -k_i a_{\text{in}}(\vec{k}). \tag{12.14}$$

Also,

$$\begin{aligned}
&[\mathbf{P}_0, a_{\text{in}}(\vec{k})] \\
&= \int d^3\vec{x}\left[f_{\vec{k}}^*(x)\left(\frac{\partial}{\partial x^0}\frac{\partial\phi_{\text{in}}(x)}{\partial x^0}\right) - \left(\frac{\partial}{\partial x^0}f_{\vec{k}}^*(x)\right)\frac{\partial\phi_{\text{in}}(x)}{\partial x^0}\right] \\
&= -\int d^3\vec{x}\left[\frac{\partial f_{\vec{k}}^*(x)}{\partial x^0}\left(\frac{\partial}{\partial x^0}\phi_{\text{in}}(x)\right) - \left(\frac{\partial}{\partial x^0}\frac{\partial f_{\vec{k}}^*(x)}{\partial x^0}\right)\phi_{\text{in}}(x)\right] \\
&= -k_0 a_{\text{in}}(\vec{k}), \tag{12.15}
\end{aligned}$$

where we have used

$$\partial_0 f_{\vec{k}}^*(x) = ik_0 f_{\vec{k}}^*(x). \tag{12.16}$$

In conclusion,

$$[\mathbf{P}_\mu, a_{\text{in}}(\vec{k})] = -k_\mu a_{\text{in}}(\vec{k}). \tag{12.17}$$

By hermitian conjugation, we get

$$[\mathbf{P}_\mu, a_{\text{in}}^\dagger(\vec{k})] = k_\mu a_{\text{in}}^\dagger(\vec{k}). \tag{12.18}$$

It follows that $a_{\text{in}}^\dagger(\vec{k})$ creates states of four-momentum $k_\mu$, whereas $a_{\text{in}}(\vec{k})$ annihilates states with momentum $k_\mu$. For instance,

$$\mathbf{P}_\mu a_{\text{in}}^\dagger(\vec{k}_1) a_{\text{in}}^\dagger(\vec{k}_2)|0\rangle = [\mathbf{P}_\mu, a_{\text{in}}^\dagger(\vec{k}_1) a_{\text{in}}^\dagger(\vec{k}_2)]|0\rangle$$
$$= (k_1 + k_2)_\mu a_{\text{in}}^\dagger(\vec{k}_1) a_{\text{in}}^\dagger(\vec{k}_2)|0\rangle, \tag{12.19}$$

$$\mathbf{P}_\mu a_{\text{in}}(\vec{k}_1) a_{\text{in}}^\dagger(\vec{k}_2)|0\rangle = [\mathbf{P}_\mu, a_{\text{in}}(\vec{k}_1) a_{\text{in}}^\dagger(\vec{k}_2)]|0\rangle$$
$$= (k_2 - k_1)_\mu a_{\text{in}}(\vec{k}_1) a_{\text{in}}^\dagger(\vec{k}_2)|0\rangle. \tag{12.20}$$

By using the condition on the spectrum of the momentum operator, we get also

$$a_{\text{in}}(\vec{k})|0\rangle = 0, \tag{12.21}$$

otherwise, this state would have a negative eigenvalue of the energy. Another obvious relation is:

$$\langle p_1, \cdots, p_M; \text{in}|k_1, \cdots, k_N; \text{in}\rangle = 0, \tag{12.22}$$

unless $M = N$ and the set $(p_1, \cdots, p_M)$ is identical to the set $(k_1, \cdots, k_N)$.

Let us now look for a relation between the in- and the $\phi$-field. The interacting field $\phi(x)$ satisfies an equation of the type

$$(\Box + m^2)\phi(x) = j(x), \tag{12.23}$$

where $j(x)$ (the scalar current) contains the self-interactions of the fields and possible interactions with other fields in the theory. We need to solve this equation with the boundary conditions at $t = \pm\infty$ relating $\phi(x)$ with the in- and out-fields. Notice that, in writing the previous equation, we have included the effects of mass renormalization, but we know that at the same time there is a wave function renormalization. Therefore, asymptotically, the field $\phi(x)$ cannot be the same as the in- and out-fields, but it will differ by the wave function renormalization factor, $\sqrt{Z}$, that, for spinor and vector fields, has been defined in eq. (10.16). Therefore, we require the following asymptotic conditions:

$$\lim_{t \to -\infty} \phi(x) = \lim_{t \to -\infty} \sqrt{Z}\,\phi_{\text{in}}(x), \quad \lim_{t \to +\infty} \phi(x) = \lim_{t \to +\infty} \sqrt{Z}\,\phi_{\text{out}}(x). \tag{12.24}$$

The solution for $\phi(x)$ is then

$$\phi(x) = \sqrt{Z}\,\phi_{\text{in}}(x) + \int d^4y \Delta_{\text{ret}}(x - y; m^2) j(y). \tag{12.25}$$

We will see in a moment the meaning of $Z$ in this context. The function $\Delta_{\text{ret}}(x; m^2)$ is the retarded Green function, defined (as in Chapter 7) by

$$(\Box + m^2)\Delta_{\text{ret}}(x; m^2) = \delta^4(x), \tag{12.26}$$

such that

$$\Delta_{\text{ret}}(x; m^2) = 0, \quad \text{for} \quad x_0 < 0. \tag{12.27}$$

In the present context, the need of the wave function renormalization comes about due to the need of having an in-field properly normalized. In fact, it must produce a single particle state, when operating on the vacuum. Notice that both $\langle 1|\phi_{\text{in}}(x)|0\rangle$ and $\langle 1|\phi(x)|0\rangle$ have the same $x$-dependence. In fact, they are proportional to $\exp(ipx)$, as it follows by taking the matrix element of eq. (12.6) between the states $|1\rangle$ and $|0\rangle$ for both types of fields. On the other hand, $\phi(x)$ may create, out of the vacuum, states other than the single particle one. Therefore, the normalization of the corresponding matrix elements cannot be the same for the two fields.

It can be shown that the asymptotic condition (12.24) cannot hold as an operator statement. This is because it is not possible to isolate the operator $\phi(x)$ from the scalar current $j(x)$ at $t = -\infty$, since this includes also the self-interactions. However, the statement can be expressed in a weak form for the matrix elements. The correct way of doing it is described in [Lehman et al. (1955)] and consists in the following procedure: first smear out the field over a space-like region

$$\phi^f(t) = i \int d^3x \left( f^*(\vec{x}, t)\partial_0 \phi(\vec{x}, t) - \partial_0 f^*(\vec{x}, t)\phi(\vec{x}, t) \right), \tag{12.28}$$

with $f(\vec{x}, t)$ an arbitrary normalizable solution of the Klein-Gordon equation

$$(\Box + m^2)f(x) = 0. \tag{12.29}$$

Then, the asymptotic condition is formulated as

$$\lim_{t \to -\infty} \langle \alpha | \phi^f(t) | \beta \rangle = \sqrt{Z} \lim_{t \to -\infty} \langle \alpha | \phi_{\text{in}}^f(t) | \beta \rangle, \tag{12.30}$$

where $|\alpha\rangle$ and $|\beta\rangle$ are any two normalizable states.

Similar considerations can be made for the out-field. In particular, the field $\phi(x)$ can also be expressed in terms of the out-field:

$$\phi(x) = \sqrt{Z}\phi_{\text{out}}(x) + \int d^4y \Delta_{\text{adv}}(x - y; m^2)j(y), \tag{12.31}$$

where $\Delta_{\text{adv}}(x; m^2)$ is the advanced Green function, defined by

$$\Delta_{\text{adv}}(x; m^2) = 0, \quad \text{for} \quad x_0 > 0, \tag{12.32}$$

and

$$(\Box + m^2)\Delta_{\text{adv}}(x; m^2) = \delta^4(x). \tag{12.33}$$

By comparing this expression with eq. (12.25) we can find a relation between the in- and out- fields

$$\sqrt{Z}\phi_{\text{in}}(x) = \sqrt{Z}\phi_{\text{out}}(x) + \int d^4y \Delta(x - y; m^2) j(y), \qquad (12.34)$$

with

$$\Delta(x - y; m^2) = \Delta_{\text{adv}}(x - y; m^2) - \Delta_{\text{ret}}(x - y; m^2). \qquad (12.35)$$

The construction of the out-creation and annihilation operators goes along the same lines discussed for the in-operators.

## 12.2   The $S$ matrix

In Section 8.2 we have introduced the scattering matrix; however, in order to reduce the calculation of its matrix elements to vacuum expectation values of $T$-products of local operators, it is convenient to use a definition based on the in- and out-states. On the other hand, recalling our previous definition, we will see that the only difference is in the terms used. In the previous Section we have shown that an initial state of $n$ non-interacting particles can be described by

$$|p_1, \cdots, p_n; \text{in}\rangle = a_{\text{in}}^\dagger(p_1) \cdots a_{\text{in}}^\dagger(p_n)|0; \text{in}\rangle \equiv |\alpha; \text{in}\rangle, \qquad (12.36)$$

where $\alpha$ stands for the collection of quantum numbers of the initial state. We will be interested in evaluating the probability amplitude for the initial state, after the scattering process, to go into a state of non-interacting particles described by the out-operators

$$|p_1', \cdots, p_n'; \text{out}\rangle = a_{\text{out}}^\dagger(p_1') \cdots a_{\text{out}}^\dagger(p_n')|0; \text{out}\rangle \equiv |\beta; \text{out}\rangle. \qquad (12.37)$$

The probability amplitude defines the $S$ matrix element

$$S_{\beta\alpha} = \langle\beta; \text{out}|\alpha; \text{in}\rangle. \qquad (12.38)$$

We can also define an operator $S$ transforming the in-states into the out-states

$$\langle\beta; \text{in}|S = \langle\beta; \text{out}|, \qquad (12.39)$$

such that

$$S_{\beta\alpha} = \langle\beta; \text{in}|S|\alpha; \text{in}\rangle. \qquad (12.40)$$

From this definition, we can derive some important properties for the operator $S$:

(1) Since the vacuum is stable and not degenerate, it follows that $|S_{00}| = 1$, and therefore

$$\langle 0; \text{in}|S = \langle 0; \text{out}| = e^{i\phi}\langle 0; \text{in}|. \qquad (12.41)$$

By choosing the phase equal to zero we get

$$|0; \text{out}\rangle = |0; \text{in}\rangle \equiv |0\rangle. \qquad (12.42)$$

(2) In an analogous way the stability of the one particle state requires

$$|p; \text{in}\rangle = |p; \text{out}\rangle. \qquad (12.43)$$

(3) The operator $S$ maps the in- to the out-fields

$$\phi_{\text{in}}(x) = S\phi_{\text{out}}(x)S^{-1}. \qquad (12.44)$$

In fact, let us consider the following matrix element

$$\langle \beta; \text{out}|\phi_{\text{out}}|\alpha; \text{in}\rangle = \langle \beta; \text{in}|S\phi_{\text{out}}|\alpha; \text{in}\rangle. \qquad (12.45)$$

But $\langle \beta; \text{out}|\phi_{\text{out}}$ is an out-state and therefore

$$\langle \beta; \text{out}|\phi_{\text{out}} = \langle \beta; \text{in}|\phi_{\text{in}}S. \qquad (12.46)$$

Then

$$\langle \beta; \text{out}|\phi_{\text{out}}(x)|\alpha; \text{in}\rangle = \langle \beta; \text{in}|S\phi_{\text{out}}(x)|\alpha; \text{in}\rangle = \langle \beta; \text{in}|\phi_{\text{in}}(x)S|\alpha; \text{in}\rangle, \qquad (12.47)$$

which proves the relation (12.44).

(4) $S$ is a unitary operator. This follows from the very definition (12.39) implying

$$S^\dagger|\alpha; \text{in}\rangle = |\alpha; \text{out}\rangle, \qquad (12.48)$$

from which

$$\langle \beta; \text{in}|SS^\dagger|\alpha; \text{in}\rangle = \langle \beta; \text{out}|\alpha; \text{out}\rangle = \delta_{\alpha\beta}, \qquad (12.49)$$

showing that

$$SS^\dagger = 1. \qquad (12.50)$$

From eq. (12.48) and the unitarity of $S$, we obtain

$$|\alpha; \text{in}\rangle = S|\alpha; \text{out}\rangle. \qquad (12.51)$$

(5) $S$ is Lorentz invariant, since it transforms in- into out-fields belonging to the same Lorentz representation. For the same reason, the $S$ matrix is invariant under any symmetry of the theory.

## 12.3   The reduction formalism

We will now introduce the reduction formalism, allowing to reduce the calculation of the $S$-matrix to vacuum expectation values of $T$-products of field operators. Let us start considering the following $S$ matrix element

$$S_{\beta\alpha p} = \langle\beta;\text{out}|\alpha,p;\text{in}\rangle, \qquad (12.52)$$

where $|\alpha,p;\text{in}\rangle$ is the state of a set of particles $\alpha$ plus an additional incoming particle of momentum $p_\mu$. Here, we will consider the case of scalar particles, but the formalism can be easily extended to other cases. We want to show that, by using the asymptotic condition, it is possible to **extract** the particle of momentum $p_\mu$ from the in-state, introducing the corresponding field operator. Let us consider the following relation

$$\langle\beta;\text{out}|\alpha,p;\text{in}\rangle = \langle\beta;\text{out}|a_{\text{in}}^\dagger(p)|\alpha;\text{in}\rangle$$

$$= \langle\beta;\text{out}|a_{\text{out}}^\dagger(p)|\alpha;\text{in}\rangle + \langle\beta;\text{out}|a_{\text{in}}^\dagger(p) - a_{\text{out}}^\dagger(p)|\alpha;\text{in}\rangle$$

$$= \langle\beta - p;\text{out}|\alpha;\text{in}\rangle$$

$$-i\int d^3\vec{x}\Big[f_{\vec{p}}(x)\left(\frac{\partial}{\partial x^0}\langle\beta;\text{out}|\phi_{\text{in}}(x) - \phi_{\text{out}}(x)|\alpha;\text{in}\rangle\right)$$

$$-\left(\frac{\partial}{\partial x^0}f_{\vec{p}}(x)\right)\langle\beta;\text{out}|\phi_{\text{in}}(x) - \phi_{\text{out}}(x)|\alpha;\text{in}\rangle\Big], \qquad (12.53)$$

where we have used eq. (12.13) and its adjoint. We have denoted $|\beta-p;\text{out}\rangle$ the out-state obtained by removing the particle with momentum $p_\mu$ from the out-state, if already present in the state. Otherwise $|\beta - p;\text{out}\rangle$ is the null vector. As we know, the integral appearing in the previous expression is time-independent, since $f_{\vec{p}}(x)$, $\phi_{\text{in}}(x)$ and $\phi_{\text{out}}(x)$ are solutions of the free Klein-Gordon equation (see Section 3.1). Therefore, we are allowed to make the following substitutions (in the weak sense, since we are taking matrix elements)

$$\lim_{x^0\to-\infty}\phi_{\text{in}}(x) = \lim_{x^0\to-\infty}\frac{1}{\sqrt{Z}}\phi(x)$$

$$\lim_{x^0\to+\infty}\phi_{\text{out}}(x) = \lim_{x^0\to+\infty}\frac{1}{\sqrt{Z}}\phi(x), \qquad (12.54)$$

obtaining

$$\langle\beta;\text{out}|\alpha,p;\text{in}\rangle = \langle\beta - p;\text{out}|\alpha;\text{in}\rangle$$

$$+\frac{i}{\sqrt{Z}}\left(\lim_{x^0\to+\infty} - \lim_{x^0\to-\infty}\right)\int d^3\vec{x}\Big[f_{\vec{p}}(x)\left(\frac{\partial}{\partial x^0}\langle\beta;\text{out}|\phi(x)|\alpha;\text{in}\rangle\right)$$

$$-\left(\frac{\partial}{\partial x^0}f_{\vec{p}}(x)\right)\langle\beta;\text{out}|\phi(x)|\alpha;\text{in}\rangle\Big]. \qquad (12.55)$$

By using the identity

$$\int_{-\infty}^{+\infty} d^4x \frac{\partial}{\partial x^0} \left[ g_1(x) \left( \frac{\partial}{\partial x^0} g_2(x) \right) - \left( \frac{\partial}{\partial x^0} g_1(x) \right) g_2(x) \right]$$

$$= \left( \lim_{x^0 \to +\infty} - \lim_{x^0 \to -\infty} \right)$$

$$\times \int d^3\vec{x} \left[ g_1(x) \left( \frac{\partial}{\partial x^0} g_2(x) \right) - \left( \frac{\partial}{\partial x^0} g_1(x) \right) g_2(x) \right]$$

$$= \int_{-\infty}^{+\infty} d^4x \left[ g_1(x)\ddot{g}_2(x) - \ddot{g}_1(x)g_2(x) \right], \tag{12.56}$$

we can write

$$\langle \beta; \mathrm{out} | \alpha, p; \mathrm{in} \rangle = \langle \beta - p; \mathrm{out} | \alpha; \mathrm{in} \rangle$$

$$+ \frac{i}{\sqrt{Z}} \int d^4x \langle \beta; \mathrm{out} | \left[ f_{\vec{p}}(x)\ddot{\phi}(x) - \ddot{f}_{\vec{p}}(x)\phi(x) \right] |\alpha; \mathrm{in} \rangle. \tag{12.57}$$

By using the Klein-Gordon equation for $f_{\vec{p}}(x)$ we get

$$\langle \beta; \mathrm{out} | \alpha, p; \mathrm{in} \rangle = \langle \beta - p; \mathrm{out} | \alpha; \mathrm{in} \rangle$$

$$+ \frac{i}{\sqrt{Z}} \int d^4x \langle \beta; \mathrm{out} | [f_{\vec{p}}(x)\ddot{\phi}(x)$$

$$- ((\nabla^2 - m^2) f_{\vec{p}}(x)) \, \phi(x)] |\alpha; \mathrm{in} \rangle. \tag{12.58}$$

Integrating by parts

$$\langle \beta; \mathrm{out} | \alpha, p; \mathrm{in} \rangle = \langle \beta - p; \mathrm{out} | \alpha; \mathrm{in} \rangle$$

$$+ \frac{i}{\sqrt{Z}} \int d^4x f_{\vec{p}}(x)(\Box + m^2) \langle \beta; \mathrm{out} | \phi(x) |\alpha; \mathrm{in} \rangle. \tag{12.59}$$

We can iterate this procedure in order to extract all the in- and out-particles from the states. Let us illustrate the case in which we want to extract from the previous matrix element an out-particle. This exercise is interesting because it shows how the $T$-product comes about. Let us suppose that the out state is of the type $\beta = (\gamma, p')$, with $p'$ the momentum of a single particle. We want to extract this particle from the state

$$\langle \gamma, p'; \mathrm{out} | \phi(x) |\alpha; \mathrm{in} \rangle = \langle \gamma; \mathrm{out} | a_{\mathrm{out}}(p')\phi(x) |\alpha; \mathrm{in} \rangle$$

$$= \langle \gamma; \mathrm{out} | [a_{\mathrm{out}}(p')\phi(x) - \phi(x)a_{\mathrm{in}}(p')] |\alpha; \mathrm{in} \rangle$$

$$+ \langle \gamma; \mathrm{out} | \phi(x)a_{\mathrm{in}}(p') |\alpha; \mathrm{in} \rangle$$

$$= \langle \gamma; \mathrm{out} | \phi(x) |\alpha - p'; \mathrm{in} \rangle$$

$$+ i \int d^3\vec{y} \Big[ f_{\vec{p}'*}(y) \left( \frac{\partial}{\partial y^0} \langle \gamma; \mathrm{out} | (\phi_{\mathrm{out}}(y)\phi(x) - \phi(x)\phi_{\mathrm{in}}(y)) |\alpha; \mathrm{in} \rangle \right)$$

$$- \left( \frac{\partial}{\partial y^0} f_{\vec{p}}^*(y) \right) \langle \gamma; \mathrm{out} | (\phi_{\mathrm{out}}(y)\phi(x) - \phi(x)\phi_{\mathrm{in}}(y)) |\alpha; \mathrm{in} \rangle \Big]. \tag{12.60}$$

Using again the time independence of the integral, we have

$$\langle \gamma; \text{out}|[\phi_{\text{out}}(y)\phi(x) - \phi(x)\phi_{\text{in}}(y)]|\alpha; \text{in}\rangle$$

$$= \frac{1}{\sqrt{Z}} \left( \lim_{y^0 \to +\infty} - \lim_{y^0 \to -\infty} \right) \langle \gamma; \text{out}|T(\phi(y)\phi(x))|\alpha; \text{in}\rangle, \quad (12.61)$$

and, proceding as in the previous case,

$$\langle \beta; \text{out}|\phi(x)|\alpha; \text{in}\rangle$$

$$= \langle \gamma; \text{out}|\phi(x)|\alpha - p'; \text{in}\rangle$$

$$+ \frac{i}{\sqrt{Z}} \int d^4 y f_{\vec{p}'}^*(y)(\Box + m^2)_y \langle \gamma; \text{out}|T(\phi(y)\phi(x))|\alpha; \text{in}\rangle. \quad (12.62)$$

Substituting into eq. (12.59)

$$\langle \beta; \text{out}|\alpha, p; \text{in}\rangle = \langle \beta - p; \text{out}|\alpha; \text{in}\rangle$$

$$+ \frac{i}{\sqrt{Z}} \int d^4 x f_{\vec{p}}(x)(\Box + m^2)_x \langle \gamma; \text{out}|\phi(x)|\alpha - p'; \text{in}\rangle$$

$$+ \left(\frac{i}{\sqrt{Z}}\right)^2 \int d^4 x d^4 y f_{\vec{p}}(x) f_{\vec{p}'}^*(y)$$

$$\times (\Box + m^2)_x (\Box + m^2)_y \langle \gamma; \text{out}|T(\phi(y)\phi(x))|\alpha; \text{in}\rangle. \quad (12.63)$$

More generally, if we want to remove all the particles in $(q_1, \cdots, q_m)$ and in $(p_1, \cdots, p_n)$ with the condition $p_i \neq q_j$, we get

$$\langle p_1, \cdots, p_m; \text{out}|q_1, \cdots, q_n; \text{in}\rangle$$

$$= \left(\frac{i}{\sqrt{Z}}\right)^{m+n} \int \prod_{i,j=1}^{m,n} d^4 x_i d^4 y_j f_{\vec{q}_i}(x_i) f_{\vec{p}_j}^*(y_j)(\Box + m^2)_{x_i}(\Box + m^2)_{y_j}$$

$$\times \langle 0|T(\phi(y_1) \cdots \phi(y_n)\phi(x_1) \cdots \phi(x_m))|0\rangle. \quad (12.64)$$

If some of the $q_i$'s are equal to some of the $p_j$'s, we can reduce further the matrix elements of the type $\langle \beta; \text{out}|\alpha - p; \text{in}\rangle$. Therefore we have shown that it is possible to evaluate all the $S$ matrix elements once one knows the vacuum expectation values of the $T$-products (called also the $n$-point Green functions). The result obtained here can be easily extended to generic fields, simply by substituting to the wave functions $f_{\vec{p}}(x)$, the corresponding solutions of the free wave equation. For instance, for spinor fields, using the corresponding wave function, that is spinors times plane waves, and to the Klein-Gordon operator the Dirac operator, $(i\partial\!\!\!/ - m)$. In the Dirac case, it is also necessary to change $i/\sqrt{Z}$ to $-i/\sqrt{Z_2}$ for fermions and to $+i/\sqrt{Z_2}$ for antifermions, where $Z_2$ is the wave function renormalization for the spinor field.

In the following, we will show how to evaluate perturbatively the vacuum expectation value of $n$ fields, in the context of the path-integral formulation for quantum fields.

## 12.4    The path integral for a free scalar field

In this Section, we will extend the path integral approach to the case of a scalar field. We will show later on how to extend the formulation to Fermi fields. Let us start with a free scalar field. In the first part of this book, we have generally denoted the scalar quantum field by $\phi(x)$. Here, we will use the symbol $\varphi(x)$ for its classical counterpart. For a neutral particle this is a real function, and its dynamics is described by the action

$$S = \int_V d^4x \frac{1}{2} \left[ \partial_\mu \varphi \partial^\mu \varphi - m^2 \varphi^2 \right] \equiv \int_t^{t'} dt \int d^3\vec{x} \frac{1}{2} \left[ \partial_\mu \varphi \partial^\mu \varphi - m^2 \varphi^2 \right].$$

(12.65)

A free field theory can be thought in terms of a continuous collection of non-interacting harmonic oscillators. For this, it is sufficient to look for the normal modes. Let us consider the Fourier transform of the field

$$\varphi(\vec{x}, t) = \frac{1}{(2\pi)^3} \int d^3\vec{k} \, e^{i\vec{k}\cdot\vec{x}} q(\vec{k}, t).$$

(12.66)

By substituting into eq. (12.65) we find

$$S = \frac{1}{2} \int_t^{t'} dt \int \frac{d^3\vec{k}}{(2\pi)^3} [\dot{q}(\vec{k}, t)\dot{q}(-\vec{k}, t)$$
$$- (|\vec{k}|^2 + m^2)q(\vec{k}, t)q(-\vec{k}, t)].$$

(12.67)

Since $\varphi(\vec{x}, t)$ is a real field

$$q^*(\vec{k}, t) = q(-\vec{k}, t),$$

(12.68)

we can write

$$S = \int_t^{t'} dt \int \frac{d^3\vec{k}}{(2\pi)^3} \frac{1}{2} \left[ |\dot{q}(\vec{k}, t)|^2 - \omega_{\vec{k}}^2 |q(\vec{k}, t)|^2 \right], \quad \omega_{\vec{k}}^2 = |\vec{k}|^2 + m^2.$$ (12.69)

Therefore the system is equivalent to a continuous set of complex oscillators $q(\vec{k}, t)$, with mass equal to one (or to pairs of real oscillators). The equations of motion are

$$\ddot{q}(\vec{k}, t) + \omega_{\vec{k}}^2 q(\vec{k}, t) = 0.$$

(12.70)

We will now evaluate the generating functional $Z[J]$, by adding an external source term to the action

$$S = \int_V d^4x \left\{ \frac{1}{2} \left[ \partial_\mu \varphi \partial^\mu \varphi - m^2 \varphi^2 \right] + J\varphi \right\}.$$

(12.71)

From the Fourier decomposition we get

$$S = \int_t^{t'} dt \int \frac{d^3\vec{k}}{(2\pi)^3} \left\{ \frac{1}{2} \left[ |\dot{q}(\vec{k},t)|^2 - \omega_{\vec{k}}^2 |q(\vec{k},t)|^2 \right] + J(-\vec{k},t)q(\vec{k},t) \right\},$$
(12.72)

where we have expanded $J(x)$ in terms of its Fourier components. Let us separate $q(\vec{k},t)$ and $J(\vec{k},t)$ into their real and imaginary parts

$$q(\vec{k},t) = x(\vec{k},t) + iy(\vec{k},t),$$
$$J(\vec{k},t) = l(\vec{k},t) + im(\vec{k},t).$$
(12.73)

It follows from the reality conditions that the real parts are even functions of $\vec{k}$, whereas the imaginary parts are odd functions:

$$x(\vec{k},t) = x(-\vec{k},t), \qquad l(\vec{k},t) = l(-\vec{k},t),$$
$$y(\vec{k},t) = -y(-\vec{k},t), \qquad m(\vec{k},t) = -m(-\vec{k},t).$$
(12.74)

Now, we can use the result found in the previous Section for a single oscillator, summing over all the oscillators. In particular we get for the exponent in eq. (11.138)

$$\frac{i}{2} \int_t^{t'} ds\, ds'\, J(s)\Delta(s - s')J(s')$$

$$\to \frac{i}{2} \int_t^{t'} ds\, ds' \int \frac{d^3\vec{k}}{(2\pi)^3} \left[ l(\vec{k},s)\Delta(s - s';\omega_{\vec{k}})l(\vec{k},s') \right.$$

$$\left. + m(\vec{k},s)\Delta(s - s';\omega_{\vec{k}})m(\vec{k},s') \right]$$

$$= \frac{i}{2} \int_t^{t'} ds\, ds' \int \frac{d^3\vec{k}}{(2\pi)^3} J(-\vec{k},s)\Delta(s - s';\omega_{\vec{k}})J(\vec{k},s'), \qquad (12.75)$$

where we have made use of eq. (12.73). $\Delta(s)$ is the propagator for an oscillator of mass equal to one, as defined in eq. (11.144):

$$\Delta(s) = -\lim_{\epsilon \to 0^+} \frac{1}{2\pi} \int_{-\infty}^{+\infty} \frac{e^{-i\nu s}}{\nu^2 - \omega^2 + i\epsilon} d\nu.$$
(12.76)

Going back to $J(\vec{x},t)$, we get

$$\frac{i}{2} \int d^4x\, d^4y\, J(x) \int \frac{d^3\vec{k}}{(2\pi)^3} \Delta(s - s';\omega_{\vec{k}}) e^{i\vec{k}\cdot(\vec{x}-\vec{y})} J(y)$$

$$\equiv \frac{i}{2} \int d^4x\, d^4y\, J(x)\Delta_F(x - y;m^2)J(y),$$
(12.77)

where we have defined $x^\mu = (\vec{x},s)$, $y^\mu = (\vec{y},s')$ and

$$\Delta_F(x,m^2) = \int \frac{d^3\vec{k}}{(2\pi)^3} \Delta(s;\omega_{\vec{k}}) e^{i\vec{k}\cdot\vec{x}} = -\lim_{\epsilon \to 0^+} \int \frac{d^4k}{(2\pi)^4} \frac{e^{-ikx}}{k^2 - m^2 + i\epsilon}.$$
(12.78)

Here $k^\mu = (\nu, \vec{k})$ and $\Delta_F$ is the Feynman propagator defined in eq. (7.10). Therefore, the generating functional is

$$Z[J] = e^{\frac{i}{2} \int d^4x d^4y J(x) \Delta_F(x-y; m^2) J(y)} Z[0]. \qquad (12.79)$$

In particular we get

$$\frac{1}{\langle 0|0 \rangle} \langle 0|T(\phi(x_1)\phi(x_2))|0\rangle = -\frac{1}{Z[0]} \frac{\delta^2 Z[J]}{\delta J(x_1)\delta J(x_2)}\Big|_{J=0}$$
$$= -i\Delta_F(x_1 - x_2; m^2). \qquad (12.80)$$

It is convenient to use the following notation for the $N$-point Green functions

$$G^{(N)}(x_1, \cdots, x_N) = \frac{\langle 0|T(\phi(x_1) \cdots \phi(x_N))|0\rangle}{\langle 0|0\rangle}. \qquad (12.81)$$

In the free case we can evaluate easily the generic $G^{(N)}$, by developing the generating functional in a series of Volterra

$$\frac{Z[J]}{Z[0]} = \sum_{n=0}^{\infty} \int d^4x_1 \cdots d^4x_n J(x_1) \cdots J(x_n)$$
$$\times \frac{(i)^n}{n!} \frac{\langle 0|T(\phi(x_1) \cdots \phi(x_n))|0\rangle}{\langle 0|0\rangle}$$
$$\equiv \sum_{n=0}^{\infty} \int d^4x_1 \cdots d^4x_n J(x_1) \cdots J(x_n)$$
$$\times \frac{(i)^n}{n!} G_0^{(n)}(x_1, \cdots, x_n), \qquad (12.82)$$

where $G_0^{(n)}$ is the Green functions for the free case. By comparison with the expansion of eq. (12.79)

$$\frac{Z[J]}{Z[0]} = 1 + \frac{i}{2} \int d^4x_1 d^4x_2 J(x_1) J(x_2) \Delta_F(x_1 - x_2)$$
$$- \frac{1}{8} \int d^4x_1 d^4x_2 J(x_1) J(x_2) \Delta_F(x_1 - x_2)$$
$$\times \int d^4x_3 d^4x_4 J(x_3) J(x_4) \Delta_F(x_3 - x_4) + \cdots. \qquad (12.83)$$

Since $Z[J]$ is even in $J$, we have

$$G_0^{(2k+1)}(x_1, \cdots, x_{2k+1}) = 0. \qquad (12.84)$$

For $G_0^{(2)}$, we get back eq. (12.80). To evaluate $G_0^{(4)}$ we have first to notice that all the $G_0^{(n)}$'s are symmetric in their arguments. Therefore, in order to extract the coefficients, we need to perform the symmetrization explicitly:

$$\int d^4x_1 \cdots d^4x_4 J(x_1) \cdots J(x_4) \Delta_F(x_1 - x_2) \Delta_F(x_3 - x_4)$$

$$= \frac{1}{3} \int d^4x_1 \cdots d^4x_4 J(x_1) \cdots J(x_4) \Big[ \Delta_F(x_1 - x_2) \Delta_F(x_3 - x_4)$$

$$+ \Delta_F(x_1 - x_3) \Delta_F(x_2 - x_4) + \Delta_F(x_1 - x_4) \Delta_F(x_2 - x_3) \Big]. \quad (12.85)$$

It follows:

$$G_0^{(4)}(x_1, \cdots, x_4) = -\Big[ \Delta_F(x_1 - x_2) \Delta_F(x_3 - x_4)$$

$$+ \Delta_F(x_1 - x_3) \Delta_F(x_2 - x_4)$$

$$+ \Delta_F(x_1 - x_4) \Delta_F(x_2 - x_3) \Big]. \quad (12.86)$$

If we represent $G_0^{(2)}(x, y)$ by a line as in Fig. 12.2, $G_0^{(4)}$ will be given by Fig. 12.3, since it is obtained in terms of products of $G_0^{(2)}$. Therefore, $G_0^{(2n)}$ is given by a combination of products of $n$ two-point functions. A diagram is disconnected if we can isolate two sub-diagrams with no connecting lines. We see that in the free case, all the non-vanishing Green functions are disconnected, except for the two point function. It is natural to introduce a functional generating only the connected Green functions. In the free case we define:

Fig. 12.2   The two-point Green's function $G_0^{(2)}(x, y)$.

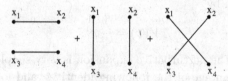

Fig. 12.3   The expansion of the four-point Green's function $G_0^{(4)}(x_1, x_2, x_3, x_4)$.

$$Z[J] = e^{iW[J]}, \quad (12.87)$$

with

$$W[J] = \frac{1}{2} \int d^4x_1 d^4x_2 J(x_1) J(x_2) \Delta_F(x_1 - x_2) + W[0], \qquad (12.88)$$

and we have

$$\frac{1}{Z[0]} \frac{\delta^2 Z[J]}{\delta J(x_1)\delta J(x_2)}\bigg|_{J=0} = i\frac{\delta^2 W[J]}{\delta J(x_1)\delta J(x_2)}\bigg|_{J=0} = i\Delta_F(x_1 - x_2). \quad (12.89)$$

Therefore, the derivatives of $W[J]$ give the connected diagrams with the proper normalization (that is divided by $Z[0]$). It is possible to show that, in general, $W[J]$ as defined through eq. (12.87), generates the connected Green functions (see, for example [Rivers (1987)]),

$$iW[J] = \sum_{n=0}^{\infty} \frac{(i)^n}{n!} \int d^4x_1 \cdots d^4x_n J(x_1) \cdots J(x_n) G_c^{(n)}(x_1, \cdots, x_n),$$
$$(12.90)$$

where

$$G_c^{(n)}(x_1, \cdots, x_n) = \frac{1}{\langle 0|0\rangle} \langle 0|T(\phi(x_1)\cdots\phi(x_n))|0\rangle_{\text{conn}}. \qquad (12.91)$$

The index "conn." denotes the connected Green functions. Notice that once these functions are evaluated, the theory is completely solved, since the generating functional is recovered by exponentiating $W[J]$.

## 12.5   The perturbative expansion for the theory $\lambda\varphi^4$

Let us consider an interacting scalar field described by the action

$$S = \int_V d^4x \left( \frac{1}{2}\left[\partial_\mu\varphi\partial^\mu\varphi - m^2\varphi^2\right] - V(\varphi) \right), \qquad (12.92)$$

where $V(\varphi)$ is the interaction potential. An example is the theory $\lambda\varphi^4$ with the potential given by

$$V(\varphi) = \frac{\lambda}{4!}\varphi^4. \qquad (12.93)$$

The derivation of the perturbative expansion is very simple in this formalism. We take advantage of the following identity, valid for any functional $F$ admitting a Volterra expansion:

$$\int \mathcal{D}(\varphi(x))F[\varphi]e^{iS+i\int d^4x J(x)\varphi(x)}$$
$$= F\left[\frac{1}{i}\frac{\delta}{\delta J}\right]\int \mathcal{D}(\varphi(x))e^{iS+i\int d^4x J(x)\varphi(x)}. \qquad (12.94)$$

We separate in eq. (12.92) the quadratic part of the action, $S_0$, from the interacting part, and then we use the previous identity

$$Z[J] = N \int \mathcal{D}(\varphi(x)) e^{iS + i \int d^4x J(x)\varphi(x)}$$

$$= N \int \mathcal{D}(\varphi(x)) e^{-i \int d^4x V(\varphi)} e^{iS_0 + i \int d^4x J(x)\varphi(x)}$$

$$= e^{-i \int d^4x V\left(\frac{1}{i}\frac{\delta}{\delta J}\right)} Z_0[J]. \tag{12.95}$$

Notice that we have suppressed the temporal limits in the expression for the generating functional. $Z_0[J]$ is the generating functional for the free case, evaluated in eq. (12.79). Therefore, we get

$$Z[J] = e^{iW[J]} = Z[0] e^{-i \int d^4x V\left(\frac{1}{i}\frac{\delta}{\delta J}\right)}$$

$$\times e^{\frac{i}{2} \int d^4x d^4y J(x) \Delta_F(x-y) J(y)}. \tag{12.96}$$

In this equation, we have suppressed the mass in the argument of $\Delta_F$. For the following calculation, it will be useful to introduce a shorthand notation

$$\int d^4x_1 \cdots d^4x_n F(x_1, \cdots, x_n) \equiv \langle F(x_1, \cdots, x_n) \rangle. \tag{12.97}$$

Then, eq. (12.96) can be rewritten in the following way

$$Z[J] = e^{iW[J]} = e^{-i\langle V\left(\frac{1}{i}\frac{\delta}{\delta J}\right)\rangle} e^{\frac{i}{2}\langle J(1)\Delta_F(1,2)J(2)\rangle} Z[0]$$

$$= e^{-i\langle V\left(\frac{1}{i}\frac{\delta}{\delta J}\right)\rangle} e^{iW_0[J]}. \tag{12.98}$$

This gives rise to the following expression for the generating functional of the connected Green functions

$$W[J] = -i \log \left[ e^{iW_0[J]} + \left( e^{-i\langle V\left(\frac{1}{i}\frac{\delta}{\delta J}\right)\rangle} - 1 \right) e^{iW_0[J]} \right]$$

$$= W_0[J] - i \log \left[ 1 + e^{-iW_0[J]} \left( e^{-i\langle V\left(\frac{1}{i}\frac{\delta}{\delta J}\right)\rangle} - 1 \right) \right.$$

$$\times e^{iW_0[J]} \right]. \tag{12.99}$$

This expression can be easily expanded in the potential $V$. To this end, let us define

$$\delta[J] = e^{-iW_0[J]} \left( e^{-i\langle V\left(\frac{1}{i}\frac{\delta}{\delta J}\right)\rangle} - 1 \right) e^{iW_0[J]}, \tag{12.100}$$

then, we can expand $W[J]$ in a series of $\delta[J]$. Notice also that the expansion is in terms of the interaction Lagrangian. Just to show how the procedure works, we will consider the expansion at first order in the potential. For higher terms in the expansion, see, for instance [Ramond (1981)].

At first order in $\delta$ we get

$$W[J] = W_0[J] - i\delta + \cdots . \tag{12.101}$$

Let us now consider the case of $\lambda\varphi^4$. The functional $\delta[J]$ can be expanded in a series of the dimensionless coupling $\lambda$

$$\delta = \lambda\delta_1 + \lambda^2\delta_2 + \cdots . \tag{12.102}$$

It follows (at first order)

$$W[J] = W_0[J] - i\lambda\delta_1 + \cdots . \tag{12.103}$$

Then, from eq. (12.102) we get

$$\delta_1 = -\frac{i}{4!}e^{-iW_0[J]}\langle \left(\frac{1}{i}\frac{\delta}{\delta J}\right)^4 \rangle e^{iW_0[J]}. \tag{12.104}$$

By performing the functional derivatives we find

$$\frac{\delta^4}{\delta J(x)^4}e^{iW_0[J]}$$
$$= (-3\Delta_F(x,x)^2 + 6i\Delta_F(x,x)\Delta_F(x,1)\Delta_F(x,2)J(1)J(2)$$
$$+\Delta_F(x,1)\Delta_F(x,2)\Delta_F(x,3)\Delta_F(x,4)J(1)J(2)J(3)J(4))$$
$$\times e^{iW_0[J]}, \tag{12.105}$$

where the integration on the variables $1, 2, 3, 4$ is understood. We obtain

$$\delta_1 = -\frac{i}{4!}\Big[\langle\Delta_F(y,1)\Delta_F(y,2)\Delta(y,3)\Delta_F(y,4)J(1)J(2)J(3)J(4)\rangle$$
$$+ 6i\langle\Delta_F(y,y)\Delta_F(y,1)\Delta_F(y,2)J(1)J(2)\rangle - 3\langle\Delta_F^2(y,y)\rangle\Big]. \tag{12.106}$$

We can now evaluate the connected Green functions by differentiating functionally eq. (12.103) and putting $J = 0$. For instance, at this order, the two-point function is given by

$$G_c^{(2)}(x_1, x_2) = -i\frac{\delta^2 W[J]}{\delta J(x_1)\delta J(x_2)}$$
$$= -i\frac{\delta^2 W_0[J]}{\delta J(x_1)\delta J(x_2)}\Big|_{J=0} - \lambda\frac{\delta^2\delta_1[J]}{\delta J(x_1)\delta J(x_2)}\Big|_{J=0}, \tag{12.107}$$

that is,

$$G_c^{(2)}(x_1, x_2) = i\Delta_F(x_1 - x_2) - \frac{\lambda}{2}\int d^4x\,\Delta_F(x_1 - x)\Delta_F(x - x)\Delta_F(x - x_2). \tag{12.108}$$

In Fig. 12.4 we give the graphical interpretation of this expression. The graphics are obtained in terms of two elements, the propagator $\Delta_F(x - y)$,

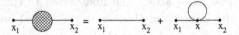

Fig. 12.4   The graphical expansion for the two-point connected Green's function at first order in $\lambda$.

corresponding to the lines and the vertex $\lambda\varphi^4$, corresponding to the point where four lines meet. In order to get $G_c^{(n)}$ we have to draw all the possible connected diagrams containing a number of interaction vertices corresponding to the considered perturbative order. The analytic expression is obtained associating a factor $i\Delta_F(x - y)$ to each line connecting the point $x$ to the point $y$, and a factor $(-i\lambda/4!)$ for each interaction vertex. The numerical factor appearing in eq. (12.108) is obtained by counting the possible ways to draw a given diagram. For instance, let us consider the diagram of Fig. 12.5

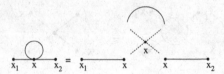

Fig. 12.5   How to get the numerical factor for the first order contribution to $G_c^{(2)}(x_1, x_2)$.

To get the numerical factor, we count the number of ways to attach the external propagators to the vertex. Then, we have to count the possibilities to attach the vertex at the internal propagator. There are four ways to attach the vertex to the first propagator and three for the second one. Therefore,

$$\left(\frac{1}{4!}\right) \cdot 4 \cdot 3 = \frac{1}{2},\tag{12.109}$$

where the factor $1/4!$ comes from the vertex. As another example, consider the diagram in Fig. 12.6.

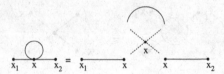

Fig. 12.6   How to get the numerical factor for one of the second order contributions to $G_c^{(2)}(x_1, x_2)$.

There are four ways to attach the first vertex to the first propagator and four for attaching the second vertex to the second propagator. Three ways for attaching the second leg of the first vertex to the second vertex and two ways of attaching the third leg of the first vertex to the second one. Therefore, we get

$$\left(\frac{1}{4!}\right)^2 \cdot 4 \cdot 4 \cdot 3 \cdot 2 = \frac{1}{6}. \tag{12.110}$$

The four-point function can be evaluated in analogous way, and the result is given in Fig. 12.7. This is obtained by functional differentiation of $W[J]$. The first order contribution is

$$G_c^{(4)}(x_1, x_2, x_3, x_4) = i\lambda \int d^4x \Delta_F(x_1 - x)\Delta_F(x_2 - x)$$
$$\times \Delta_F(x_3 - x)\Delta_F(x_4 - x), \tag{12.111}$$

where the numerical coefficient is one, since there are $4 \times 3 \times 2$ ways of attaching the vertex to the four external legs.

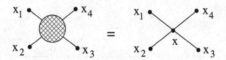

Fig. 12.7  The graphical expansion up to the second order of $G_c^{(4)}(x_1, x_2, x_3, x_4)$.

## 12.6  The Feynman rules in momentum space

The rules for the perturbative expansion of Green's functions in space-time, that we got in the previous Section, can be easily translated in momentum space. Here we will simply give the result (see, the exercise (2) in this Chapter)

- For each propagator draw a line with associated momentum $p$ (see Fig. 12.8).
- For each factor $(-i\lambda/4!)$ draw a vertex with the convention that the momentum flux is zero (see Fig. 12.9).
- To get $G_c^{(n)}$ draw all the topological inequivalent diagrams after having fixed the external legs. The number of ways of drawing a given diagram is its topological weight. The contribution of such a diagram is multiplied by its topological weight.

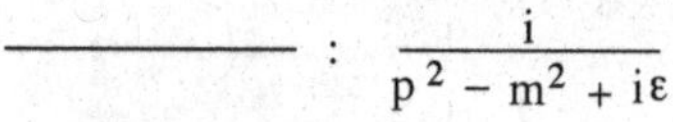

$$\underline{\phantom{xxxxxxxxx}} \quad : \quad \frac{i}{p^2 - m^2 + i\varepsilon}$$

Fig. 12.8   The propagator in momentum space.

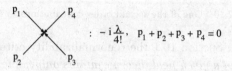

$$: \quad -i\frac{\lambda}{4!}, \quad p_1 + p_2 + p_3 + p_4 = 0$$

Fig. 12.9   The vertex in momentum space.

- After the requirement of the conservation of the four-momentum at each vertex, we need to integrate over all the independent internal four-momenta, with weight

$$\int \frac{d^4 q}{(2\pi)^4}. \tag{12.112}$$

A more systematic way is to associate a factor

$$-\frac{i\lambda}{4!}(2\pi)^4 \delta^4 \left(\sum_{i=1}^{4} p_i\right) \tag{12.113}$$

to each vertex and integrate over all the momenta of the internal lines. This gives automatically a factor

$$(2\pi)^4 \delta^4 \left(\sum_{i=1}^{n} p_i\right), \quad (n = \text{number of external lines}), \tag{12.114}$$

corresponding to the conservation of the total four-momentum of the process.

By using these rules, we can easily evaluate the second order contribution to the two-point function, corresponding to Fig. 12.10. We get

$$4 \times 4 \times 3 \times 2 \times \left(-\frac{i\lambda}{4!}\right)^2 \frac{i}{p_2^2 - m^2 + i\epsilon} \int \left(\prod_{i=1}^{3} \frac{d^4 q_i}{(2\pi)^4} \frac{i}{q_i^2 - m^2 + i\epsilon}\right)$$

$$\times (2\pi)^4 \delta^4(p_1 - q_1 - q_2 - q_3)(2\pi)^4 \delta^4(q_1 + q_2 + q_3 - p_2)\frac{i}{p_1^2 - m^2 + i\epsilon}$$

$$= \frac{\lambda^2}{6} \frac{1}{(p_1^2 - m^2 + i\epsilon)^2}(2\pi)^4 \delta^4(p_1 + p_2)$$

$$\times \int \left(\prod_{i=1}^{3} \frac{d^4 q_i}{(2\pi)^4} \frac{i}{q_i^2 - m^2 + i\epsilon}\right) (2\pi)^4 \delta^4(p_1 - q_1 - q_2 - q_3). \tag{12.115}$$

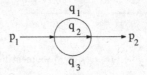

Fig. 12.10   One of the second order contributions to $G_c^{(2)}$.

We recall from Section 10.1 the renormalizability criterium: *if the Lagrangian density of a given field theory contains only monomial in the fields with mass dimension smaller than or equal to the number of space-time dimensions, then, the theory is renormalizable.* It follows that the scalar field theory with interaction $\lambda\varphi^4$ is renormalizable in four space-time dimensions.

## 12.7   Regularization in $\lambda\varphi^4$

In this Section, we will evaluate the one-loop contribution to the two-point Green function, by using the dimensional regularization. For a complete one-loop renormalization of the theory, we should consider also the four-point function. This is left to the reader in the exercise (4) of this Chapter.

Then, let us consider the action in dimensions $d = 2\omega$

$$S_{2\omega} = \int d^{2\omega}x \left[ \frac{1}{2}\partial_\mu\varphi\partial^\mu\varphi + \frac{1}{2}m^2\varphi^2 + \frac{\lambda}{4!}\varphi^4 \right]. \qquad (12.116)$$

In our unit system ($\hbar = c = 1$), the action $S_{2\omega}$ should be dimensionless. As discussed in Section 10.1, from the previous requirement we can evaluate the mass dimensions of the scalar field, by looking at the kinetic term. It follows:

$$dim[\varphi] = \frac{d}{2} - 1 = \omega - 1. \qquad (12.117)$$

From this, it follows at once that

$$dim[m] = 1, \quad dim[\lambda] = 4 - 2\omega. \qquad (12.118)$$

As we did in QED for the electric charge, it is convenient to introduce a dimensionful parameter $\mu$ ($dim[\mu] = 1$), in order to be able to define a dimensionless coupling

$$\lambda_{\text{new}} = \lambda_{\text{old}}(\mu^2)^{\omega-2}. \qquad (12.119)$$

In terms of the new coupling ($\lambda = \lambda_{\text{new}}$):

$$S_{2\omega} = \int d^{2\omega}x \left[ \frac{1}{2}\partial_\mu\varphi\partial^\mu\varphi + \frac{1}{2}m^2\varphi^2 + \frac{\lambda}{4!}(\mu^2)^{2-\omega}\varphi^4 \right]. \qquad (12.120)$$

The Feynman rules are slightly modified:

- The scalar product among two four-vectors becomes a sum over $2\omega$ components.
- In the loop integrals we will have a factor

$$\int \frac{d^{2\omega}p}{(2\pi)^{2\omega}}, \qquad (12.121)$$

and the $\delta$ function in $2\omega$ dimensions is defined by

$$\int d^{2\omega}p\, \delta^{(2\omega)}\left(\sum_i p_i\right) = 1. \qquad (12.122)$$

- The coupling $\lambda$ goes into $\lambda(\mu^2)^{2-\omega}$.

Also, we will do all the calculations in the Euclidean version. Remember that in Section 10.3 we gave the relevant integrals both in the Euclidean and in the Minkowskian version. Furthermore, the Feynman rules in the Euclidean space are quite simple.

We start by going to the Euclidean formulation. Let us denote the time by $t_M$. We need only to express the propagator in the Euclidean space. This can be done by a Wick rotation, through the relation $t_E = it_M$, between the Euclidean and Minkowskian times. The result is that the propagator is given by

$$i\Delta_F(x_M) = \int \frac{d^4 p_E}{(2\pi)^4} \frac{e^{-ip_E x_E}}{p_E^2 + m^2}, \qquad (12.123)$$

where we have defined $x_M = (t_M, \vec{x})$, $x_E = (t_E, \vec{x})$) and $p_E = (E_E, \vec{p})$ (see exercise (3) in this Chapter).

From this, we get the following modifications of Feynman's rules in the Euclidean version:

- Associate to each line a propagator

$$\frac{1}{p^2 + m^2}. \qquad (12.124)$$

- To each vertex associate a factor

$$-\frac{\lambda}{4!}. \qquad (12.125)$$

This rule follows because the weight in the Euclidean path-integral is given by $e^{-S_E}$ instead of $e^{iS}$, as shown in Section 11.6.

The other rules of the previous Section are unchanged.

We will calculate the first order contribution to the two-point function, corresponding to the diagram (tadpole) of Fig. 12.11. This is given by

$$(2\pi)^{2\omega}\delta^{(2\omega)}(p_1 + p_2)\frac{1}{p_1^2 + m^2}T_2\frac{1}{p_1^2 + m^2}, \tag{12.126}$$

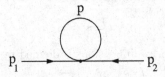

Fig. 12.11   The one-loop contribution to the two-point function in the $\lambda\varphi^4$ theory.

with

$$
\begin{aligned}
T_2 &= \frac{1}{2}(-\lambda)(\mu^2)^{2-\omega}\int\frac{d^{2\omega}p}{(2\pi)^{2\omega}}\frac{1}{p^2 + m^2}\\
&= -\frac{\lambda}{2}(\mu^2)^{2-\omega}\frac{\pi^\omega}{(2\pi)^{2\omega}}\Gamma(1-\omega)\frac{1}{(m^2)^{1-\omega}}\\
&= -\frac{\lambda}{2}\frac{m^2}{(4\pi)^2}\left(\frac{4\pi\mu^2}{m^2}\right)^{2-\omega}\Gamma(1-\omega).
\end{aligned}
\tag{12.127}
$$

For $\omega \to 2$ we get

$$
\begin{aligned}
T_2 &\approx -\frac{\lambda}{2}\frac{m^2}{(4\pi)^2}\left[1 + (2-\omega)\log\frac{4\pi\mu^2}{m^2} + \cdots\right]\left[-\frac{1}{2-\omega} - \psi(2) + \cdots\right]\\
&= \frac{\lambda}{32\pi^2}m^2\left[\frac{1}{2-\omega} + \psi(2) + \log\frac{4\pi\mu^2}{m^2} + \cdots\right],
\end{aligned}
\tag{12.128}
$$

where $\psi(n)$ was defined in eq. (10.70).

## 12.8   Renormalization in $\lambda\varphi^4$

In order to renormalize up to one-loop order the two-point function, we proceed as in QED. That is we split the original Lagrangian, expressed in terms of bare couplings and bare fields in a part

$$\mathcal{L}_p = \frac{1}{2}\partial_\mu\varphi\partial^\mu\varphi + \frac{1}{2}m^2\varphi^2 + \frac{\lambda}{4!}\mu^{4-2\omega}\varphi^4, \tag{12.129}$$

depending on the renormalized parameters $m$ and $\lambda$ and field $\varphi$, and in the counterterm contribution

$$\mathcal{L}_{ct} = \frac{1}{2}\delta Z \partial_\mu \varphi \partial^\mu \varphi + \frac{1}{2}\delta m^2 \varphi^2 + \frac{\delta\lambda}{4!}\mu^{4-2\omega}\varphi^4. \tag{12.130}$$

The sum of these two expressions,

$$\mathcal{L} = \frac{1}{2}(1+\delta Z)\partial_\mu\varphi\partial^\mu\varphi + \frac{1}{2}(m^2+\delta m^2)\varphi^2 + \frac{(\lambda+\delta\lambda)}{4!}\mu^{4-2\omega}\varphi^4, \tag{12.131}$$

reproduces the original Lagrangian, if we redefine couplings and fields as follows

$$\varphi_B = (1+\delta Z)^{1/2}\varphi \equiv Z_\varphi^{1/2}\varphi, \tag{12.132}$$

$$m_B^2 = \frac{m^2+\delta m^2}{1+\delta Z} = (m^2+\delta m^2)Z_\varphi^{-1}, \tag{12.133}$$

$$\lambda_B = \mu^{4-2\omega}\frac{\lambda+\delta\lambda}{(1+\delta Z)^2} = \mu^{4-2\omega}(\lambda+\delta\lambda)Z_\varphi^{-2}. \tag{12.134}$$

In fact,

$$\mathcal{L} = \mathcal{L}_p + \mathcal{L}_{ct} = \frac{1}{2}\partial_\mu\varphi_B\partial^\mu\varphi_B + \frac{1}{2}m_B^2\varphi_B^2 + \frac{\lambda_0}{4!}\varphi_B^4. \tag{12.135}$$

The counterterms are then evaluated by the requirement of removing the divergences arising in the limit $\omega \to 2$, except for a finite part which is fixed by the renormalization conditions.

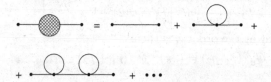

Fig. 12.12   The contributions to the two-point connected Green function.

From the results of the previous Section, by summing up all the contributions (see Fig. 12.12) we get the following expression for the two-point connected Green function

$$G_c^{(2)}(p_1,p_2) = (2\pi)^{2\omega}\delta^{2\omega}(p_1+p_2)\left[\frac{1}{p_1^2+m^2} + \frac{1}{p_1^2+m^2}T_2\frac{1}{p_1^2+m^2} + \cdots\right]$$

$$\approx (2\pi)^{2\omega}\delta^{2\omega}(p_1+p_2)\frac{1}{p_1^2+m^2-T_2}. \tag{12.136}$$

Therefore, the effect of the first order contribution is to produce a mass correction determined by $T_2$. The divergent part can be compensated by the counterterm $\delta m^2$, which produces an additional contribution to the one-loop diagram, represented in Fig. 12.13. We choose the counterterm as

$$\frac{1}{2}\delta m^2 \varphi^2 = \frac{\lambda}{64\pi^2} m^2 \left[ \frac{2}{\epsilon} + F_1 \left( \omega, \frac{m^2}{\mu^2} \right) \right] \varphi^2, \tag{12.137}$$

with $\epsilon = 4 - 2\omega$ and $F_1$ an arbitrary dimensionless function.

$$\bullet\!\!\!-\!\!\!-\!\!\!\times\!\!\!-\!\!\!-\!\!\!\bullet \;\; = \; -\,\delta m^2$$

Fig. 12.13   The Feynman rule for the counterterm $\delta m^2$.

Adding the contribution of the counterterm to the previous result we get

$$G_c^{(2)}(p_1, p_2) = (2\pi)^{2\omega} \delta^{2\omega}(p_1 + p_2) \Big[ \frac{1}{p_1^2 + m^2 - T_2}$$
$$+ \frac{1}{p_1^2 + m^2}(-\delta m^2) \frac{1}{p_1^2 + m^2} \Big]. \tag{12.138}$$

At the order $\lambda$ we have

$$G_c^{(2)}(p_1, p_2) = (2\pi)^{2\omega} \delta^{2\omega}(p_1 + p_2) \frac{1}{p_1^2 + m^2 + \delta m^2 - T_2}. \tag{12.139}$$

Of course, the same result could have been obtained by noticing that the counterterm modifies the propagator by shifting $m^2$ into $m^2 + \delta m^2$. Now the combination $\delta m^2 - T_2$ is finite and given by

$$\delta m^2 - T_2 = \frac{\lambda}{32\pi^2} m^2 \left[ F_1 - \psi(2) + \log \frac{m^2}{4\pi\mu^2} \right]. \tag{12.140}$$

The finite result at the order $\lambda$ is then

$$G_c^{(2)}(p_1, p_2) = (2\pi)^{2\omega} \delta^{2\omega}(p_1 + p_2)$$
$$\times \frac{1}{p_1^2 + m^2 \left( 1 + \frac{\lambda}{32\pi^2} \left[ F_1 - \psi(2) + \log \frac{m^2}{4\pi\mu^2} \right] \right)}. \tag{12.141}$$

This expression has a pole in Minkowski space, and we can define the renormalization condition by requiring that the inverse propagator at the physical mass be (in Euclidean space):

$$G_c^{(2)^{-1}} = p^2 + m_{phys}^2, \tag{12.142}$$

where $m_{phys}$ can be read in eq. (12.141). This choice determines uniquely the $F_1$ term. Notice that, at one-loop there is no wave function renormalization. In fact, this comes about only at two-loop level (see, for instance, [Ramond (1981)]).

## 12.9 Fermionic oscillators and Grassmann algebras

The Feynman formulation of path integral can be extended to the case of fermions. However, since the path integral is strictly related to the classical description of the theory, it should be nice to have such a "classical" description for the fermionic systems. In fact, such a description is discussed in [Casalbuoni (1976a)] and [Casalbuoni (1976b)]. The description is based on the use of the elements of a Grassmann algebra ("anticommuting numbers") as dynamical variables. After quantization, the elements of a Grassmann algebra lead naturally to anticommutation relations among the corresponding operators. Before proceeding further, let us introduce some concepts about Grassmann algebras, that will be useful in the following.

First of all, we recall that an algebra $V$ is nothing but a vector space where a bilinear mapping, $V \times V \to V$ (the algebra product) is defined. A Grassmann algebra $\mathcal{G}_n$ is a vector space of dimensions $2^n$ and it can be described in terms of $n$ generators $\theta_i$ $(i = 1, \cdots, n)$ satisfying the relations

$$\theta_i \theta_j + \theta_j \theta_i \equiv [\theta_i, \theta_j]_+ = 0, \quad i, j = 1, \cdots, n. \tag{12.143}$$

Forming all the possible products among the generators, one gets a basis of $\mathcal{G}_n$:

$$1, \ \theta_i, \ \theta_i \theta_j, \ \theta_i \theta_j \theta_k, \cdots, \theta_1 \theta_2 \cdots \theta_n. \tag{12.144}$$

The number of these elements is easily evaluated:

$$1 + n + \frac{n(n-1)}{2} + \cdots = \sum_{k=0}^{n} \frac{n!}{k!(n-k)!} = 2^n. \tag{12.145}$$

Notice, that there are no other possible elements since, from eq. (12.143), it follows

$$\theta_i^2 = 0. \tag{12.146}$$

The subset of elements of $\mathcal{G}_n$, generated by the product of an even number of generators, is called the even part of the algebra, $\mathcal{G}_n^{(0)}$, whereas the subset generated by the product of an odd number of generators is the odd part, $\mathcal{G}_n^{(1)}$. It follows

$$\mathcal{G}_n = \mathcal{G}_n^{(0)} \oplus \mathcal{G}_n^{(1)}. \tag{12.147}$$

A decomposition of this kind is called a **grading** of the vector space, if it is possible to assign a grade to each element of the subspaces. In this case the grading is

(1) If $x \in \mathcal{G}_n^{(0)}$, $\deg(x) = +1$.
(2) If $x \in \mathcal{G}_n^{(1)}$, $\deg(x) = -1$.

Notice that this grading is also consistent with the algebra product, since if $x, y$ are monomials of $\mathcal{G}_n$, then

$$\deg(xy) = \deg(x)\deg(y). \tag{12.148}$$

Since we have no space to enter into the details of the classical mechanics based on Grassmann variables, we will formulate the path integral starting directly from quantum mechanics. Let us recall that, in the Feynman formulation, one introduces states which are eigenstates of a complete set of observables. In the case of bosonic fields, one considers eigenstates of the field operators in the Heisenberg representation

$$\phi(\vec{x}, t)|\varphi(\vec{x}), t\rangle = \varphi(\vec{x})|\varphi(\vec{x}), t\rangle. \tag{12.149}$$

But this is possible since the fields commute at equal times

$$[\phi(\vec{x}, t), \phi(\vec{y}, t)] = 0, \tag{12.150}$$

and we can diagonalize simultaneously the set of operators $\phi(\vec{x}, t)$. In the case of fermionic fields, this is not possible, since they anticommute at equal times. Therefore, we cannot define simultaneous eigenstates of fermionic fields. To gain a better understanding of this point, let us consider free fermionic fields, and let us recall that this is nothing but a set of harmonic oscillators of Fermi type, that is they are described by creation and annihilation operators satisfying the anticommutation relations

$$[a_i, a_j]_+ = \left[a_i^\dagger, a_j^\dagger\right]_+ = 0, \quad \left[a_i, a_j^\dagger\right]_+ = \delta_{ij}. \tag{12.151}$$

For simplicity, consider the case of a single Fermi oscillator, and let us try to diagonalize the corresponding annihilation operator

$$a|\theta\rangle = \theta|\theta\rangle. \tag{12.152}$$

Then, $a^2 = 0$ implies $\theta^2 = 0$. Therefore, the eigenvalue $\theta$ cannot be an ordinary number, but it should be identified with an odd element of a Grassmann algebra. Consider now $n$ Fermi oscillators, and require

$$a_i|\theta_1, \cdots, \theta_n\rangle = \theta_i|\theta_1, \cdots, \theta_n\rangle, \tag{12.153}$$

we get

$$a_i\left(a_j|\theta_1, \cdots, \theta_n\rangle\right) = \theta_j\left(a_i|\theta_1, \cdots, \theta_n\rangle\right) = \theta_j\theta_i|\theta_1, \cdots, \theta_n\rangle. \tag{12.154}$$

But $a_i a_j = -a_j a_i$ and therefore

$$[\theta_i, \theta_j]_+ = 0. \tag{12.155}$$

The conclusion is that the equations (12.153) are meaningful, only if we identify the eigenvalues $\theta_i$ with the elements of a Grassmann algebra $\mathcal{G}_n$. In this case, the eigenstate in eq. (12.152) can be easily evaluated. We find:

$$|\theta\rangle = e^{\theta a^\dagger}|0\rangle, \tag{12.156}$$

where $|0\rangle$ is the ground state of the oscillator. In fact,

$$a|\theta\rangle = a(1 + \theta a^\dagger)|0\rangle = \theta a a^\dagger|0\rangle = \theta(1 - a^\dagger a)|0\rangle$$
$$= \theta|0\rangle = \theta(1 + \theta a^\dagger|0\rangle = \theta|\theta\rangle, \tag{12.157}$$

where we have used

$$e^{\theta a^\dagger} = 1 + \theta a^\dagger, \tag{12.158}$$

since $\theta^2 = a^{\dagger\,2} = 0$. We also have

$$a^\dagger|\theta\rangle = a^\dagger(1 + \theta a^\dagger)|0\rangle = a^\dagger|0\rangle = \frac{\partial}{\partial\theta}(1 + \theta a^\dagger)|0\rangle = \frac{\partial}{\partial\theta}|\theta\rangle. \tag{12.159}$$

Here we have introduced the left-derivative on the algebra $\mathcal{G}_n$, defined by

$$\frac{\partial}{\partial\theta_i}\theta_{i_1}\theta_{i_2}\cdots\theta_{i_k} = \delta_{i_1,i}\theta_{i_2}\theta_{i_3}\cdots\theta_{i_k} - \delta_{i_2,i}\theta_{i_1}\theta_{i_3}\cdots\theta_{i_k}$$
$$+ \cdots + (-1)^{k-1}\delta_{i_k,i}\theta_{i_1}\theta_{i_3}\cdots\theta_{i_{k-1}}. \tag{12.160}$$

We see that the canonical anticommutation relation

$$[a, a^\dagger]_+ = 1 \tag{12.161}$$

is realized on the Grassmann algebra, through the identification

$$a \approx \theta, \qquad a^\dagger \approx \frac{\partial}{\partial\theta}. \tag{12.162}$$

We want to define a scalar product in the vector space of the eigenstates of the Fermi annihilation operators[1]. To this end, we extend the algebra $\mathcal{G}_1$ generated by $\theta$ to a complex Grassmann algebra, $\mathcal{G}_2$, generated by $\theta$ and $\theta^*$. The operation "*" is defined as the complex conjugation on the ordinary numbers and as an involution of the algebra. This is an automorphism of the algebra with square one. More precisely, the conjugation maps $\theta$ in $\theta^*$

---

[1]The vector spaces we are dealing with, are not defined on real or complex numbers, but rather on a Grassmann algebra. A space of this sort is called a modulus over the algebra by mathematicians,

and $\theta^*$ in $\theta$. In the case of $\mathcal{G}_n$ we extend it to $\mathcal{G}_{2n}$, requiring the following property for the conjugation:

$$(\theta_{i_1}\theta_{i_2}\cdots\theta_{i_k})^* = \theta_{i_k}^*\cdots\theta_{i_2}^*\theta_{i_1}^*. \tag{12.163}$$

In the algebra $\mathcal{G}_2$ we can define normalized eigenstates

$$|\theta\rangle = e^{\theta a^\dagger}|0\rangle N(\theta, \theta^*), \tag{12.164}$$

where $N(\theta, \theta^*) \in \mathcal{G}_2^{(0)}$, with the phase fixed by the convention

$$N^*(\theta, \theta^*) = N(\theta, \theta^*). \tag{12.165}$$

We define the corresponding bra as

$$\langle\theta| = N(\theta, \theta^*)\langle 0|e^{\theta^* a}. \tag{12.166}$$

The normalization factor $N(\theta, \theta^*)$ is determined by the condition

$$1 = \langle\theta|\theta\rangle = N^2(\theta, \theta^*)\langle 0|e^{\theta^* a}e^{\theta a^\dagger}|0\rangle =$$
$$= N^2(\theta, \theta^*)(1 + \theta^*\theta) = N^2(\theta, \theta^*)e^{-\theta\theta^*}. \tag{12.167}$$

It follows

$$N(\theta, \theta^*) = e^{\theta\theta^*/2}. \tag{12.168}$$

Let us recall some simple property of a Fermi oscillator. The occupation number operator has eigenvalues 1 and 0. In fact,

$$(a^\dagger a)^2 = a^\dagger a a^\dagger a = a^\dagger(1 - a^\dagger a)a = a^\dagger a, \tag{12.169}$$

from which

$$a^\dagger a(a^\dagger a - 1) = 0. \tag{12.170}$$

Starting from the state with zero eigenvalue for $a^\dagger a$, we can build up only the eigenstate with eigenvalue 1, $a^\dagger|0\rangle$, since $a^{\dagger 2}|0\rangle = 0$. Therefore, the space of the eigenstates of $a^\dagger a$ is a two-dimensional space. By defining the ground state as

$$|0\rangle = \begin{bmatrix} 0 \\ 1 \end{bmatrix}. \tag{12.171}$$

It is easily seen that (see the exercise (4) in this Chapter)

$$a = \begin{bmatrix} 0 & 0 \\ 1 & 0 \end{bmatrix}, \quad a^\dagger = \begin{bmatrix} 0 & 1 \\ 0 & 0 \end{bmatrix}. \tag{12.172}$$

The state with occupation number equal to 1 is

$$|1\rangle = \begin{bmatrix} 1 \\ 0 \end{bmatrix}. \tag{12.173}$$

On this basis, the normalized state $|\theta\rangle$ is given by

$$|\theta\rangle = e^{\theta\theta^*/2}e^{\theta a^\dagger}|0\rangle = e^{\theta\theta^*/2}(1 + \theta a^\dagger)|0\rangle =$$
$$= e^{\theta\theta^*/2}(|0\rangle + \theta|1\rangle) = e^{\theta\theta^*/2}\begin{bmatrix}\theta\\1\end{bmatrix}, \qquad (12.174)$$

whereas

$$\langle\theta| = e^{\theta\theta^*/2}\left[\theta^*, 1\right]. \qquad (12.175)$$

Let us now come to the definition of the integral over a Grassmann algebra. To this end, we recall that the crucial element, in order to derive the path integral formalism from quantum mechanics, was the completeness relation for the eigenstates of the position operator

$$\int dq|q\rangle\langle q| = 1. \qquad (12.176)$$

We will consider the annihilation operator, $a$, as a generalized position operator, and we will require that its eigenstate satisfies a completeness relation. Therefore, we will "define" the integration over the Grassmann algebra $\mathcal{G}_2$ by requiring that the states defined in eqs. (12.174) and (12.175) satisfy

$$\int d\theta^* d\theta|\theta\rangle\langle\theta| = \begin{pmatrix}1 & 0\\0 & 1\end{pmatrix} \equiv 1_2, \qquad (12.177)$$

with the right-hand side the identity on the two-dimensional vector space. Using the explicit representation for the states

$$\int d\theta^* d\theta|\theta\rangle\langle\theta| = \int d\theta^* d\theta e^{\theta\theta^*}\begin{pmatrix}\theta\\1\end{pmatrix}\begin{pmatrix}\theta^* & 1\end{pmatrix}$$
$$= \int d\theta^* d\theta e^{\theta\theta^*}\begin{pmatrix}\theta\theta^* & \theta\\\theta^* & 1\end{pmatrix}$$
$$= \int d\theta^* d\theta\begin{pmatrix}\theta\theta^* & \theta\\\theta^* & 1+\theta\theta^*\end{pmatrix} = \begin{pmatrix}1 & 0\\0 & 1\end{pmatrix}. \qquad (12.178)$$

Therefore, the integration must satisfy

$$\int d\theta^* d\theta \cdot \theta\theta^* = 1, \qquad (12.179)$$

$$\int d\theta^* d\theta \cdot \theta = \int d\theta^* d\theta \cdot \theta^* = \int d\theta^* d\theta \cdot 1 = 0. \qquad (12.180)$$

These relations determine the integration in a unique way, since they fix the value of the integral over all the basis elements of $\mathcal{G}_2$, and the integral over a generic element of the algebra is defined by linearity. The previous integration rules can be obtained as integration over the algebra $\mathcal{G}_1$ requiring

$$[d\theta, d\theta^*]_+ = [\theta, d\theta^*]_+ = [\theta, d\theta]_+ = [\theta^*, d\theta]_+ = [\theta^*, d\theta^*]_+ = 0. \quad (12.181)$$

The rules (12.179) and (12.180) are then obtained requiring that

$$\int d\theta \cdot \theta = 1, \quad \int d\theta^* \cdot \theta^* = 1, \quad \int d\theta \cdot 1 = 0, \quad \int d\theta^* \cdot 1 = 0. \quad (12.182)$$

For $n$ Fermi oscillators we get

$$|\theta_1, \cdots, \theta_n\rangle = e^{\sum_{i=1}^n \theta_i \theta_i^*/2} e^{\sum_{i=1}^n \theta_i a_i^\dagger} |0\rangle, \quad (12.183)$$

and we require

$$\int \left( \prod_{i=1}^n d\theta_i^* d\theta_i \right) |\theta_1, \cdots, \theta_n\rangle\langle\theta_1, \cdots, \theta_n| = 1. \quad (12.184)$$

This relation is satisfied assuming

$$\int d\theta_i \theta_j = \delta_{ij} \quad (12.185)$$

and

$$[d\theta_i, \theta_j]_+ = [d\theta_i, d\theta_j]_+ = 0, \quad (12.186)$$

and analogous relations for the conjugated variables. Notice that the measure $d\theta$ is translationally invariant. In fact, if $f(\theta) = a + b\theta$ and $\alpha \in \mathcal{G}_1^{(1)}$

$$\int d\theta f(\theta) = \int d\theta f(\theta + \alpha), \quad (12.187)$$

since

$$\int d\theta\, b\alpha = 0. \quad (12.188)$$

## 12.10   Properties of the integration over Grassmann variables

Let us consider the Grassmann algebra $\mathcal{G}_n$. Its generic element can be expanded over the basis elements as

$$f(\theta_1, \cdots, \theta_n) = f_0 + \sum_{i_1} f_1(i_1)\theta_{i_1} + \sum_{i_1, l_2} f_2(i_1, i_2)\theta_{i_1}\theta_{i_2} + \cdots$$

$$+ \sum_{i_1, \cdots, i_n} f_n(i_1, \cdots, i_n)\theta_{i_1} \cdots \theta_{i_n}. \quad (12.189)$$

Of course, the coefficients of the expansion are antisymmetric. If we integrate $f$ over the algebra $\mathcal{G}_n$, only the last element of the expansion gives a non-vanishing contribution. Notice that, due to the antisymmetry, we have

$$\sum_{i_1,\cdots,i_n} f_n(i_1,\cdots,i_n)\theta_{i_1}\cdots\theta_{i_n} = n! f_n(1,\cdots,n)\theta_1\cdots\theta_n. \qquad (12.190)$$

It follows

$$\int d\theta_n \cdots d\theta_1 f(\theta_1,\cdots,\theta_n) = n! f_n(1,\cdots,n). \qquad (12.191)$$

An interesting property of the integration rules for Grassmann variables is the integration by parts formula. This can be seen either by the observation that a derivative destroys a factor $\theta_i$ in each element of the basis and therefore

$$\int d\theta_n \cdots d\theta_1 \frac{\partial}{\partial\theta_j} f(\theta_1,\cdots,\theta_n) = 0, \qquad (12.192)$$

either by the property that the integral is invariant under translations (see the previous Section and Section 11.5). Let us now consider the transformation properties of the measure under a change of variables, and consider an invertible linear transformation

$$\theta_i = \sum_k a_{ik}\theta'_k. \qquad (12.193)$$

The eq. (12.185) must be required to hold in each basis, since it can be seen easily that the previous transformation leaves invariant the multiplication rules. In fact,

$$\theta'_i\theta'_j = c_{ik}c_{jh}\theta_k\theta_h = -c_{ik}c_{jh}\theta_h\theta_k = -c_{jh}c_{ik}\theta_h\theta_k = -\theta'_j\theta'_i, \qquad (12.194)$$

where $c_{ij}$ is the inverse of the matrix $a_{ij}$. Therefore, we require

$$\int d\theta_i\theta_j = \int d\theta'_i\theta'_j = \delta_{ij}. \qquad (12.195)$$

Defining

$$d\theta_i = \sum_k d\theta'_k b_{ki}, \qquad (12.196)$$

we obtain

$$\int d\theta_i\theta_j = \int \sum_{h,k} d\theta'_k b_{ki} a_{jh}\theta'_h = \sum_{h,k} b_{ki} a_{jh}\delta_{hk} = (ab)_{ji} = \delta_{ij}, \qquad (12.197)$$

from which

$$b_{ij} = (a^{-1})_{ij}. \tag{12.198}$$

The transformation property of the measure is

$$d\theta_n \cdots d\theta_1 = \det|a^{-1}|d\theta'_n \cdots d\theta'_1, \tag{12.199}$$

where we have used the anticommutation properties of $d\theta_i$'s among themselves. We see also that

$$\det|a| = \det\left|\frac{\partial\theta}{\partial\theta'}\right| \tag{12.200}$$

and

$$d\theta'_n \cdots d\theta'_1 = \det^{-1}\left|\frac{\partial\theta'}{\partial\theta}\right|d\theta_n \cdots d\theta_1. \tag{12.201}$$

Therefore the rule for a linear change of variables is

$$\int d\theta'_n \cdots d\theta'_1 f(\theta'_i) = \int d\theta_n \cdots d\theta_1 \det^{-1}\left|\frac{\partial\theta'}{\partial\theta}\right| f(\theta'_i(\theta_i)). \tag{12.202}$$

This rule should be compared with the one for commuting variables

$$\int dx'_1 \cdots dx'_n f(x'_i) = \int dx_1 \cdots dx_n \det\left|\frac{\partial x'}{\partial x}\right| f(x'_i(x_i)). \tag{12.203}$$

In conclusion, the Grassmann measure transforms according to the inverse Jacobian rather than with the Jacobian itself.

As in the commuting case, the more important integral to be evaluated, with regard to the path integral approach, is the Gaussian one

$$I = \int d\theta_n \cdots d\theta_1 e^{\sum_{i,j} A_{ij}\theta_i\theta_j/2} \equiv \int d\theta_n \cdots d\theta_1 e^{\theta^T A\theta/2}, \tag{12.204}$$

where the matrix $A$ is supposed to be real and antisymmetric

$$A^T = -A. \tag{12.205}$$

As shown in the Appendix, a real, antisymmetric matrix can be reduced to the form

$$A_s = \begin{bmatrix} 0 & \lambda_1 & 0 & 0 & \cdots \\ -\lambda_1 & 0 & 0 & 0 & \cdots \\ 0 & 0 & 0 & \lambda_2 & \cdots \\ 0 & 0 & -\lambda_2 & 0 & \cdots \\ \cdot & \cdot & \cdot & \cdot & \cdots \\ \cdot & \cdot & \cdot & \cdot & \cdots \\ \cdot & \cdot & \cdot & \cdot & \cdots \end{bmatrix}, \tag{12.206}$$

by means of an orthogonal transformation $S$. Then, performing the following change of variables

$$\theta_i = \sum_j (S^T)_{ij} \theta_j', \tag{12.207}$$

we get

$$\theta^T A \theta = (S^T \theta')^T A S^T \theta' = \theta'^T S A S^T \theta' = \theta'^T A_s \theta', \tag{12.208}$$

and using $\det|S| = 1$

$$I = \int d\theta_n' \cdots d\theta_1' e^{\theta'^T A_s \theta'/2}. \tag{12.209}$$

The expression in the exponent is given by

$$\frac{1}{2} \theta'^T A_s \theta' = \lambda_1 \theta_1' \theta_2' + \cdots + \lambda_{\frac{n}{2}} \theta_{n-1}' \theta_n', \quad \text{for } n \text{ even}$$

$$\frac{1}{2} \theta'^T A_s \theta' = \lambda_1 \theta_1' \theta_2' + \cdots + \lambda_{\frac{n-1}{2}} \theta_{n-2}' \theta_{n-1}', \text{ for } n \text{ odd.} \tag{12.210}$$

From eq. (12.191) we get a contribution to the integral, only from the term in the expansion of the exponential $(\theta'^T A_s \theta')^{\frac{n}{2}}$ for $n$ even. In the case of $n$ odd we get $I = 0$, since $\theta'^T A_s \theta'$ does not contain $\theta_n'$. Therefore, for $n$ even

$$I = \int d\theta_n' \cdots d\theta_1' e^{\theta'^T A_s \theta'/2} = \int d\theta_n' \cdots d\theta_1' \frac{1}{(\frac{n}{2})!} \left( \frac{1}{2} \theta'^T A_s \theta' \right)^{\frac{n}{2}}$$

$$= \int d\theta_n' \cdots d\theta_1' \frac{1}{(\frac{n}{2})!} (\frac{n}{2})! \lambda_1 \cdots \lambda_{\frac{n}{2}} \theta_1' \cdots \theta_n'$$

$$= \lambda_1 \cdots \lambda_{\frac{n}{2}} = \sqrt{\det A}, \tag{12.211}$$

where we have made use of $\det|A_s| = \det|A|$. Then,

$$\int d\theta_n \cdots d\theta_1 e^{\theta^T A \theta/2} = \sqrt{\det A}. \tag{12.212}$$

This result is valid also for $n$ odd, since in this case $\det|A| = 0$. A similar result holds for commuting variables. In fact, let us define

$$J = \int dx_1 \cdots dx_n e^{-x^T A x/2}, \tag{12.213}$$

with $A$ a real symmetric matrix. Then, $A$ can be diagonalized by an orthogonal transformation with the result

$$J = \int dx_1' \cdots dx_n' e^{-x'^T A_d x'/2} = (2\pi)^{\frac{n}{2}} \frac{1}{\sqrt{\det A}}. \tag{12.214}$$

Again, the result for fermionic variables is opposite to the one of the commuting case. Another useful integral is

$$I(A;\chi) = \int d\theta_n \cdots d\theta_1 e^{\theta^T A\theta/2 + \chi^T \theta}, \qquad (12.215)$$

where the $\chi_i$'s are Grassmann variables anticommuting with the $\theta_i$'s. To evaluate this integral, let us put

$$\theta = \theta' + A^{-1}\chi. \qquad (12.216)$$

We get

$$\frac{1}{2}\theta^T A\theta + \chi^T \theta = \frac{1}{2}(\theta'^T - \chi^T A^{-1})A(\theta' + A^{-1}\chi) + \chi^T(\theta' + A^{-1}\chi)$$

$$= \frac{1}{2}\theta'^T A\theta' + \frac{1}{2}\chi^T A^{-1}\chi, \qquad (12.217)$$

where we have used the antisymmetry of $A$ and $\chi^T \theta' = -\theta'^T \chi$. Since the measure is invariant under translations, we find

$$I(A;\chi) = \int d\theta'_n \cdots d\theta'_1 e^{\theta'^T A\theta'/2 + \chi^T A^{-1}\chi/2}$$

$$= e^{\chi^T A^{-1}\chi/2}\sqrt{\det A}, \qquad (12.218)$$

again to be compared with the result for commuting variables

$$I(A;J) = \int dx_1 \cdots dx_n e^{-x^T Ax/2 + J^T x}$$

$$= e^{J^T A^{-1}J/2}\frac{(2\pi)^{\frac{n}{2}}}{\sqrt{\det A}}. \qquad (12.219)$$

The previous equations can be generalized to complex variables

$$\int d\theta_1 d\theta_1^* \cdots d\theta_n d\theta_n^* e^{\theta^\dagger A\theta} = \det|A| \qquad (12.220)$$

and

$$\int d\theta_1 d\theta_1^* \cdots d\theta_n d\theta_n^* e^{\theta^\dagger A\theta + \chi^\dagger \theta + \theta^\dagger \chi} = e^{-\chi^\dagger A^{-1}\chi}\det|A|. \qquad (12.221)$$

The second of these equations follows from the first one as in the real case.

## 12.11   The path integral for the fermionic theories

In this Section, we will discuss the path integral formulation for a system described by anticommuting variables. Considering a Fermi oscillator, one could start from the ordinary quantum mechanical description and then, by using the completeness in terms of Grassmann variables, it would be possible to derive the corresponding path integral formulation. We will give here only the result of this analysis. It turns out that the amplitude among eigenstates of the Fermi annihilation operator is given by (see, for example, [Popov (1991)])

$$\langle \theta', t' | \theta, t \rangle = \int_{\theta(t)=\theta,t}^{\theta^*(t')=\theta'^*,t'} \mathcal{D}(\theta, \theta^*) e^{\theta'^* \theta(t')/2 + \theta^*(t)\theta/2 + iS}, \qquad (12.222)$$

where

$$\mathcal{D}(\theta, \theta^*) = \prod d\theta d\theta^* \qquad (12.223)$$

and

$$S = \int_t^{t'} \left[ \frac{i}{2}(\theta^* \dot{\theta} - \dot{\theta}^* \theta) - H(\theta, \theta^*) \right] dt. \qquad (12.224)$$

Here, in the case of a Fermi oscillator, $H = \theta^* \theta$. The expression is very similar to the one obtained in the commuting case, if we realize that the previous formula is the analogue in the phase space using complex coordinates of the type $q \pm ip$ (see, for example, [Popov (1991)]). It should be noticed that the particular boundary conditions for the $\theta$ variables originate from the fact that the equations of motion are of the first order. Correspondingly, we can assign only the coordinates at a particular time. In field theories, we will be interested in the generating functional and the boundary conditions are arbitrary. For instance, we can take the fields vanishing at $t = \pm\infty$. Furthermore, the boundary conditions do not affect the dependence of the generating functional on the external sources. The extension to the Fermi fields is done exactly as we have done in the bosonic case, and correspondingly the generating functional for the free Dirac theory will be

$$Z[\eta, \bar{\eta}] = \int \mathcal{D}(\psi, \bar{\psi}) e^{i \int [\bar{\psi}(i\hat{\partial} - m)\psi + \bar{\eta}\psi + \bar{\psi}\eta]}, \qquad (12.225)$$

where $\eta$ and $\bar{\eta}$ are the external Grassmann sources. Using the integration formula (12.218) we get formally

$$Z[\eta, \bar{\eta}] = Z[0] e^{-(i\bar{\eta})\frac{-i}{ii\hat{\partial} - m}(i\eta)} = Z[0] e^{-i\bar{\eta}\frac{1}{i\hat{\partial} - m}\eta}. \qquad (12.226)$$

Here, the determinant of the Dirac operator has been included in $Z[0]$. In this expression, we have to specify how to define the inverse of the Dirac operator. We know that this is nothing but the Dirac propagator defined by

$$(i\hat{\partial} - m)S_F(x - y) = \delta^4(x - y). \tag{12.227}$$

It follows

$$Z[\eta, \bar{\eta}] = Z[0]e^{-i\int d^4x d^4y \bar{\eta}(x)S_F(x-y)\eta(y)}, \tag{12.228}$$

with

$$\begin{aligned}
S_F(x) &= \lim_{\epsilon \to 0^+} \int \frac{d^4k}{(2\pi)^4} \frac{1}{\hat{k} - m + i\epsilon} e^{-ikx} \\
&= \lim_{\epsilon \to 0^+} \int \frac{d^4k}{(2\pi)^4} \frac{\hat{k} + m}{k^2 - m^2 + i\epsilon} e^{-ikx}.
\end{aligned} \tag{12.229}$$

The Green functions are obtained by differentiating the generating functional with respect to the external sources (Grassmann left-differentiation). In particular, the two-point function is given by

$$\frac{1}{Z[0]} \frac{\delta Z}{\delta\bar{\eta}(x)\delta\eta(y)}\Big|_{\eta=\bar{\eta}=0} = \frac{1}{\langle 0|0\rangle} \langle 0|T(\psi(x)\bar{\psi}(y))|0\rangle = iS_F(x - y). \tag{12.230}$$

The $T$-product in this expression is the one for the Fermi fields (see Section 7.1), as it follows from the derivation made in the bosonic case, taking into account that in this case, the "classical fields" are anticommuting.

We will not insist further on this discussion, but all we have done in the bosonic case can be easily generalized to the case of Fermi fields.

## 12.12    Exercises

(1) Evaluate the two-point and the four-point connected Green's functions in configuration space in the $\lambda\phi^4$ theory.
(2) Express the previous results in momentum space.
(3) Evaluate and renormalize, at one-loop level, the four-point function of the $\lambda\varphi^4$ theory.
(4) Prove eq. (12.172).
(5) Prove eqs. (12.220) and (12.221). Hint: make use of the results of the real case given in eqs. (12.211) and (12.218) and define complex Grassmann variables. For example, define $\theta_1 = (\eta + \eta^*)/\sqrt{2}$, $\theta_2 = -i(\eta - \eta^*)/\sqrt{2}$, and analogous relations for the other pairs of real variables.

(6) Following what we did for a single bosonic degree of freedom, derive eq. (12.222).

# Chapter 13

# The quantization of the gauge fields

## 13.1 Path integral quantization of the gauge theories

As we have seen in Chapter 5, the quantization of the electromagnetic field is far from being trivial. The difficulty being related to the gauge invariance of the theory. In this Section, we would like to show that analogous difficulties arise in the path integral quantization. To understand this point, we will consider the quadratic part of the action. Therefore, the evaluation of the path integral reduces to calculate a Gaussian integral of the type given in eqs. (12.218) and (12.219) for the fermionic and the bosonic case respectively. For instance, in the case of bosons, we need the following expression:

$$I(A; J) = \int dx_1 \cdots dx_n e^{-x^T A x/2 + J^T x} = e^{J^T A^{-1} J/2} \frac{(2\pi)^{\frac{n}{2}}}{\sqrt{\det A}}. \qquad (13.1)$$

The operator $A$ appearing in the Gaussian factor is the one giving rise to the equations of motion (wave operator) for the field, which in the previous expression has been discretized and represented by the variables $x_i$. The result of the integration is meaningful only if the operator $A$ is non-singular (it needs to have an inverse), otherwise the integral is not well defined. This is precisely the point where we encounter a problem in the case of a gauge theory. Remember, from Section 5.1 (see eq. (5.3)), that the wave operator for the electromagnetic field, in momentum space, is given by

$$(-k^2 g^{\mu\nu} + k^\mu k^\nu) A_\nu(k) = 0. \qquad (13.2)$$

But this operator has a null eigenvector $k_\nu$, since

$$(-k^2 g^{\mu\nu} + k^\mu k^\nu) k_\nu = 0, \qquad (13.3)$$

and, therefore, it is a singular operator. This is strictly related to the gauge invariance, which, in momentum space, reads as

$$A_\mu(k) \to A_\mu(k) + k_\mu \Lambda(k). \qquad (13.4)$$

In fact, we see that the equations of motion are invariant under a gauge transformation, precisely because the wave operator has $k_\mu$ as a null eigenvector. Another way of seeing this problem is the following: in order to get physical results as, for instance, to evaluate the $S$ matrix elements, it is necessary to integrate over gauge-invariant functionals

$$\int \mathcal{D}(A_\mu) F[A_\mu] e^{iS}, \tag{13.5}$$

with $F[A_\mu] = F[A_\mu^\Omega]$ (here $A_\mu^\Omega$ is the gauge transformed of $A_\mu$). But, if we insist on the gauge invariance of both the action and the functional $F$, then, it follows that the integrand is invariant along the gauge group orbits. The gauge group orbits are defined as the sets of points, in the field space, which can be reached by a given $A_\mu$ via a gauge transformation. In other words, given $A_\mu$, its orbit is given by all the fields $A_\mu^\Omega$ obtained by varying the parameters of the gauge group (that is varying $\Omega$), see Fig. 13.1. For instance, in the abelian case

$$A_\mu^\Omega = A_\mu + \partial_\mu \Omega. \tag{13.6}$$

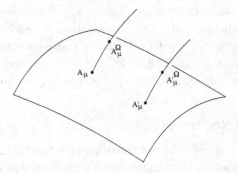

Fig. 13.1　The orbits of the gauge group in field space.

Since the orbits define equivalence classes (or cosets), we can imagine the field space as the set of all the orbits. Correspondingly, we can divide our functional integral in a part parallel, and in a part perpendicular to the orbits. Then, the reason why our functional integral is not well defined depends on the fact that the integrand along the direction parallel to the orbits is invariant, implying that the integral is infinite. But, this observation suggests a simple solution to the problem. We could define the integral by dividing it by the integral along the orbits, which, as we shall see, is the volume of the gauge group. In order to gain a clear understanding about this point, let us consider the following very simple example, in two dimensions

$$Z = \int_{-\infty}^{+\infty} dx\,dy\, e^{-\frac{1}{2}a^2(x-y)^2}. \tag{13.7}$$

The two points $x$ and $y$ simulate the possible values of the field in two points. At the same time the exponent can be seen as a Euclidean action with the (gauge) symmetry

$$x^{\Omega} = x + \Omega, \quad y^{\Omega} = y + \Omega. \tag{13.8}$$

The integral (13.7) is infinite due to this invariance. In fact, this implies that the integrand depends only on $x-y$, and therefore the integration over the remaining variable gives rise to an infinite result. In order to proceed along the lines we have oulined above, let us partition the field space, that is the two-dimensional space $(x, y)$, in the gauge group orbits. Given a point $(x_0, y_0)$, the group orbit is given by the set of points $(x^{\Omega} = x+\Omega, y^{\Omega} = y+\Omega)$ corresponding to the line $x - y = x_0 - y_0$, as shown in Fig. 13.2.

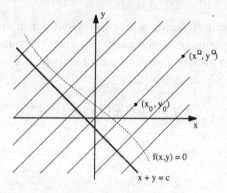

Fig. 13.2 The field space $(x, y)$ partitioned in the orbits of the gauge group. We show also the line $x + y = c$ and the arbitrary line $f(x, y) = 0$ (dashed).

We can perform the integral by integrating along the perpendicular direction, that is along the line $x + y = c$, with $c$ a constant, and then integrate along the orbits. It is this last integral which gives a divergent result, since we are adding up the same contribution an infinite number of times. Therefore, we introduce new variables $x - y$ and $x + y$, integrating only on $x - y$. Or, said in a different way, we perform both integrals, but we divide the result by the integral over $x+y$, which cancels the divergence at the numerator. This can be done in a more systematic way, by using the following identity

$$1 = \int_{-\infty}^{+\infty} d\Omega \delta \left( \frac{1}{2}(x+y) + \Omega \right). \tag{13.9}$$

Then, we multiply $Z$ by this factor obtaining

$$Z = \int d\Omega \left[ \int dx dy \delta \left( \frac{1}{2}(x+y) + \Omega \right) e^{-\frac{1}{2}a^2(x-y)^2} \right]. \tag{13.10}$$

The integral inside the parentheses is what we are looking for, since it is evaluated along the line $x + y = constant$. Of course, the result should not depend on the choice of this constant (it corresponds to translate the integration line in a way parallel to the orbits). In fact, by doing the change of variables

$$x \to x + \Omega, \qquad y \to y + \Omega, \tag{13.11}$$

we get

$$Z = \int d\Omega \left[ \int dx dy \delta \left( \frac{1}{2}(x+y) \right) e^{-\frac{1}{2}a^2(x-y)^2} \right]. \tag{13.12}$$

As promised, the infinite factor is now factorized out, and we see that this is nothing but the volume of the gauge group ($\Omega$ parametrize the gauge group which, in the present case is isomorphic to $R^1$). Then, we define our integral as

$$Z' = \frac{Z}{V_G} = \int dx dy \delta \left( \frac{1}{2}(x+y) \right) e^{-\frac{1}{2}a^2(x-y)^2}, \tag{13.13}$$

with

$$V_G = \int d\Omega. \tag{13.14}$$

In this way, the integral is defined by integrating over the particular surface $x+y = 0$. Of course, we would get the same result integrating over any other line (or surface in the general case), having the property of intersecting each orbit only once. For instance, let us choose the line (see Fig. 13.2)

$$f(x,y) = 0. \tag{13.15}$$

Then, proceeding as in eq. (13.9), we consider the following identity, which allows us to define a gauge invariant function $\Delta_f(x,y)$

$$1 = \Delta_f(x,y) \int d\Omega \delta \left[ f(x^\Omega, y^\Omega) \right] = \Delta_f(x,y) \int d\Omega \delta \left[ f(x+\Omega, y+\Omega) \right]. \tag{13.16}$$

In the previous case we had $f(x, y) = (x + y)/2 = 0$ giving rise to $f(x^\Omega, y^\Omega) = (x + y)/2 + \Omega = 0$, from which $\Delta_f(x, y) = 1$. Multiplying $Z$ by the expression (13.16), we get

$$Z = \int d\Omega \left[ \int dx dy \Delta_f(x, y) \delta \left[ f(x + \Omega, y + \Omega) \right] e^{-\frac{1}{2} a^2 (x-y)^2} \right]. \qquad (13.17)$$

Let us show that $\Delta_f(x, y)$ is gauge invariant. In fact, by the transformation $x \to x + \Omega'$, $y \to y + \Omega'$ we obtain

$$1 = \Delta_f(x + \Omega', y + \Omega') \int d\Omega \delta \left[ f(x + \Omega + \Omega', y + \Omega + \Omega') \right], \qquad (13.18)$$

and changing the integration variable $\Omega \to \Omega + \Omega'$

$$1 = \Delta_f(x + \Omega', y + \Omega') \int d\Omega \delta \left[ f(x + \Omega, y + \Omega) \right]. \qquad (13.19)$$

Comparison with eq. (13.16) gives

$$\Delta_f(x, y) = \Delta_f(x + \Omega', y + \Omega'). \qquad (13.20)$$

The integral, inside the parentheses in eq. (13.17), does not depend on $\Omega$, therefore it is gauge invariant. This can be shown exactly, as we did before, in the case $f(x, y) = (x + y)/2$. Therefore, we can factorize out the integration over $\Omega$ (the volume of the gauge group)

$$Z = \int d\Omega \left[ \int dx dy \Delta_f(x, y) \delta \left[ f(x, y) \right] e^{-\frac{1}{2} a^2 (x-y)^2} \right], \qquad (13.21)$$

and define

$$Z' = \frac{Z}{V_G} = \int dx dy \Delta_f(x, y) \delta \left[ f(x, y) \right] e^{-\frac{1}{2} a^2 (x-y)^2}. \qquad (13.22)$$

The function $\Delta_f(x, y)$ (called the Faddeev-Popov determinant) [Faddeev and Popov (1967)] can be calculated directly from its definition. To this end, notice that in (13.21) $\Delta_f(x, y)$ appears multiplied by $\delta[f(x, y)]$. As a consequence, we need to know $\Delta_f$ only at points very close at the surface $f(x, y) = 0$. But in this case, the equation $f(x^\Omega, y^\Omega) = 0$ has the only solution $\Omega = 0$, since, by hypothesis, the surface $f(x, y)$ crosses the gauge orbits only once. Therefore, we can write

$$\delta[f(x^\Omega, y^\Omega)] = \frac{1}{\left[ \dfrac{\partial f^\Omega}{\partial \Omega} \right]_{\Omega=0}} \delta[\Omega], \qquad (13.23)$$

giving rise to

$$\Delta_f(x, y) = \frac{\partial f^\Omega}{\partial \Omega} \bigg|_{\Omega=0}, \qquad (13.24)$$

where $f^\Omega = f(x^\Omega, y^\Omega)$. Clearly, $\Delta_f$ represents the answer of the gauge fixing, $f = 0$, to an infinitesimal gauge transformation. As a last remark, notice that our original integral can be written as

$$Z = \int d^2 x e^{-\frac{1}{2} x_i A_{ij} x_j},$$ (13.25)

with

$$A = a^2 \begin{bmatrix} 1 & -1 \\ -1 & 1 \end{bmatrix},$$ (13.26)

a singular matrix. As we have already noticed, the singularity of $A$ is related to the gauge invariance, and the solution, we have given here, solves also the problem of the singularity of $A$. In fact, integrating over $x - y$, means to integrate only on the eigenvector of $A$ corresponding to the non-zero eigenvalue.

Since the solution of our problem is of geometrical type, it is very easy to generalize it to the case of a gauge field theory, abelian or not. We start by choosing a surface in the gauge fields space given by

$$f_B(A_B^\mu) = 0, \qquad B = 1, \cdots, n,$$ (13.27)

where $n$ is the number of parameters of Lie $G$. Also, this surface should intersect the gauge orbits only once. To write down the analogue of eq. (13.16), we need also a measure on the gauge group. This can be obtained as follows. We have seen that we are interested in the solutions of

$$f_B(A_\mu^\Omega) = 0,$$ (13.28)

around $f_B(A_\mu) = 0$. Therefore, it is enough to know the measure around the identity of the gauge group. Then, the transformation $\Omega$ can be written as

$$\Omega \approx 1 + i\alpha_A(x) T^A.$$ (13.29)

At first order in the group parameters, we have

$$\Omega \Omega' \approx 1 + i(\alpha_A + \alpha'_A) T^A.$$ (13.30)

Therefore, the invariant measure around the identity is given by

$$d\mu(\Omega) = \prod_{A,x} d\alpha_A(x).$$ (13.31)

It follows that for infinitesimal transformations

$$d\mu(\Omega \Omega') = d\mu(\Omega' \Omega) = d\mu(\Omega).$$ (13.32)

Next, we define the Faddeev-Popov determinant as in (13.16):

$$1 = \Delta_f[A_\mu] \int d\mu(\Omega)\delta[f_A(A_\mu^\Omega)]. \tag{13.33}$$

In this expression we have introduced a functional delta-function, defined by the identity

$$1 = \int d\mu(\Omega)\delta[\Omega], \tag{13.34}$$

having the usual properties extended to the functional case. We can see that $\Delta_f$ is a gauge invariant functional, since

$$\Delta_f^{-1}[A_\mu^{\Omega'^{-1}}] = \int d\mu(\Omega)\delta[f_A(A_\mu^{(\Omega'^{-1}\Omega)})], \tag{13.35}$$

and changing variables $\Omega'^{-1}\Omega = \Omega''$ we get

$$\Delta_f^{-1}(a_\mu^{\Omega'^{-1}}) = \int d\mu(\Omega'\Omega'')\delta[f_A(A_\mu^{\Omega''})] = \int d\mu(\Omega)\delta[f_A(A_\mu^\Omega)]. \tag{13.36}$$

Consider now the naive path integral

$$Z = \int \mathcal{D}(A_\mu)F[A_\mu]e^{i\int \mathcal{L}_A d^4x}, \tag{13.37}$$

where $F$ is a gauge invariant functional. Since in the following $F$ will not play any particular role, we will discuss only the case $F = 1$, but all the following considerations are valid for an arbitrary gauge invariant $F$. We multiply $Z$ by the identity (13.33)

$$Z = \int d\mu(\Omega)\left[\int \mathcal{D}(A_\mu)e^{i\int \mathcal{L}_A d^4x}\delta[f_A(A_\mu^\Omega)]\Delta_f[A_\mu]\right]. \tag{13.38}$$

The functional measure is invariant under gauge transformations

$$A_\mu \to \Omega A_\mu \Omega^{-1} + \frac{i}{g}(\partial_\mu\Omega)\Omega^{-1}, \tag{13.39}$$

since the homogeneous part has determinant one[1], and the inhomogeneous part gives rise to a translation. Then, by performing the change of variables $A_\mu \to A_\mu^\Omega$, and using the gauge invariance of $\mathcal{D}(A_\mu)$, $\mathcal{L}_A$ and $\Delta_f[A_\mu]$, we get

$$Z = \int d\mu(\Omega)\left[\int \mathcal{D}(A_\mu)e^{i\int \mathcal{L}_A d^4x}\delta[f_A(A_\mu)]\Delta_f[A_\mu]\right]. \tag{13.40}$$

---

[1]However, there are cases in which the measure is not invariant. See the last Section of this Chapter.

Again, we define the functional integral as

$$Z \to \frac{Z}{V_G} = \frac{Z}{\int d\mu(\Omega)} = \int \mathcal{D}(A_\mu) e^{i \int \mathcal{L}_A d^4 x} \delta[f_A(A_\mu)] \Delta_f[A_\mu]. \quad (13.41)$$

The evaluation of $\Delta_f$ goes as before. Since in the previous equation it appears multiplied by $\delta[f_A(A_\mu)]$, the only solution of $f_A(A_\mu^\Omega) = 0$ in this neighborhood is $\Omega = 1$, or $\alpha_A = 0$. Therefore,

$$\delta[f_A(A_\mu^\Omega)] = \det{}^{-1} \left\| \frac{\delta f_A(A_\mu^\Omega)}{\delta \alpha_B} \bigg|_{\alpha_A = 0} \right\| \delta[\alpha_A], \quad (13.42)$$

from which

$$\Delta_f[A_\mu] = \det \left\| \frac{\delta f_A(A_\mu^\Omega)}{\delta \alpha_B} \bigg|_{\alpha_A = 0} \right\|. \quad (13.43)$$

Once again, the Faddeev-Popov determinant $\Delta_f[A_\mu]$ is the answer of the gauge-fixing to an infinitesimal gauge transformation. The final expression that we get for $Z$ is

$$Z = \int \mathcal{D}(A_\mu) \delta[f_A(A_\mu)] \det \left\| \frac{\delta f_A(A_\mu^\Omega)}{\delta \alpha_B} \bigg|_{\alpha_A = 0} \right\| e^{i \int \mathcal{L}_A d^4 x}. \quad (13.44)$$

This expression gives the right measure to be used in the path integral quantization of the gauge fields. In particular, the vacuum expectation value (v.e.v.) of a gauge field functional will be defined as

$$\langle 0 | T^*(F[\mathbf{A}]) | 0 \rangle_f = \int \mathcal{D}(A_\mu) \delta[f] \Delta_f e^{iS} F[A], \quad (13.45)$$

where the index $f$ specifies the gauge in which we evaluate the v.e.v. of $F$. About this point, we can prove an important result, that is: if $F[A]$ is a gauge invariant functional

$$F[A^\Omega] = F[A], \quad (13.46)$$

then, its v.e.v. does not depend on the gauge fixing $f_A = 0$. We will prove this result by showing the relation between the v.e.v.'s of $F[A]$ in two different gauges. Consider the identity

$$\langle 0 | T^*(F[\mathbf{A}]) | 0 \rangle_{f_1} = \int \mathcal{D}(A_\mu) \delta[f_1] \Delta_{f_1} e^{iS} F[A] \int d\mu(\Omega) \delta[f_2^\Omega] \Delta_{f_2}, \quad (13.47)$$

where we have used eq. (13.33) and defined $f_2^\Omega \equiv f_2[A^\Omega]$. By the change of variable $A_\mu \to A_\mu^\Omega$, we get

$$\langle 0 | T^*(F[\mathbf{A}]) | 0 \rangle_{f_1}$$

$$= \int d\mu(\Omega) \int \mathcal{D}(A_\mu) \delta[f_1^{\Omega^{-1}}] \Delta_{f_1} e^{iS} F[A^{\Omega^{-1}}] \delta[f_2] \Delta_{f_2}$$

$$= \int d\mu(\Omega) \int \mathcal{D}(A_\mu) e^{iS} \delta[f_2] \Delta_{f_2} \left( \delta[f_1^\Omega] \Delta_{f_1} F[A^\Omega] \right), \quad (13.48)$$

where in the second line we have made a further change of variables $\Omega \to \Omega^{-1}$ and used the invariance of $d\mu(\Omega)$. Then,

$$\langle 0|T^*(F[\mathbf{A}])|0\rangle_{f_1} = \int d\mu(\Omega)\langle 0|T^*(\delta[f_1^\Omega]\Delta_{f_1}F[\mathbf{A}^\Omega])|0\rangle_{f_2}. \qquad (13.49)$$

This is the relation we were looking for. If $F$ is gauge invariant, we can use the definition of $\Delta_{f_1}$ on the right-hand side of the previous equation, obtaining

$$\langle 0|T^*(F[\mathbf{A}])|0\rangle_{f_1} = \langle 0|T^*(F[\mathbf{A}])|0\rangle_{f_2}. \qquad (13.50)$$

In the following, we will define a generating functional as

$$Z_f[\eta_\mu^A] = \int \mathcal{D}(A_\mu)\delta[f]\Delta_f e^{iS} e^{i\int d^4x\eta_\mu^A A_A^\mu}, \qquad (13.51)$$

which is not gauge invariant, but in terms of which we can easily construct the v.e.v.'s of gauge invariant quantities, since it is built up in terms of the correct functional measure.

In perturbation theory, it is convenient to write the integrand of eq. (13.51) in the form $\exp(iS_{\text{eff}})$, where $S_{\text{eff}}$ is an effective action taking into account the gauge fixing, $f = 0$, and the Faddeev-Popov determinant. We will discuss later the exponentiation of $\Delta_f$. As far as $\delta[f]$ is concerned, we can exponentiate it by using its Fourier representation or, more conveniently, by choosing the gauge condition in the form $f_A(A) - B_A(x) = 0$, with $B_A(x)$ a set of arbitrary functions independent on the gauge fields. Then,

$$\begin{aligned}\Delta_{(f-B)}[A] &= \det\left\|\frac{\delta(f_A[A^\Omega] - B_A)}{\delta\alpha_B}\bigg|_{\alpha_A=0}\right\| \\ &= \det\left\|\frac{\delta f_A[A^\Omega]}{\delta\alpha_B}\bigg|_{\alpha_A=0}\right\| = \Delta_f[A], \qquad (13.52)\end{aligned}$$

and therefore,

$$1 = \Delta_f[A]\int d\mu(\Omega)\delta[f_A^\Omega - B_A]. \qquad (13.53)$$

We can further multiply this identity by the following expression, which does not depend on the gauge fields

$$\text{constant} = \int \mathcal{D}(B)e^{-\frac{i}{2\beta}\int d^4x\sum_A B_A^2(x)}, \qquad (13.54)$$

obtaining

$$\begin{aligned}\text{constant} &= \Delta_f[A]\int d\mu(\Omega)\mathcal{D}(B)e^{-\frac{i}{2\beta}\int d^4x\sum_A(f_A^\Omega(x))^2}\delta(f_A - B_A) \\ &= \Delta_f[A]\int d\mu(\Omega)e^{-\frac{i}{2\beta}\int d^4x\sum_A(f_A^\Omega(x))^2}. \qquad (13.55)\end{aligned}$$

Multiplying this expression by the functional (13.37), we can show, as before, that the gauge volume factorizes out, allowing us to define a class of generating functionals

$$Z_\beta[\eta_\mu^A] = \int \mathcal{D}(A_\mu) \Delta_f[A] e^{i \int d^4x (\mathcal{L}_A - \frac{1}{2\beta} \sum_A (f_A)^2)} e^{i \int d^4x \eta_\mu^A A_A^\mu}. \tag{13.56}$$

Finally, defining

$$S_f = -\frac{1}{2\beta} \int d^4x \sum_A (f_A(x))^2, \tag{13.57}$$

we find the analogue of eq. (13.49), that is

$$\langle 0|T^*(F[\mathbf{A}])|0\rangle_{f_1} = \int d\mu(\Omega) \langle 0|T^*(\Delta_{f_1} e^{iS_{f_1}^\Omega} F[\mathbf{A}^\Omega])|0\rangle_{f_2} \tag{13.58}$$

showing again that the v.e.v. of a gauge invariant functional gives rise to a gauge invariant expression.

## 13.2   Path integral quantization of QED

Let us now discuss the path integral quantization of QED in a covariant gauge. The Lagrangian density is

$$\mathcal{L} = \bar\psi \left[ i\gamma^\mu (\partial_\mu + ieA_\mu) - m \right] \psi - \frac{1}{4} F_{\mu\nu} F^{\mu\nu}, \tag{13.59}$$

with

$$F_{\mu\nu} = \partial_\mu A_\nu - \partial_\nu A_\mu \tag{13.60}$$

and invariant under

$$\psi(x) \to \psi'(x) = e^{-ie\alpha(x)} \psi(x),$$
$$A_\mu(x) \to A'_\mu(x) = A_\mu(x) + \frac{1}{e} \partial_\mu \alpha(x). \tag{13.61}$$

The Lorentz gauge is defined by

$$f(A_\mu) = \partial^\mu A_\mu(x) = 0 \tag{13.62}$$

and

$$f(A_\mu^\Omega) = \partial^\mu A_\mu(x) + \frac{1}{e} \Box \alpha(x), \tag{13.63}$$

with

$$\left. \frac{\delta f(A_\mu^\Omega)}{\delta \alpha(y)} \right|_{\alpha=0} = \frac{1}{e} \Box_x \delta^4(x-y). \tag{13.64}$$

Then, the Faddeev-Popov determinant is field-independent, and it can be absorbed into the normalization of the generating functional, which is given by

$$Z[\eta, \bar{\eta}, \eta_\mu] = N \int \mathcal{D}(\psi, \bar{\psi}) \mathcal{D}(A_\mu) e^{i \int d^4 x \mathcal{L}} \delta[\partial_\mu A^\mu]$$
$$\times e^{i \int [\bar{\eta}\psi + \bar{\psi}\eta + \eta_\mu A^\mu] d^4 x}. \tag{13.65}$$

We extract the interaction term in the usual way

$$Z[\eta, \bar{\eta}, \eta_\mu] = e^{-ie \int d^4 x (-\frac{1}{i}\frac{\delta}{\delta\bar{\eta}})\gamma^\mu(\frac{1}{i}\frac{\delta}{\delta A^\mu})(\frac{1}{i}\frac{\delta}{\delta\eta})} Z_0[\eta, \bar{\eta}, \eta_\mu], \tag{13.66}$$

with

$$Z_0[\eta, \bar{\eta}, \eta_\mu] = N \int \mathcal{D}(\psi, \bar{\psi}) \mathcal{D}(A_\mu) e^{i \int d^4 x (\mathcal{L}_0 + \mathcal{L}_\psi)} \delta[\partial_\mu A^\mu]$$
$$\times e^{i \int [\bar{\eta}\psi + \bar{\psi}\eta + \eta_\mu A^\mu] d^4 x}, \tag{13.67}$$

where

$$\int d^4 x \mathcal{L}_0 = -\frac{1}{4} \int d^4 x F_{\mu\nu} F^{\mu\nu} = \frac{1}{2} \int d^4 x \, A_\mu (g^{\mu\nu}\Box - \partial^\mu\partial^\nu) A_\nu \tag{13.68}$$

and

$$\int d^4 x \mathcal{L}_\psi = \int d^4 x \bar{\psi}(i\gamma^\mu\partial_\mu - m)\psi. \tag{13.69}$$

Then, using $\delta[\partial_\mu A^\mu]$ and integrating over the Fermi fields, we get

$$Z_0[\eta, \bar{\eta}, \eta_\mu] = e^{-i\langle\bar{\eta}(x) S_F(x-y)\eta(y)\rangle}$$
$$\times N \int \mathcal{D}(A_\mu) \delta[\partial_\mu A^\mu] e^{\frac{i}{2}\langle A_\mu(x)\Box A^\mu(x)\rangle} e^{i\langle\eta_\mu(x) A^\mu(x)\rangle}, \tag{13.70}$$

where $N$ is an irrelevant normalization factor. We can exponentiate the term $\delta[\partial_\mu A^\mu]$ through its Fourier transform

$$\delta[\partial_\mu A^\mu] = \int \mathcal{D}(C) e^{i\langle C(x)\partial_\mu A^\mu(x)\rangle} = \int \mathcal{D}(C) e^{-i\langle\partial_\mu C(x) A^\mu(x)\rangle}, \tag{13.71}$$

obtaining the following expression for $Z_0$

$$Z_0[\eta, \bar{\eta}, \eta_\mu] = N e^{-i\langle\bar{\eta}(x) S_F(x-y)\eta(y)\rangle}$$
$$\times \int \mathcal{D}(A_\mu)\mathcal{D}(C) e^{i\langle\frac{1}{2} A_\mu(x)\Box A^\mu(x)\rangle}$$
$$\times e^{i\langle(\eta_\mu(x) - \partial_\mu C(x)) A^\mu(x)\rangle}. \tag{13.72}$$

Performing the Gaussian integration over $A_\mu$ we get

$$Z_0[\eta, \bar\eta, \eta_\mu] = N e^{-i\langle \bar\eta(x) S_F(x-y)\eta(y)\rangle}$$
$$\times \int \mathcal{D}(C) e^{\frac{1}{2}\langle i(\eta_\mu(x) - \partial_\mu C(x))(\frac{i}{\Box})_{x,y} i(\eta^\mu(y) - \partial^\mu C(y))\rangle}$$
$$= e^{-i\langle \bar\eta(x) S_F(x-y)\eta(y)\rangle} e^{-\frac{i}{2}\langle \eta_\mu \frac{1}{\Box}\eta^\mu\rangle}$$
$$\times \int \mathcal{D}(C) e^{i\langle \partial_\mu C \frac{1}{\Box}\eta^\mu\rangle - \frac{i}{2}\langle \partial_\mu C \frac{1}{\Box}\partial^\mu C\rangle}, \tag{13.73}$$

where, the inverse of the D'Alembertian, $1/\Box$, will be defined more precisely later on. After integrating by parts, we can rewrite the integral over $C(x)$ as

$$\int \mathcal{D}(C) e^{-i\langle C \frac{1}{\Box}\partial_\mu \eta^\mu\rangle + \frac{i}{2}\langle CC\rangle}. \tag{13.74}$$

Integrating once more

$$e^{\frac{1}{2}\langle(-\frac{i}{\Box}\partial_\mu\eta^\mu)(+i)(-\frac{i}{\Box}\partial_\nu\eta^\nu)\rangle} = e^{\frac{i}{2}\langle \eta^\mu \frac{\partial_\mu \partial_\nu}{\Box^2}\eta^\nu\rangle}. \tag{13.75}$$

The final result is

$$Z_0[\eta, \bar\eta, \eta_\mu] = Z[0] e^{-i\langle \bar\eta(x) S_F(x-y)\eta(y)\rangle} e^{-\frac{i}{2}\langle \eta^\mu \left(\frac{g_{\mu\nu}}{\Box} - \frac{\partial_\mu \partial_\nu}{\Box^2}\right)\eta^\nu\rangle}. \tag{13.76}$$

Introducing the function

$$G_{\mu\nu} = \lim_{\epsilon \to 0^+} \int \frac{d^4 k}{(2\pi)^4}\left[-\frac{g_{\mu\nu}}{k^2 + i\epsilon} + \frac{k_\mu k_\nu}{(k^2 + i\epsilon)^2}\right] e^{-ikx}, \tag{13.77}$$

where we have defined the inverse of $\Box$ with the usual Feynman prescription, we can write

$$Z_0[\eta, \bar\eta, \eta_\mu] = Z[0] e^{-i\langle \bar\eta(x) S_F(x-y)\eta(y)\rangle} e^{-\frac{i}{2}\langle \eta^\mu(x) G_{\mu\nu}(x-y)\eta^\nu(y)\rangle}. \tag{13.78}$$

The photon propagator in this gauge is given by

$$\frac{1}{\langle 0|0\rangle}\langle 0|T(A_\mu(x) A_\nu(x))|0\rangle = iG_{\mu\nu}(x-y). \tag{13.79}$$

Notice that this propagator is transverse, that is, it satisfies the Lorentz condition

$$\partial^\mu G_{\mu\nu}(x) = 0. \tag{13.80}$$

The Feynman rules in this gauge are the same discussed in Section 8.5, except for the photon propagator which, in this gauge, is

$$-i\left[\frac{g_{\mu\nu}}{k^2 + i\epsilon} - \frac{k_\mu k_\nu}{(k^2 + i\epsilon)^2}\right]. \tag{13.81}$$

Of course, the additional term in the propagator, proportional to $k_\mu k_\nu$ does not affect the physics, since the photon is coupled to a conserved current, satisfying $k_\mu j^\mu = 0$.

We can perform an analogous calculation, but using the gauge fixing in the form $f - B$, discussed at the end of the previous Section, that is, using the generating functional (13.56)

$$Z_\beta[\eta, \bar{\eta}, \eta_\mu] = N \int \mathcal{D}(\psi, \bar{\psi}) \mathcal{D}(A_\mu) e^{i \int d^4 x [\mathcal{L} - \frac{1}{2\beta}(\partial_\mu A^\mu)^2]}$$

$$\times e^{\int d^4 x [\bar{\eta}\psi + \bar{\psi}\eta + \eta_\mu A^\mu]}. \tag{13.82}$$

In this case, the only effect of the gauge fixing is to change the Lagrangian $\mathcal{L}_0$ in $\mathcal{L}_0' = \mathcal{L}_0 - (\partial_\mu A^\mu)^2/(2\beta)$. Of course, the addition of this term removes the problem of the singular wave operator. In fact we have

$$\int d^4 x \mathcal{L}_0' = \frac{1}{2} \int d^4 x A^\mu \left( g_{\mu\nu}\Box - (1 - \frac{1}{\beta})\partial_\mu \partial_\nu \right) A^\nu. \tag{13.83}$$

The functional integral, to be evaluated, is

$$Z_0[\eta, \bar{\eta}, \eta_\mu] = N e^{-i\langle \bar{\eta}(x) S_F(x-y)\eta(y)\rangle}$$

$$\times \int \mathcal{D}(A_\mu) e^{i \int d^4 x (\frac{1}{2} A^\mu K_{\mu\nu} A^\nu + \eta_\mu A^\mu)}, \tag{13.84}$$

with

$$K_{\mu\nu} = g_{\mu\nu}\Box - \frac{\beta - 1}{\beta}\partial_\mu \partial_\nu. \tag{13.85}$$

We get

$$Z_\beta[\eta, \bar{\eta}, \eta_\mu] = Z[0]e^{-i\langle \bar{\eta}(x) S_F(x-y)\eta(y)\rangle}e^{-\frac{i}{2}\langle \eta_\mu(x) G^{\mu\nu}(x-y)\eta^\nu(y)\rangle}, \tag{13.86}$$

where

$$K_{\mu\nu}G^{\nu\rho}(x) = \delta_\mu^\rho \delta^4(x). \tag{13.87}$$

We can see that $K_{\mu\nu}$ is non-singular for $\beta \neq \infty$. Equation (13.87) can be solved easily in momentum space

$$\left(-g_{\mu\nu}k^2 + \frac{\beta - 1}{\beta}k_\mu k_\nu\right) G^{\nu\rho}(k) = \delta_\nu^\rho. \tag{13.88}$$

It is enough to write the most general second rank symmetric tensor function of $k_\mu$:

$$G_{\mu\nu}(k) = Ag_{\mu\nu} + Bk_\mu k_\nu. \tag{13.89}$$

We find

$$-Ak^2 = 1, \qquad \frac{\beta - 1}{\beta}A - \frac{1}{\beta}Bk^2 = 0. \tag{13.90}$$

Therefore,

$$G_{\mu\nu}(k) = -\frac{g_{\mu\nu}}{k^2} + (1 - \beta)\frac{k_\mu k_\nu}{k^4}. \tag{13.91}$$

The singularities are defined as usual, shifting $k^2$ by $i\epsilon$, obtaining

$$G_{\mu\nu}(x) = \lim_{\epsilon \to 0^+} \int \frac{d^4k}{(2\pi)^4} \left[ -\frac{g_{\mu\nu}}{k^2 + i\epsilon} + (1 - \beta)\frac{k_\mu k_\nu}{(k^2 + i\epsilon)^2} \right] e^{-ikx}. \tag{13.92}$$

At the beginning of this Section, we have made a choice of gauge corresponding to $\beta = 0$ (the so-called Landau gauge). In fact, in this case,

$$\lim_{\beta \to 0} e^{-\frac{i}{2\beta} \int d^4x (\partial_\mu A^\mu)^2} \to \delta[\partial_\mu A^\mu]. \tag{13.93}$$

On the other hand, the gauge used in Section 5.2 corresponds to $\beta = 1$. In fact, we have

$$G_{\mu\nu}(x) = -g_{\mu\nu}\Delta_F(x; m^2 = 0). \tag{13.94}$$

This is called the Feynman gauge. As we see, the parameter $\beta$ selects different covariant gauges. As already stressed, the physics does not depend on $\beta$, since the photon couples to the fermionic current $\bar{\psi}\gamma^\mu\psi$, which is conserved. As a consequence, the $\beta$ dependent part of the propagator, being proportional to $k_\mu k_\nu$, does not contribute. The situation is quite different in the non-abelian case, where the fermionic current is not conserved, since the gauge fields are not neutral under gauge transformations. Therefore, we need further contributions in order to cancel the $\beta$ dependent terms. Such contributions arise from the Faddeev-Popov determinant.

### 13.3   Path integral for the non-abelian gauge theories

In the case of a non-abelian gauge theory, there are several differences with respect to QED. The pure gauge Lagrangian is not quadratic, it contains three-linear and four-linear interaction terms and, furthermore, the Faddeev-Popov determinant depends on the gauge fields. In this Section, we will consider the pure gauge case, since the fermionic coupling is completely analogous to the abelian case. Separating the Lagrangian in the quadratic part plus the rest, we get

$$\mathcal{L}_A = -\frac{1}{4}\sum_C F_{\mu\nu C}F_C^{\mu\nu} = \mathcal{L}_A^{(2)} + \mathcal{L}_A^I, \tag{13.95}$$

where $\mathcal{L}_A^{(2)}$ is the quadratic part,

$$\mathcal{L}_A^{(2)} = -\frac{1}{4}(\partial_\mu A_{\nu C} - \partial_\nu A_{\mu C})(\partial^\mu A_C^\nu - \partial^\nu A_C^\mu), \qquad (13.96)$$

and $\mathcal{L}_A^{(I)}$ the interacting one,

$$\mathcal{L}_A^{(I)} = g f_C^{AB} A_{\mu A} A_{\nu B} \partial^\mu A_C^\nu - \frac{1}{4} g^2 f_C^{AB} f_C^{DE} A_A^\mu A_B^\nu A_{\mu D} A_{\nu E}. \qquad (13.97)$$

The interaction terms can be extracted by functional integration with the usual trick

$$Z[\eta_\mu] = e^{i S_A^{(I)} \left[ \frac{1}{i} \frac{\delta}{\delta \eta_\mu^A} \right]} Z_0[\eta_\mu], \qquad (13.98)$$

with $S_A^{(I)} = \int d^4 x \mathcal{L}_A^{(I)}$ and

$$Z_0[\eta_\mu] = N \int \mathcal{D}(A_\mu) e^{i \int d^4 x \left[ \mathcal{L}_A^{(2)} - \frac{1}{2\beta} \sum_A (\partial_\mu A_A^\mu)^2 \right]}$$

$$\times e^{i \int d^4 x \eta_A^\mu A_\mu^A} \Delta_f[A_\mu]. \qquad (13.99)$$

Consider now $\Delta_f[A_\mu]$. In the Lorentz gauge

$$f_A[A_\mu] = \partial_\mu A_A^\mu. \qquad (13.100)$$

Under an infinitesimal gauge transformation (for later convenience we have changed the definition of the gauge parameters, sending $\alpha_A \to g\alpha_A$)

$$\delta A_C^\mu = g f_C^{AB} \alpha_A A_B^\mu + \partial^\mu \alpha_C, \qquad (13.101)$$

we have

$$f_C[A_\mu^\Omega] = \partial_\mu A_C^\mu + g f_C^{AB} \partial_\mu(\alpha_A A_B^\mu) + \Box \alpha_C. \qquad (13.102)$$

Then

$$\frac{\delta f_C[A_\mu^\Omega(x)]}{\delta \alpha_B(y)} \bigg|_{\alpha_A = 0} = \delta_{BC} \Box_x^2 \delta^4(x - y) + g f_C^{BA} \partial_\mu(\delta^4(x - y) A_A^\mu)$$

$$\equiv M_{CB}(x - y). \qquad (13.103)$$

Since $\Delta_f[A]$ is the determinant of this expression (see eq. (13.43)), it depends explicitly on the gauge fields. In order to get a convenient expression for the generating functional, it is useful to use an exponential form. This can be done, recalling from eq. (12.221) that a determinant can be written as a Gaussian integral over Grassmann variables.

$$\Delta_f[A_\mu] = \det \left\| \frac{\delta f_A[A_\mu^\Omega(x)]}{\delta \alpha_B(y)} \bigg|_{\alpha_A = 0} \right\|$$

$$= N \int \mathcal{D}(c, c^*) e^{i \int c_A^*(x) M_{AB}(x-y) c_B(y) d^4 x d^4 y}. \qquad (13.104)$$

The Grassmann fields, $c_A(x)$, are called ghost fields, or Faddeev-Popov ghosts. They carry zero spin and therefore violate the spin-statistics theorem. However, this has no physical consequences, since our generating functional does not generate amplitudes with external ghosts. The action for the ghost fields can be read from eqs. (13.103) and (13.104), obtaining

$$S_{FPG} = \int d^4x \, c_A^*(x) \left[ \delta^{AB} \Box - g f_B^{AC} \partial_\mu A_C^\mu \right] c_B(x). \qquad (13.105)$$

The ghosts are zero mass particles, interacting with the gauge fields through the interaction term

$$S_{FPG}^{(I)} = -g \int d^4x \, c_A^* f_B^{AC} \partial_\mu A_C^\mu c_B. \qquad (13.106)$$

It is convenient to modify the generating functional, introducing external sources for the ghost fields

$$Z[\eta, \eta*, \eta_\mu] = e^{iS_A^{(I)} \left[ \frac{1}{i} \frac{\delta}{\delta \eta_\mu} \right]} e^{iS_{FPG}^{(I)} \left[ -\frac{1}{i} \frac{\delta}{\delta \eta}, \frac{1}{i} \frac{\delta}{\delta \eta^*} \right]} Z_0[\eta, \eta^*, \eta_\mu], \qquad (13.107)$$

with

$$Z_0[\eta, \eta^*, \eta_\mu] = \int \mathcal{D}(A_\mu) \mathcal{D}(c, c^*)$$
$$\times e^{i \int d^4x \left[ \mathcal{L}_A^{(2)} - \frac{1}{2\beta} \sum_A (\partial_\mu A_A^\mu)^2 + \eta_\mu^A A_A^\mu \right]}$$
$$\times e^{i \int d^4x \left[ c_A^* \Box c_A + c_A^* \eta^A + \eta^{A*} c_A \right]}. \qquad (13.108)$$

The integration over $A_A^\mu$ can be done as in the abelian case. For the ghost fields we use eq. (12.221) with the result

$$\int \mathcal{D}(c, c^*) e^{i \langle c_A^* \Box c_A + c_A^* \eta^A + \eta^{A*} c_A \rangle}$$
$$= e^{-\langle (i\eta^{A*})(-\frac{i}{\Box})(i\eta^A) \rangle} = e^{-i \langle \eta^{A*} \frac{1}{\Box} \eta^A \rangle}. \qquad (13.109)$$

Defining

$$\Delta_{AB}(x) = -\delta_{AB} \lim_{\epsilon \to 0^+} \int \frac{d^4k}{(2\pi)^4} \frac{e^{-ikx}}{k^2 + i\epsilon}, \qquad (13.110)$$

we get

$$Z_0[\eta, \eta^*, \eta_\mu] = e^{-\frac{i}{2} \langle \eta_\mu^A(x) G_{AB}^{\mu\nu}(x-y) \eta_{\nu B}(y) \rangle}$$
$$\times e^{-i \langle \eta^{A*}(x) \Delta_{AB}(x-y) \eta^B(y) \rangle}, \qquad (13.111)$$

with

$$G_{AB}^{\mu\nu} = \delta_{AB} G^{\mu\nu}(x-y). \qquad (13.112)$$

It is then easy to read the Feynman rules. For the propagators we get

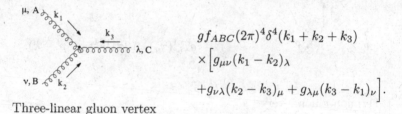

k

μ, A        ν, B        $-i\delta_{AB}\Big(g_{\mu\nu} - (1-\beta)\dfrac{k_\mu k_\nu}{k^2 + i\epsilon}\Big)\dfrac{1}{k^2 + i\epsilon},$

Gauge field propagator

k

A        B        $i\delta_{AB}\dfrac{1}{k^2 + i\epsilon}.$

Ghost field propagator

The Feynman rules for the vertices are obtained by taking the Fourier transform of the interaction terms. Notice that here we have written $f_C^{AB} \equiv f_{ABC}$, since we are dealing with compact Lie groups.

μ, A  $k_1$

$k_3$  λ, C

ν, B  $k_2$

$$gf_{ABC}(2\pi)^4\delta^4(k_1 + k_2 + k_3)$$
$$\times\Big[g_{\mu\nu}(k_1 - k_2)_\lambda$$
$$+g_{\nu\lambda}(k_2 - k_3)_\mu + g_{\lambda\mu}(k_3 - k_1)_\nu\Big].$$

Three-linear gluon vertex

μ, A  $k_1$        ρ, D

$k_4$

$k_3$

ν, B  $k_2$        λ, C

$$-ig^2(2\pi)^4\delta^4(\sum_{i=1}^{4}k_i)$$
$$\times\Big[f_{ABE}f_{CDE}(g_{\mu\lambda}g_{\nu\rho} - g_{\nu\lambda}g_{\mu\rho})$$
$$+f_{ACE}f_{BDE}(g_{\mu\nu}g_{\lambda\rho} - g_{\lambda\nu}g_{\mu\rho})$$
$$+f_{ADE}f_{CBE}(g_{\mu\lambda}g_{\rho\nu} - g_{\rho\lambda}g_{\mu\nu})\Big].$$

Four-linear gluon vertex

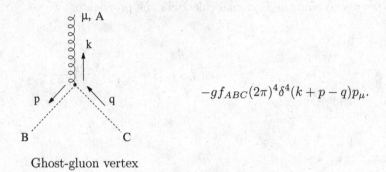

$$-gf_{ABC}(2\pi)^4\delta^4(k+p-q)p_\mu.$$

Ghost-gluon vertex

When we have fermions coupled to the gauge fields, we need to add the rule for the vertex. By a simple comparison with the coupling in QED, we obtain the following rule

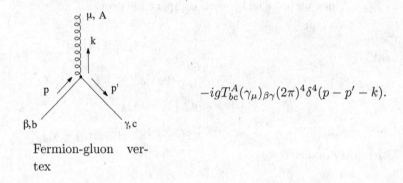

$$-igT^A_{bc}(\gamma_\mu)_{\beta\gamma}(2\pi)^4\delta^4(p-p'-k).$$

Fermion-gluon vertex

## 13.4   Ward-Takahashi identities and anomalies

So far, we have assumed that the symmetries of the classical Lagrangian are still valid at the quantum level. This is not always the case, and the motivations can be easily understood using the path integral approach. In fact, it is not enough that the Lagrangian is invariant under some transformation, but also the functional measure should be invariant. In fact, we will see that there are situations in which the measure is not invariant and, therefore, the symmetry is broken at the quantum level. Before dealing with this problem, we will derive the Ward-Takahashi identities (see [Ward (1950)] and [Takahashi (1957)]), that are the expression of Noether's theorem in the quantum context.

Let us consider a field theory with a Lagrangian invariant under the following global transformation

$$\phi_i \to \phi_i + \delta\phi_i, \tag{13.113}$$

with

$$\delta\phi_i(x) = -i\epsilon_A(T^A)_{ij}\phi_j(x). \tag{13.114}$$

The change of the action under the same transformation but with a parameter space-time dependent, that is $\epsilon_A \to \epsilon_A(x)$, is given by

$$\delta S = \int d^4x \left[ \frac{\partial\mathcal{L}}{\partial\phi_i}\delta\phi_i + \frac{\partial\mathcal{L}}{\partial\phi_{i,\mu}}\delta\phi_{i,\mu} \right], \tag{13.115}$$

where

$$\delta\phi_{i,\mu} = -i\epsilon_A(T^A)_{ij}\partial_\mu\phi_j - i(\partial_\mu\epsilon_A)(T^A)_{ij}\phi_j. \tag{13.116}$$

Substituting we find

$$\delta S = \int d^4x \Big[ -i\frac{\partial\mathcal{L}}{\partial\phi_i}\epsilon_A(T^A)_{ij}\phi_j - i\frac{\partial\mathcal{L}}{\partial\phi_{i,\mu}}\epsilon_A(T^A)_{ij}\partial_\mu\phi_j$$
$$-i\frac{\partial\mathcal{L}}{\partial\phi_{i,\mu}}(T^A)_{ij}\phi_j\partial_\mu\epsilon_A \Big]. \tag{13.117}$$

Since, by hypothesis, for $\epsilon_A$ independent on $x$, we have $\delta S = 0$, it follows that the first two terms must cancel each other. Therefore, we are left with

$$\delta S = \int d^4x \, (\partial^\mu\epsilon_A)j_\mu^A, \tag{13.118}$$

where

$$j_\mu^A = -i\frac{\partial\mathcal{L}}{\partial\phi_{i,\mu}}(T^A)_{ij}\phi_j \tag{13.119}$$

are the Noether currents, after factorizing the infinitesimal parameters $\epsilon_A$. However, if we take $\epsilon_A(x)$ to describe a variation around the classical solutions, that is, if the fields inside the currents satisfy the equations of motion, then we must have $\delta S = 0$. It follows,, integrating by parts,

$$\delta S = -\int d^4x \, \epsilon_A \partial_\mu j_A^\mu, \tag{13.120}$$

from which

$$\partial_\mu j_A^\mu = 0. \tag{13.121}$$

Let us now consider the generating functional associated to this theory

$$Z[\eta] = \int \mathcal{D}(\phi) e^{iS[\phi] + i \int d^4 x \, \eta_i \phi_i}. \tag{13.122}$$

If we perform the change of variables

$$\phi_i \to \phi_i + \delta\phi_i = \phi_i - i\epsilon_A(x)(T^A)_{ij}\phi_j, \tag{13.123}$$

and assume that the integration measure is invariant, we get

$$Z[\eta] = \int \mathcal{D}(\phi) e^{iS[\phi] + i \int d^4 x \, \eta_i \phi_i} e^{\int d^4 x (i\partial^\mu \epsilon_A j_\mu^A + \epsilon_A \eta_i (T^A)_{ij}\phi_j)}. \tag{13.124}$$

By expanding this expression at the first order in $\epsilon_A(x)$, and by comparison with eq. (13.122), we get

$$0 = \int \mathcal{D}(\phi) e^{iS[\phi] + i \int d^4 x \, \eta_i \phi_i} \left[ -i\partial^\mu j_\mu^A + \eta_i (T^A)_{ij}\phi_j \right]. \tag{13.125}$$

We can generate the Ward-Takahashi identities through functional differentiation with respect to the sources and, then, taking $\eta_i = 0$. For instance, for $\eta_i = 0$, we get

$$\partial^\mu \langle 0 | j_\mu^A(x) | 0 \rangle = 0. \tag{13.126}$$

Differentiating once (remember that one gets the $T^*$ product) we have

$$\partial_x^\mu \langle 0 | T(j_\mu^A(x)\phi_i(y)) | 0 \rangle = -\delta^4(x - y)\langle 0 | (T^A)_{ij}\phi_j(y) | 0 \rangle, \tag{13.127}$$

and, differentiating $N$ times

$$\partial_x^\mu \langle 0 | T(j_\mu^A(x)\phi_{i_1}(x_1) \cdots \phi_{i_N}(x_N)) | 0 \rangle$$
$$= \sum_{p=1}^N \delta^4(x - x_p)$$
$$\times \langle 0 | T(\phi_{i_1}(x_1) \cdots (-(T^A)_{i_p j}\phi_j(x_p)) \cdots \phi_{i_N}(x_N)) | 0 \rangle. \tag{13.128}$$

This derivation is correct only if the integration measure is invariant under the change of variables (13.123). We will illustrate a simple situation where the measure is not invariant. Consider a massless fermion coupled to a gauge field. Since we are interested in the transformation properties of the functional measure, we will take the generating functional at zero sources:

$$Z = \int \mathcal{D}\psi \mathcal{D}\bar{\psi} e^{i \int d^4 x \, \bar{\psi} i \hat{D} \psi}, \tag{13.129}$$

where

$$D_\mu = \partial_\mu + ig A_\mu. \tag{13.130}$$

For the sake of simplicity, we have assumed the gauge field to be an external one. If $A_\mu$ is quantized, we have to insert a further functional integration in $A_\mu$, but this will not play any role in the following. The Lagrangian is invariant under the global transformation

$$\psi \to e^{i\alpha\gamma_5}\psi, \qquad (13.131)$$

giving rise to the classically conserved current

$$J_\mu = \bar{\psi}\gamma_\mu\gamma_5\psi. \qquad (13.132)$$

We will show that the quantum corrections destroy the conservation law. To this end, we proceed as in the previous discussion, performing the following change of variables inside $Z$

$$\psi(x) \to \psi'(x) = (1 + i\alpha(x)\gamma_5)\psi(x),$$
$$\bar{\psi}(x) \to \bar{\psi}'(x) = \bar{\psi}(x)(1 + i\alpha(x)\gamma_5). \qquad (13.133)$$

The variation of the action is given by

$$\delta S = \int d^4x\, \alpha(x)\partial_\mu(\bar{\psi}\gamma^\mu\gamma_5\psi). \qquad (13.134)$$

In order to evaluate the change of the integration measure, it is convenient to expand the fermion field in a basis of eigenvectors of $i\hat{D}$

$$i\hat{D}\phi_m(x) = \lambda_m\phi_m(x),$$
$$-iD_\mu\tilde{\phi}_m(x)\gamma^\mu = \lambda_m\tilde{\phi}_m(x), \qquad (13.135)$$

where we have used a discrete notation for the eigenvalues. In the case of a zero gauge field $A_\mu$, the eigenvalues $\lambda_m$ are given by

$$\lambda_m^2 = k_0^2 - |\vec{k}|^2. \qquad (13.136)$$

Notice that they become negative definite after being Wick rotated

$$\lambda_m^2 \to -k_4^2 - |\vec{k}|^2 = -k_E^2. \qquad (13.137)$$

By hypothesis, the eigenvectors $\phi_m$ span an orthonormal basis, therefore, we can expand $\psi$ and $\bar{\psi}$ as

$$\psi(x) = \sum_m a_m\phi_m(x), \qquad \bar{\psi}(x) = \sum_m b_m\tilde{\phi}_m(x), \qquad (13.138)$$

where $a_m$ and $b_m$ are Grassmann coefficients. Then, except for a possible unessential constant, we have

$$\mathcal{D}\psi\mathcal{D}\bar{\psi} = \prod_m da_m db_m. \qquad (13.139)$$

We can evaluate the effect of the change of variables on the Grassmann coefficients, starting from

$$\psi'(x) = (1 + i\alpha(x)\gamma_5)\psi(x) = \sum_m a'_m \phi_m(x). \qquad (13.140)$$

From the orthogonality relation

$$\int d^4x\, \phi_m^\dagger \phi_n = \delta_{mn}, \qquad (13.141)$$

we get

$$a'_m = \int d^4x\, \phi_m^\dagger(x)\psi'(x) = a_m + \sum_n \int d^4x\, \phi_m^\dagger(x)i\alpha(x)\gamma_5\phi_n(x)a_n$$

$$= a_m + \sum_n C_{mn}a_n, \qquad (13.142)$$

with

$$C_{mn} = \int d^4x\, \phi_m^\dagger(x)i\alpha(x)\gamma_5\phi_n(x). \qquad (13.143)$$

A similar transformation holds for $b_m$. Therefore, from the transformation rules for the Grassmann measure, we get (see eq. (12.199))

$$\mathcal{D}\psi'\mathcal{D}\bar\psi' = \frac{1}{det|\mathcal{I}|^2}\mathcal{D}\psi\mathcal{D}\bar\psi, \qquad (13.144)$$

where $\mathcal{I} = 1 + C$ is the Jacobian of the transformation. Since we are interested in the infinitesimal transformation, we can write

$$det|\mathcal{I}| = e^{Tr\,\log(1+C)} \approx e^{Tr\,C}, \qquad (13.145)$$

that is

$$\log det|\mathcal{I}| \approx Tr\,C = i\int d^4x\,\alpha(x)\sum_n \phi_n^\dagger(x)\gamma_5\phi_n(x). \qquad (13.146)$$

The coefficient of $\alpha(x)$ is the trace of $\gamma_5$ taken over all the Hilbert space. Although the trace of $\gamma_5$ is zero on the spinor space, it is multiplied by an infinite factor arising from the trace over the eigenfuctions $\phi_n(x)$. Therefore, this expression is ill defined and it needs to be regularized. Furthermore, in order to maintain the gauge invariance, we need a gauge invariant regularization. A natural choice, for the regularization of the trace, is the following

$$\sum_n \phi_n^\dagger(x)\gamma_5\phi_n(x) = \lim_{M\to\infty}\sum_n \phi_n^\dagger(x)\gamma_5\phi_n(x)e^{\lambda_n^2/M^2}, \qquad (13.147)$$

where we have introduced a factor which is convergent in the euclidean space. The previous expression can be rewritten as follows:

$$\sum_n \phi_n^\dagger(x)\gamma_5\phi_n(x) = \lim_{M\to\infty} \sum_n \phi_n^\dagger(x)\gamma_5 e^{(i\hat{D})^2/M^2}\phi_n(x)$$

$$= \lim_{M\to\infty} \langle x|tr\left[\gamma_5 e^{(i\hat{D})^2/M^2}\right]|x\rangle, \qquad (13.148)$$

where $tr$ is the trace over the Dirac matrices. The evaluation of $(\hat{D})^2$ gives

$$(\hat{D})^2 = \gamma_\mu\gamma_\nu D^\mu D^\nu = \frac{1}{2}[\gamma_\mu,\gamma_\nu]_+ D^\mu D^\nu + \frac{1}{2}[\gamma_\mu,\gamma_\nu]_- D^\mu D^\nu$$

$$= D^2 - \frac{i}{2}\sigma_{\mu\nu}[D^\mu, D^\nu]_- = D^2 + \frac{g}{2}\sigma_{\mu\nu}F^{\mu\nu}. \qquad (13.149)$$

In order to get a contribution from the trace over the Dirac indices, we need at least four $\gamma$ matrices. Therefore, the leading term is obtained by expanding the exponential up to the order $(\sigma_{\mu\nu}F^{\mu\nu})^2$ and neglecting $A_\mu$ in all the other places. We get

$$\sum_n \phi_n^\dagger(x)\gamma_5\phi_n(x) = \lim_{M\to\infty} tr\left[\gamma_5 \frac{1}{2!}\left(-\frac{g}{2M^2}\sigma_{\mu\nu}F^{\mu\nu}\right)^2\right]\langle x|e^{-\Box/M^2}|x\rangle.$$

$$(13.150)$$

The matrix element can be easily evaluated by inserting momentum eigenstates and performing a Wick's rotation

$$\langle x|e^{-\Box/M^2}|x\rangle = \lim_{x\to y}\int \frac{d^4k}{(2\pi)^4}e^{k^2/M^2}e^{-ik(x-y)}$$

$$= i\int \frac{d^4k_E}{(2\pi)^4}e^{-k_E^2/M^2} = i\frac{M^4}{16\pi^2}. \qquad (13.151)$$

It follows

$$\sum_n \phi_n^\dagger\gamma_5\phi_n = \frac{ig^2}{8\cdot 16\pi^2}tr[\gamma_5\sigma_{\mu\nu}\sigma_{\rho\lambda}]F^{\mu\nu}F^{\rho\lambda}. \qquad (13.152)$$

In our definitions

$$\gamma_5 = i\gamma^0\gamma^1\gamma^2\gamma^3. \qquad (13.153)$$

The trace is easily evaluated by calculating first

$$tr[\gamma_5\sigma_{01}\sigma_{23}] = -itr[\gamma^0\gamma^1\gamma^2\gamma^3\gamma_0\gamma_1\gamma_2\gamma_3] = -4i, \qquad (13.154)$$

and using $\epsilon^{0123} = 1$ we get

$$tr[\gamma_5\sigma_{\mu\nu}\sigma_{\rho\lambda}] = 4i\epsilon_{\mu\nu\rho\lambda}. \qquad (13.155)$$

Then

$$\sum_n \phi_n^\dagger \gamma_5 \phi_n = -\frac{g^2}{32\pi^2} \epsilon_{\mu\nu\rho\lambda} F^{\mu\nu} F^{\rho\lambda}. \tag{13.156}$$

Therefore, the determinant we are looking for is given by

$$\det |\mathcal{I}| = e^{-i \int d^4x\, \alpha(x) \frac{g^2}{32\pi^2} \epsilon_{\mu\nu\rho\lambda} F^{\mu\nu} F^{\rho\lambda}}, \tag{13.157}$$

and after the change of variables the generating functional is given by

$$Z = \int \mathcal{D}\psi \mathcal{D}\bar{\psi} e^{i \int d^4x\, \bar{\psi} i \hat{D} \psi}$$

$$\times e^{i \int d^4x\, \alpha(x)(\partial_\mu(\bar{\psi}\gamma^\mu\gamma_5\psi) + \frac{g^2}{16\pi^2}\epsilon_{\mu\nu\rho\lambda} F^{\mu\nu} F^{\rho\lambda})}, \tag{13.158}$$

leading to the anomalous conservation law

$$\partial_\mu j_5^\mu = -\frac{g^2}{16\pi^2} \epsilon_{\mu\nu\rho\lambda} F^{\mu\nu} F^{\rho\lambda}. \tag{13.159}$$

This is called the Adler-Bell-Jackiw (ABJ) anomaly [Adler (1969)], [Bell and Jackiw (1969)] (see [Fujikawa (1979)] for the first derivation of the anomaly using the path integral). The previous result can be extended to the case of non-abelian chiral symmetries as, for instance, to the standard model, where the $SU(2)$ symmetry acts only on the left-handed fermions and the corresponding transformations involve $\gamma_5$. The ABJ anomaly is modified by a factor

$$Tr[T^A[T^B, T^C]_+], \tag{13.160}$$

where the term $T^A$ comes in the chiral transformation in conjunction with $\gamma_5$, whereas the other two matrices arise from the regulator. It is a crucial point in the renormalization of the gauge theories, that the gauge symmetry is preserved at the quantum level. It follows that the standard model is consistent only if the anomaly vanishes. One has to check that the factor in eq. (13.160) vanishes for all the generators of $SU(2) \otimes U(1)$. It is not difficult to show that this is indeed the case if the sum of the electric charges of the fermions is zero

$$\sum_{\text{fermions}} Q = 0. \tag{13.161}$$

This condition is indeed satisfied for any generation of quarks and leptons, in fact

$$Q_{\text{electron}} + Q_{\text{neutrino}} + 3(Q_{\text{up}} + Q_{\text{down}}) = -1 + 0 + 3\left(\frac{2}{3} - \frac{1}{3}\right) = 0. \tag{13.162}$$

## 13.5   Exercises

(1) Derive the Feynman rules for non-abelian gauge fields, given in Section 13.4.

(2) Derive the Ward-Takahashi identities in the case of a charged Dirac field, interacting with an electromagnetic field and with charge $Q$. Use the definition of the conserved current given in eq. (13.119).

(3) Consider the v.e.v. in the same field theory of the previous exercise:

$$\langle 0|T(j_\mu(x)\psi(x_1)\bar\psi(x_2)|0\rangle. \tag{13.163}$$

Define its Fourier transform as

$$\int d^4x\, d^4x_1\, d^4x_2\, e^{(ikx+ip'x_1-ipx_2)}\langle 0|T(j_\mu(x)\psi(x_1)\bar\psi(x_2)|0\rangle$$

$$= iS_F(p')(-iQ\Gamma^\mu(p',p))(iS_F(p))\delta^4(p'+k-p), \tag{13.164}$$

where $\Gamma^\mu(p',p)$ is the vertex function (see (10.170)). Using the Ward-Takahashi identities derived in the previous exercise, show that

$$(p'-p)^\mu\Gamma_\mu(p',p) = \Gamma^{(2)}(p') - \Gamma^{(2)}(p), \tag{13.165}$$

where $\Gamma^{(2)}(p)$ was defined in (10.170). Using this relation, prove (10.169).

# Appendix A

Given a real and antisymmetric matrix $n \times n$, $A$, we want to show that it is possible to put it in the form (12.206) by means of an orthogonal transformation. To this end, let us notice that $iA$ is a hermitian matrix and, as such, it can be diagonalized through a unitary transformation $U$:

$$U(iA)U^\dagger = A_d, \tag{A.1}$$

where $A_d$ is diagonal and real. The eigenvalues of $iA$ satisfy the equation

$$\det|iA - \lambda \cdot 1| = 0. \tag{A.2}$$

Since $A^T = -A$ it follows

$$\det|iA - \lambda \cdot 1| = \det|(iA - \lambda \cdot 1)^T| = \det|-iA - \lambda \cdot 1| = 0. \tag{A.3}$$

Therefore, both $\lambda$ and $-\lambda$ are eigenvalues of $iA$. It follows that $A_d$ can be written in the form

$$A_d = \begin{bmatrix} \lambda_1 & 0 & 0 & 0 & \cdots \\ 0 & -\lambda_1 & 0 & 0 & \cdots \\ 0 & 0 & \lambda_2 & 0 & \cdots \\ 0 & 0 & 0 & -\lambda_2 & \cdots \\ \cdot & \cdot & \cdot & \cdot & \cdots \\ \cdot & \cdot & \cdot & \cdot & \cdots \\ \cdot & \cdot & \cdot & \cdot & \cdots \end{bmatrix}. \tag{A.4}$$

If $n$ is odd, $A_d$ has necessarily a zero eigenvalue, that we will write as $(A_d)_{nn}$. We have also defined the $\lambda_i$ to be positive definite. Each submatrix of the type

$$\begin{bmatrix} \lambda_i & 0 \\ 0 & -\lambda_i \end{bmatrix}, \tag{A.5}$$

can be put in the form

$$\begin{bmatrix} 0 & i\lambda_i \\ -i\lambda_i & 0 \end{bmatrix},$$   (A.6)

using the $2 \times 2$ unitary matrix

$$V_2 = \frac{1}{\sqrt{2}} \begin{bmatrix} i & 1 \\ 1 & i \end{bmatrix}.$$   (A.7)

In fact

$$V_2 \begin{bmatrix} \lambda_i & 0 \\ 0 & -\lambda_i \end{bmatrix} V_2^\dagger = \begin{bmatrix} 0 & i\lambda_i \\ -i\lambda_i & 0 \end{bmatrix}.$$   (A.8)

Defining

$$V = \begin{bmatrix} V_2 & 0 & \cdots \\ 0 & V_2 & \cdots \\ \cdot & \cdot & \cdots \\ \cdot & \cdot & \cdots \\ \cdot & \cdot & \cdots \end{bmatrix},$$   (A.9)

with $V_{nn} = 0$ for $n$ odd, we get

$$VA_dV^\dagger = \begin{bmatrix} 0 & i\lambda_1 & 0 & 0 & \cdots \\ -i\lambda_1 & 0 & 0 & 0 & \cdots \\ 0 & 0 & 0 & i\lambda_2 0 & \cdots \\ 0 & 0 & -i\lambda_1 & 0 & \cdots \\ \cdot & \cdot & \cdot & \cdot & \cdots \\ \cdot & \cdot & \cdot & \cdot & \cdots \\ \cdot & \cdot & \cdot & \cdot & \cdots \end{bmatrix} = iA_s,$$   (A.10)

with $A_s$ defined in (12.206). Therefore

$$(VU)(iA)(VU)^\dagger = iA_s$$   (A.11)

or

$$(VU)(A)(VU)^\dagger = A_s.$$   (A.12)

$A$ and $A_s$ are real matrices, therefore $VU$ must also be real. However, $VU$ being real and unitary it must be orthogonal.

# Bibliography

Adler, S. L. (1969). *Phys. Rev.* **177**, p. 2426.

Anderson, C. D. (1932). *Science* **76**, p. 238.

Bell J. S. and Jackiw R. (1969) *Nuovo Cim.* *A* **60**, p. 47.

Bethe, H. A. (1947). *Phys. Rev.* **72**, p. 339.

Born, M., Heisenberg, W. and Jordan, P. (1925). *Z. f. Phys.* **35**, p. 557.

Born, M. and Jordan, P. (1925). *Z. f. Phys.* **34**, p. 858.

Bose, S. N. (1924). *Z. f. Phys.* **26**, p. 178.

Casalbuoni R. (1976a) *Nuovo Cim.* *A* **33**, p. 115.

Casalbuoni R. (1976b) *Nuovo Cim.* *A* **33**, p. 389.

Casimir, H. B. G. (1948). *Proc. K. Ned. Akad. Wet.* **51**, p. 797.

Davisson, C. J. and Germer, L. H. (1927). *Nature* **119**, p. 558.

De Broglie, L. (1924). *Recherches sur la thorie des quanta*, Ph.D. thesis, Paris University, Paris, France.

Dirac, P. A. M. (1926). *Proc. Roy. Soc.* **A112**, p. 661.

Dirac, P. A. M. (1928a). *Proc. Roy. Soc.* **A117**, p. 610.

Dirac, P. A. M. (1928b). *Proc. Roy. Soc.* **A118**, p. 351.

Dirac, P. A. M. (1931). *Proc. Roy. Soc.* **A133**, p. 60.

Dirac, P. A. M. (1934). *Rapports du Septième conseil de physique* (Gauthier-Villars, Paris).

Dirac, P. A. M. (2001). *Lecture on Quantum Mechanics* (Dover Publications Inc., Mineola, New York).

Dyson, F. J. (1949a). *Phys. Rev.* **75**, p. 486.

Dyson, F. J. (1949b). *Phys. Rev.* **75**, p. 1736.

Einstein, A. (1924). *Sitz. Ber, Preuss. Ak. Wiss.* , p. 61.

Faddeev L . D. and Popov V. N. (1967). *Phys. Lett.* **25B**, p. 29.

Fermi, E. (1926). *Z. f. Phys.* **36**, p. 902.

Feynman, R. P. (1948a). *Rev. Mod. Phys.* **20**, p. 367.

Feynman, R. P. (1948b). *Phys. Rev.* **74**, p. 939.

Feynman, R. P. (1948c). *Phys. Rev.* **74**, p. 1430.

Feynman, R. P. (1949a). *Phys. Rev.* **76**, p. 749.

Feynman, R. P. (1949b). *Phys. Rev.* **76**, p. 769.

Feynman, R. P. (1949c). *Phys. Rev.* **76**, p. 769.

Feynman, R. P. (1950). *Phys. Rev.* **80**, p. 440.

Feynman, R. P. and Hibbs, A. R. (2010). *Quantum Mechanics and Path Integrals* (Dover Publications Inc., Mineola, New York).

Foldy, L. L. and Wouthuysen, S. A. (1950). *Phys. Rev.* **78**, p. 29.

Fujikawa K. (1979). *Phys. Rev. Lett.* **42**, p. 1195.

Gordon, W. (1926). *Z. f. Phis.* **40**, p. 117.

Heisenberg, W. (1925). *Z. f. Phys.* **33**, p. 879.

Heisenberg, W. (1934). *Z. f. Phys.* **90**, p. 209.

Heisenberg, W. and Pauli, W. (1929). *Z. f. Phys.* **56**, p. 1.

Heisenberg, W. and Pauli, W. (1930). *Z. f. Phys.* **59**, p. 168.

Ito, D., Koba, Z. and Tomonaga, S. (1948). *Prog. Theor. Phys.* **3**, p. 276.

Itzykson, C. and Zuber, J. B. (1980). *Quantum Field Theory* (McGraw-Hill, New York).

Jauch, J. M. and Rohrlich, F. (1980). *The Theory of Photons and Electrons* (Springer-Verlag).

Jordan, P. and Klein, O. (1927). *Z. f. Phys.* **45**, p. 751.

Jordan, P. and Wigner, E. (1928). *Z. f. Phys.* **47**, p. 631.

Kanesawa, S. and Tomonaga, S. (1948a). *Prog. Theor. Phys.* **3**, p. 1.

Kanesawa, S. and Tomonaga, S. (1948b). *Prog. Theor. Phys.* **3**, p. 101.

Klein, O. (1926). *Z. f. Phis.* **37**, p. 895.

Klein, O. and Nishina, Y. (1929). *Z. f. Phys.* **52**, p. 853.

Koba, Z., Tati, T. and Tomonaga, S. (1947). *Rev. Mod. Phys.* **2**, p. 101.

Koba, Z. and Tomonaga, S. (1948). *Prog. Theor. Phys.* **3**, p. 290.

Lamb, W. E. and Retherford, R. C. (1947). *Phys. Rev.* **72**, p. 241.

Lehman H., Simazik K. and Zimmermann W. (1955). *Nuovo Cimento* **1**, p. 1425.

Nielsen, H. B. and Chada, S. (1976). *Nucl. Phys.* **B105**, p. 445.

Noether, E. (1918). *Kgl. Ges. d. Wiss., Nachrichten, Gottingen, Math. Phys. KI.* , p. 235.

Pais, A. (1986). *Inward Bound* (Clarendon Press, Oxford).

Pauli, W. (1925). *Z. f. Phys.* **31**, p. 373.

Pauli, W. (1940). *Phys. Rev.* **58**, p. 716.

Peskin, M. E. and Schroeder, D. V. (1995). *An Introduction to Quantum Field Theory* (Addison-Wesley, Oxford).

Pickering, A. (1984). *Constructing Quarks* (The University of Chicago Press, Chicago).

Popov, V. N. (1991). *Functional Integrals and Collective Excitations* (Cambridge University Press).

Ramond, P. (1981). *Field Theory, A Modern Primer* (Benjamin Cummings, Reading, MA).

Rivers, R. J. (1987). *Path Integral Methods in Quantum Field Theory* (Cmbridge University Press, Cambridge).

Schroedinger, E. (1926a). *Ann. der Phys.* **79**, p. 361.

Schroedinger, E. (1926b). *Ann. der Phis.* **81**, p. 109.

Schwinger, J. (1948a). *Phys. Rev.* **73**, p. 416.

Schwinger, J. (1948b). *Phys. Rev.* **74**, p. 1439.

Schwinger, J. (1949a). *Phys. Rev.* **75**, p. 651.

Schwinger, J. (1949b). *Phys. Rev.* **76**, p. 790.

Schwinger, J. (1951a). *Phys. Rev.* **82**, p. 664.

Schwinger, J. (1951b). *Phys. Rev.* **82**, p. 914.

Schwinger, J. (1951c). *Phys. Rev.* **91**, p. 713.

Schwinger, J. (1951d). *Proc. Nat. Acad. Sci.* **37**, p. 452.

Serber, R. (1936). *Phys. Rev.* **49**, p. 545.

Sparnaay, M. J. (1958). *Physica* **24**, p. 751.

Stueckelberg, E. C. G. (1942). *Helv. Phys. Acta* **15**, p. 23.

Takahashi, Y. (1957). *Nuovo Cimento* **Ser 10, 6**, p. 370.

Tomonaga, S. (1946). *Prog. Theor. Phys.* **1**, p. 27.

Tomonaga, S. (1948). *Phys. Rev.* **74**, p. 224.

Ward, J. (1950). *Phys. Rev.* **78**, p. 182.

Weinberg, S. (1995a). *The Quantum Theory of Fields, Vol. I* (Cambridge University Press, Cambridge).

Weinberg, S. (1995b). *The Quantum Theory of Fields, Vol. II* (Cambridge University Press, Cambridge).

Weyl, H. (1929a). *Proc. Nat. Ac. Sci.* **15**, p. 232.

Weyl, H. (1929b). *Z. f. Phys.* **56**, p. 330.

Yang, C. N. and Mills, R. L. (1954). *Phys. Rev.* **96**, p. 191.

# Index